T0179839

Ultra-Wideband Radio Technologies: Advanced Concepts and Applications

Ultra-Wideband Radio Technologies: Advanced Concepts and Applications

Edited by **Kevin Merriman**

New Jersey

Published by Clanrye International,
55 Van Reypen Street,
Jersey City, NJ 07306, USA
www.clanryeinternational.com

Ultra-Wideband Radio Technologies:
Advanced Concepts and Applications
Edited by Kevin Merriman

International Standard Book Number: 978-1-63240-505-0 (Hardback)

Printed in the United States of America.

Contents

Preface

Ultra-Wideband Radio has both traditional and nascent applications. Ultra-Wideband Radio, referred to as UWB, is a characteristic of a new radio access philosophy and utilizes several GHz of bandwidth. It promises increased data rate communication over small distances, as well as modern radar sensing and localization applications with extraordinary resolution. This book analyzes algorithms, hardware and application hassles in this field of radio technology for interactions, localization and sensing. It includes contributions from renowned experts and will be beneficial for readers interested in this field.

The information shared in this book is based on empirical researches made by veterans in this field of study. The elaborative information provided in this book will help the readers further their scope of knowledge leading to advancements in this field.

Finally, I would like to thank my fellow researchers who gave constructive feedback and my family members who supported me at every step of my research.

Editor

MIRA – Physical Layer Optimisation for the Multiband Impulse Radio UWB Architecture

Rainer Moorfeld, Adolf Finger, Hanns-Ulrich Dehner, Holger Jäkel, Martin Braun and Friedrich K. Jondral

Additional information is available at the end of the chapter

1. Introduction

Future wireless communication systems have to be realised in a simple and energy efficient manner while guaranteeing sufficient performance. Furthermore, the available frequency resources have to be used flexibly and efficiently. In this context two different approaches have been considered in recent years: On one hand OFDM-based overlay systems in which a primary user dynamically allocates unused frequencies to one or more secondary users [57] and on the other hand unlicensed, easy-to-realise and low-cost ultra-wideband (UWB) systems. This underlying technology operates with an extremely low transmission power over a wide frequency range and does not interfere with existing licensed systems [15].

In order to establish UWB on the consumer market it has to get along with some challenges. Such challenges are, e.g., the realisation of practical, low-complex and energy-efficient transceiver architectures, the investigation of methods for accurate synchronization and channel estimation or the handling of high sample rates. To meet these requirements this chapter considers a non-coherent multiband impulse radio UWB (MIR-UWB) system [11, 45, 46]. The MIR-UWB system focuses on short-range high data rate communication applications. The MIR-UWB system is an alternative to the architectures Multiband OFDM UWB [2] and Direct Sequence UWB [16] which have been proposed within the IEEE 802.15.3a standardization process.

The chapter is organised as follows: Section 2 gives a short introduction into the physical layer architecture of the non-coherent MIR-UWB system. In the following section 3 the performance of the energy detection receiver is analysed with respect to different aspects. In contrast section 4 deals with interference investigations for the non-coherent MIR-UWB system aiming at an efficient and intelligent interference handling. The chapter concludes with section 5 in which a summary is given.

2. Multiband impulse radio

The idea of the MIR-UWB architecture is based on [45, 46]. The architecture proposed there comprises a transmitter using multiple bands and impulse radio within the bands to transmit data and a receiver, which detects only the energy of the transmitted impulses. The combination of energy detection receiver and multiband enables a flexible high data rate system with low power consumption.

2.1. Transmitter

The MIR-UWB transmitter is based on a multiband pulse generation followed by a modulator. The multiband pulse generator generates a pulse with a specified bandwidth for every subband. Subbands can be activated or deactivated using the bandplan. Different possibilities to generate theses pulses are shown in [30]. Each subband pulse will be modulated with different data, all subband pulses are summed up to a multiband pulse, amplified and transmitted. Figure 1 shows a transmitter based on an *oscillator bank* pulse generator.

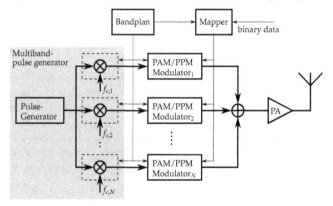

Figure 1. MIR-UWB Transmitter based on *oscillator bank* pulse generation

2.2. Receiver

The MIR-UWB receiver is based on N parallel energy detection receivers. A filter bank separates the individual subband pulses and an energy detector measures the energy in every subband. Based on the measured energy, the demodulator makes his decision. For pulse amplitude modulation (PAM) and its special case of on-off-keying (OOK) the demodulation process needs to know the SNR in each subband. This can be estimated using a preamble [46]. The channel state information can be used for Detect and Avoid (DAA) algorithms [34] and to increase the performance of the multiband system [28]. Pulse position modulation (PPM) and transmit reference (TR) do not need any channel state information.

3. Energy detection

The MIR-UWB architecture is based on energy detection. Thus the receiver detects *only* the energy of the received signal in a specified window. The disadvantage in performance is

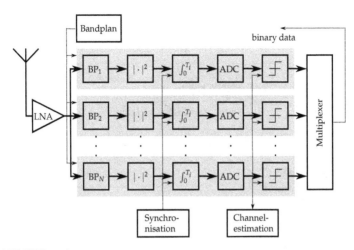

Figure 2. MIR-UWB receiver

accompanied by a very simple receiver design [36, 38, 54]. The performance measure is based on the average symbol error probability (SEP) or bit error probability (BEP) and will be derived in the following section.

3.1. Demodulation

In the additive white gaussian noise (AWGN) channel the received signal R is the sum of the transmitted signal s and white Gaussian noise W with the power spectral density of $N_0/2$:

$$R(t) = s(t) + W(t) = a_m p(t) + W(t). \tag{1}$$

The transmitted signal s is a weighted pulse p. The amplitude of the pulse is a_m. Thus, the energy of a transmitted signal is:

$$E_S = \int_{t_0}^{t_0+T_i} s^2(t)\, dt = a_m^2 \int_{t_0}^{t_0+T_i} p^2(t)\, dt$$

$$= \sum_{i=1}^{2D} s_i = a_m^2 \sum_{i=1}^{2D} p_i,$$

where $s_i := s(i/(2B))$ and $p_i := p(i/(2B))$ and B denoting the bandwidth. The received energy can be approximated by a finite sum of $2D = 2T_i B$ samples [56]. The event $\{A = a_m\}$ with the range $\mathcal{A} = \{a_0, \dots, a_{M-1}\}$ describes a transmitted symbol with the amplitude a_m. Without loss of generality the integration starts at $t_0 = 0$. In order to measure the energy, the detector squares the received signal $R(t)$ and integrates the result over the time interval T_i. The received energy, normalized by the power spectral density $N_0/2$, is:

$$Y = \frac{2}{N_0} \int_0^{T_i} R^2(t)\, dt. \tag{2}$$

A time discrete representation of the received energy Y is:

$$Y = \frac{1}{N_0 B} \sum_{i=1}^{2D} (s_i + W_i)^2,$$ (3)

where $W_i := W(i/(2B))$. If the symbol energy is $E_S = 0$, the received energy Y will be χ^2 distributed with the degree of freedom of $2D$. If the symbol energy $E_S > 0$, the received energy will be noncentral χ^2 distributed with the degree of freedom of $2D$ and the noncentrality parameter μ:

$$\mu = \sum_{i=1}^{2D} s_i^2 = \frac{2}{N_0} \int_0^T s^2(t)\, dt =: \frac{2E_s}{N_0} = 2\gamma,$$ (4)

where $\gamma = E_S/N_0$ is the SNR at the receiver. Thus, the distribution of the received energy Y depends on the symbol energy E_S:

$$Y \sim \begin{cases} \chi_{2D}^2 & \text{for } E_S = 0 \\ \chi_{2D}^2(2\gamma) & \text{for } E_S > 0. \end{cases}$$

The conditional probability density function $f_{Y|A}(\cdot|a_0)$ with $a_0 = 0$ and $E_S = 0$ of the received energy Y is:

$$f_{Y|A}(y|a_0) = \frac{1}{2^D \Gamma(D)} y^{D-1} \exp\left(-\frac{y}{2}\right).$$ (5)

The conditional probability density function $f_{Y|A}(\cdot|a_m)$ with $a_m > 0$ and $E_S > 0$ of the received energy Y is:

$$f_{Y|A}(y|a_m) = \frac{1}{2}\left(\frac{y}{2\gamma}\right)^{\frac{D-1}{2}} \exp\left(-\frac{2\gamma + y}{2}\right) I_{D-1}\left(\sqrt{2\gamma y}\right),$$ (6)

where Γ is the gamma function [18, eq. 8.310.1] and I_n is the modified Bessel function of the first kind of order n [1, eq. 9.6.3].

3.2. AWGN channel

First we calculate the bit error probability of the energy detection receiver in the AWGN channel (1). This receiver detects only the energy of the received signal (2), (3).

3.2.1. Pulse amplitude modulation

The M-PAM modulated signal is:

$$s(t) = \sum_{k=-\infty}^{\infty} a_{m,k} p(t - kT_r)$$

and transmits $\log_2(M)$ bit per symbol. The energy of the mth symbol is:

$$E_{S_m} = \underbrace{\int_0^{T_i} s^2(t)\, dt}_{=\sum\limits_{i=1}^{2D} s_i^2} = a_m^2 \underbrace{\int_0^{T_i} p^2(t)\, dt}_{=\sum\limits_{i=1}^{2D} p_i^2} = a_m^2 E_p,$$

where E_p is the energy of an unmodulated pulse p. The demodulator has to decide, which symbol m with the energy E_{S_m} and the amplitude a_m has been transmitted, based on the observation of the random variable Y. The optimal receiver, i. e. the receiver with the lowest probability to make a wrong decision, makes the decision for the symbol that has been sent most likely, given a certain energy y at the receiver. Thus, the receiver makes the decision for the symbol m with the amplitude a_m, when [25]:

$$\mathbb{P}\{A = a_m | Y = y\} \geq \mathbb{P}\{A = a_k | Y = y\}, \quad \forall\, m \neq k. \tag{7}$$

This is the maximum a posteriori probability (MAP) decision rule. If all transmitted symbols are equal probable, it can be reduced to the maximum-likelihood (ML) decision rule:

$$f_{Y|A}(y|a_m) \geq f_{Y|A}(y|a_k), \quad \forall\, m \neq k,$$

using the Bayes theorem:

$$\mathbb{P}(A = a_m | Y = y) = \frac{f_{Y|A}(y|a_m)\mathbb{P}(A = a_m)}{f_Y(y)} = \frac{f_{Y|A}(y|a_m)1/M}{f_Y(y)},$$

because $\mathbb{P}(A = a_m) = 1/M$ for all $m \in \{0, 1, \dots, M-1\}$ and $f_Y(y)$ are independent of m. For the M-PAM modulated signal, we use the ML receiver with multiple hypothesis testing:

$$m = \underset{k \in [0, M-1]}{\arg\max}\ f_{Y|A}(y|a_k), \tag{8}$$

with the conditional probability density function $f_{Y|A}$ based on (5) and (6).

The SEP P_e for the energy detection receiver in the AWGN channel with M-PAM signals can be calculated as:

$$P_e(\overline{\gamma}, \mathbf{a}, \boldsymbol{\rho}, D) = 1 - P_c(\overline{\gamma}, \mathbf{a}, \boldsymbol{\rho}, D)$$
$$= 1 - \sum_{m=0}^{M-1} \mathbb{P}(\rho_m \leq Y < \rho_{m+1} | A = a_m)\mathbb{P}(A = a_m), \tag{9}$$

where P_c is the probability of a correct decision and $\mathbb{P}(\rho_m < Y \leq \rho_{m+1} | A = a_m)$ is the conditional probability, that the received energy Y is in the interval $[\rho_m, \rho_{m+1})$, with the optimal interval thresholds ρ. Thus, the decision has been made using the ML decision rule (8). $\mathbb{P}(A = a_m)$ is the a priori probability, that the symbol m has been sent and $\mathbb{P}(A = a_m) = 1/M$ for all $m \in \{0, 1, \dots, M-1\}$. The conditional probability $\mathbb{P}(\rho_m \leq Y < \rho_{m+1} | A = a_m)$ is:

$$\mathbb{P}(\rho_m < y \leq \rho_{m+1} | A = a_m) = \int_{\rho_m}^{\rho_{m+1}} f_{Y|A}(y|a_m)\,\mathrm{d}y$$
$$= F_{Y|A}(\rho_{m+1}|a_m) - F_{Y|A}(\rho_m|a_m). \tag{10}$$

The related distribution function $F_{Y|A}(\cdot|0)$ can be calculated in closed form:

$$F_{Y|A}(y|0) = \int_0^y \frac{1}{2^D \Gamma(D)} u^{D-1} \exp\left(-\frac{u}{2}\right)\,\mathrm{d}u = \frac{\Gamma\left(D, \frac{y}{2}\right)}{\Gamma(D)}, \tag{11}$$

where $\Gamma(\cdot)$ is the Gamma function [18, eq. 8.310.1] and $\Gamma(\cdot,\cdot)$ is the incomplete Gamma function [18, eq. 8.350.1]. The distribution function (11) can be also displayed with the help of the Marcum-$\mathcal{Q}$ function:

$$\mathcal{Q}_m(a,b) = \int_b^\infty x \left(\frac{x}{a}\right)^{m-1} \exp\left(-\frac{x^2+a^2}{2}\right) I_{m-1}(ax)\, dx, \tag{12}$$

where I_n is the modified Bessel function of the first kind of order n [1, eq. 9.6.3]. Thus, an alternative representation for the distribution function (11) is for $D \in \mathbb{Z}^+$ with [49, eq. 2.1.124] and [52, eq. 4.71]:

$$F_{Y|A}(y|0) = 1 - \mathcal{Q}_D(0, \sqrt{y}). \tag{13}$$

If a symbol with an amplitude $a_m > 0$ has been sent, the conditional probability density function $f_{Y|A}(\cdot|a_m)$ has the form (6). There does not exist a closed form for the distribution function in general. But for $D \in \mathbb{Z}^+$ it can also be solved in closed form with the help of the Marcum-$\mathcal{Q}$ function (12):

$$\begin{aligned} F_{Y|A}(y|a_m) &= \int_0^y \frac{1}{2}\left(\frac{u}{2a_m^2\gamma}\right)^{\frac{D-1}{2}} \exp\left(-\frac{2a_m^2\gamma+u}{2}\right) I_{D-1}\left(\sqrt{2a_m^2\gamma u}\right) du \\ &= 1 - \mathcal{Q}_D\left(a_m\sqrt{2\gamma}, \sqrt{y}\right). \end{aligned} \tag{14}$$

Combining (13) and (14) with (9), the SEP P_e for an energy detection receiver with M-PAM for an SNR of $\gamma = E_p/N_0$ is [35]:

$$P_e(\gamma, \mathbf{a}, \boldsymbol{\rho}, M, D) = 1 - \frac{1}{M}\left[\sum_{m=0}^{M-1} \mathcal{Q}_D\left(a_m\sqrt{2\gamma}, \sqrt{\rho_{m+1}}\right) - \mathcal{Q}_D\left(a_m\sqrt{2\gamma}, \sqrt{\rho_m}\right)\right] \tag{15}$$

with the M symbol amplitudes $\mathbf{a} = (a_0, a_1, \ldots, a_{M-2}, a_{M-1})$, $M+1$ interval thresholds $\boldsymbol{\rho} = (\rho_0, \rho_1, \ldots, \rho_{M-1}, \rho_M)$ and the degree of freedom $2D$. Applying the interval thresholds $\rho_0 = 0$ und $\rho_M \to \infty$, we get $\mathcal{Q}_D\left(a_m\sqrt{2\gamma}, \sqrt{\rho_0}\right) = 0$ and $\mathcal{Q}_D\left(a_m\sqrt{2\gamma}, \sqrt{\rho_M}\right) = 1$. Combining this with (15), it reduces to:

$$P_e(\gamma, \mathbf{a}, \boldsymbol{\rho}, M, D) = \frac{1}{M}\left[M-1+\sum_{m=1}^{M-1} \mathcal{Q}_D\left(a_m\sqrt{2\gamma}, \sqrt{\rho_m}\right) - \sum_{m=0}^{M-2} \mathcal{Q}_D\left(a_m\sqrt{2\gamma}, \sqrt{\rho_{m+1}}\right)\right]. \tag{16}$$

Figure 3 shows the influence of different degrees of freedom on the BEP of an energy detection receiver with OOK and 2-PPM. For large degrees of freedom, a higher SNR is necessary to achieve the same BEP. This is due to an increasing amount of noise at the detector. OOK shows a slightly better performance than 2-PPM. Figure 4 shows the influence of higher order modulation on the BEP.

3.2.1.1. Optimal interval thresholds

The optimal interval thresholds to minimise the SEP have to fulfil the following optimisation problem:

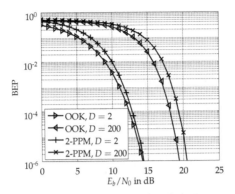

Figure 3. BEP for OOK and 2-PPM with different degrees of freedom

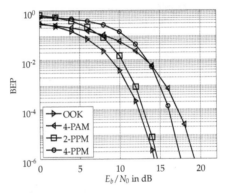

Figure 4. BEP for multilevel M-PAM and M-PPM for $D = 2$

$$\underset{\rho}{\text{minimise}} \quad P_e(\gamma, \boldsymbol{a}, \boldsymbol{\rho}, M, D)$$

$$\text{subject to} \quad \frac{1}{M} \sum_{m=0}^{M-1} a_m^2 = 1,$$

Thus, the optimal interval threshold ρ_{opt} between the two symbol amplitudes a_m and a_{m+1} has to fulfil the following equation:

$$f_{Y|A}(\rho_{\text{opt}}|a_m) = f_{Y|A}(\rho_{\text{opt}}|a_{m+1}), \tag{17}$$

where $f_{Y|A}$ are the conditional probability density functions based on (5) and (6). With these optimal interval thresholds, the symbol decision is based on the ML-criteria (8). Unfortunately, there is no closed form solution for determining the optimal interval thresholds. Thus, they have to be calculated numerically. Figure 5 shows the conditional probability density functions with equidistant symbol amplitudes a_m and optimal interval thresholds ρ_1, ρ_2 and ρ_3 with an SNR of $\gamma = 10\,\text{dB}$.

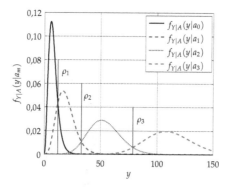

Figure 5. Optimal interval thresholds for 4-PAM ($\gamma = 10\,\text{dB}$)

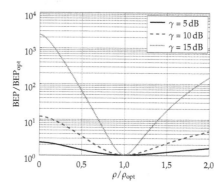

Figure 6. Sensitivity of the BEP related to the interval threshold ρ_1 for OOK

Figure 6 shows the influence of a non optimal interval threshold ρ_1 on the SEP for OOK ($M = 2$). In such a case, the SEP gets more sensitive for high SNR.

3.2.1.2. Optimal amplitudes

Now we try to minimise the SEP by optimising the symbol amplitudes a. The optimisation problem is now [33, 37]:

$$\underset{a}{\text{minimise}} \quad P_e(\gamma, a, \rho, M, D)$$

$$\text{subject to} \quad \frac{1}{M} \sum_{m=0}^{M-1} a_m^2 = 1,$$

This optimisation problem can only be solved numerically, because the optimal interval thresholds ρ are based on the amplitudes a and there exists no closed form solution for the optimal interval thresholds ρ (17). For OOK ($M = 2$) the optimal amplitudes are $a_{\text{opt}} = (0, 2)$. In this case they are independent of the SNR γ. For $M > 2$ it is possible to calculate a set of optimal amplitudes a_{opt} for every SNR γ. Figure 7 shows the SEP for 4-PAM for different symbol amplitudes a_1 and a_2 for a SNR of $16\,\text{dB}$. For figure 7 the amplitudes $a_0 = 0$ and

$a_3 = 1$ are set. The minimal SEP has been reached for $a = (0, 0.35, 0.67, 1)$. Figure 8 shows

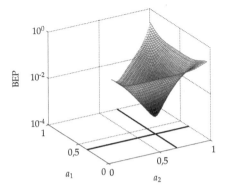

Figure 7. SEP for 4-PAM with different interval thresholds

the gain for 4-PAM with optimal amplitudes for different degrees of freedom. The results show impressive gains for large degrees of freedom. Figure 9 shows the optimal amplitudes for different degrees of freedom. For $D = 2$ the amplitudes are almost equidistant but for $D = 200$ the amplitudes are adjusted and not equidistant any more.

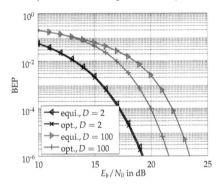

Figure 8. BEP for 4-PAM with equidistant and optimal amplitudes

3.3. Flat fading channel

To analyse the performance of an energy detection receiver we need a channel model that enables a good approximation of the energy at the receiver. Investigations of the IEEE channel model (802.15.3a) show that the energy at the receiver can be approximated by a random variable which is constant for one symbol.

Figure 10 compares the channel's magnitude (denoted as CIR) to a moving average of width 100 MHz and 1 GHz of the energy in the IEEE channel model. Figure 11 shows the magnitude at the receiver for a detector with 100 MHz and 1 GHz bandwidth. Thus we can use the

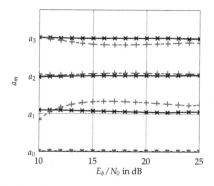

Figure 9. Optimal amplitudes for different degrees of freedom ($D = 2$, $D = 200$)

flat fading channel model to model the energy at the receiver in a frequency selective fading channel.

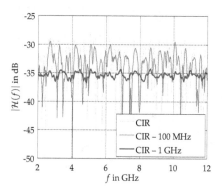

Figure 10. Moving average of the energy at the receiver (IEEE 802.15.3a, CM1)

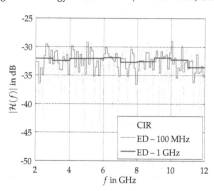

Figure 11. Energy at the receiver for different bandwidth (IEEE 802.15.3a, CM1)

In a flat fading channel, the random path attenuation H is assumed to be constant for the duration of a symbol. Thus, the received signal $R(t)$ in the interval $0 \leq t \leq T_S$ is:

$$R(t) = Hs(t) + W(t),$$

where $W(t)$ denotes the additive white Gaussian noise (AWGN) with a spectral power density of $N_0/2$.

The instantaneous SNR γ in the flat fading channel is:

$$\gamma = h^2 \frac{E_S}{N_0},$$

where h denotes the instantaneous path attenuation and E_S denotes the symbol energy. Since the SNR γ is random, γ is a realization of a a random variable Γ. The average SNR $\overline{\gamma}$ can be calculated using the expectation of the random variable H^2:

$$\overline{\gamma} = \Omega \frac{\overline{E_S}}{N_0} = \int_0^\infty \gamma f_\Gamma(\gamma) \, \mathrm{d}\gamma$$

with $\Omega = \mathbb{E}(H^2)$. Introducing a change of variable, the probability density function of the random variable Γ is [52, eq. 2.3]:

$$f_\Gamma(\gamma) = \frac{f_H\left(\sqrt{\frac{\Omega\gamma}{\overline{\gamma}}}\right)}{2\sqrt{\frac{\gamma\overline{\gamma}}{\Omega}}}. \tag{18}$$

To calculate the average SEP in a flat fading channel, we have to solve the following integral [52, eq. 1.8]:

$$\overline{P}_e(\overline{\gamma}) = \int_0^\infty P_{e,AWGN}(\gamma) f_\Gamma(\gamma) \, \mathrm{d}\gamma. \tag{19}$$

3.3.1. Pulse amplitude modulation

Using the SEP in the AWGN channel (16) and the probability density function of the random SNR Γ in the flat fading channel, the average SEP for M-PAM is:

$$\overline{P}_e(\overline{\gamma}, \boldsymbol{a}, \boldsymbol{\rho}, M, D) = \frac{1}{M} \Bigg[M - 1 - \mathcal{Q}_D\left(0, \sqrt{\rho_1}\right)$$

$$+ \sum_{m=1}^{M-1} \int_0^\infty \mathcal{Q}_D\left(a_m\sqrt{2\gamma}, \sqrt{\rho_m}\right) f_\Gamma(\gamma) \, \mathrm{d}\gamma$$

$$- \sum_{m=1}^{M-2} \int_0^\infty \mathcal{Q}_D\left(a_m\sqrt{2\gamma}, \sqrt{\rho_{m+1}}\right) f_\Gamma(\gamma) \, \mathrm{d}\gamma \Bigg], \tag{20}$$

with the symbol amplitudes $\boldsymbol{a} = (a_0, a_1, \dots, a_{M-2}, a_{M-1})$ and the interval thresholds $\boldsymbol{\rho} = (\rho_0, \rho_1, \dots, \rho_{M-1}, \rho_M)$. (20) is a general solution for the average SEP of an energy detection receiver in a flat fading channel with M-PAM.

3.3.2. Rayleigh fading

Rayleigh distributed path gains are used to model fading channels with no line-of-sight (NLOS) [49, 50, 55]. Thus, the random variable H is *Rayleigh* distributed:

$$f_H(h) = \frac{2h}{\Omega} \exp\left(-\frac{h^2}{\Omega}\right), \quad h \geq 0. \tag{21}$$

In UWB channels with a large bandwidth and a corresponding high temporal resolution, it is questionable, if the central limit theorem is applicable [4, 32, 59]. Nevertheless, some UWB channel measurements show a good fit to the *Rayleigh* distribution [17, 24, 51]. Using (18), the probability density function of the random SNR is:

$$f_\Gamma(\gamma) = \frac{1}{\overline{\gamma}} \exp\left(-\frac{\gamma}{\overline{\gamma}}\right), \quad \gamma \geq 0. \tag{22}$$

Combining (20) and (22) we get an integral of the form:

$$Y_D(\overline{\gamma}, a_m, \rho_m) = \frac{1}{\overline{\gamma}} \int_0^\infty Q_D\left(a_m\sqrt{2\gamma}, \sqrt{\rho_m}\right) \exp\left(-\frac{\gamma}{\overline{\gamma}}\right) \, \mathrm{d}\gamma. \tag{23}$$

The integral in (23) can be solved using [40, eq. 12], given that the interval thresholds do not depend on the instantaneous SNR γ but on the average SNR $\overline{\gamma}$. This is the case for an energy detection receiver with limited channel state information (only knowledge of the average SNR). The closed form solution for Y_D is then:

$$Y_D(\overline{\gamma}, a_m, \rho_m) = \exp\left(-\frac{\rho_m}{2}\right) \left\{ \left(\frac{1/\overline{\gamma} + a_m^2}{a_m^2}\right)^{D-1} \right.$$

$$\left. \cdot \left[\exp\left(\frac{\rho_m}{2}\frac{a_m^2}{a_m^2 + 1/\overline{\gamma}}\right) - \sum_{d=0}^{D-2} \frac{1}{d!}\left(\frac{\rho_m}{2}\frac{a_m^2}{a_m^2 + 1/\overline{\gamma}}\right)^d\right] + \sum_{d=0}^{D-2} \frac{1}{d!}\left(\frac{\rho_m}{2}\right)^d \right\}. \tag{24}$$

Combining (20) and (24) yields to the closed form solution for the energy detection receiver in a *Rayleigh* fading channel with M-PAM:

$$\overline{P}_{e,ray}(\overline{\gamma}, \mathbf{a}, \boldsymbol{\rho}, M, D) = \frac{1}{M}\left[M - 1 - Q_D\left(0, \sqrt{\rho_1}\right)\right.$$

$$\left. + \sum_{m=1}^{M-1} Y_D\left(\overline{\gamma}, a_m, \rho_m\right) - \sum_{m=1}^{M-2} Y_D\left(\overline{\gamma}, a_m, \rho_{m+1}\right)\right]. \tag{25}$$

3.3.3. Rician fading

Rice distributed path gains are used to model line-of-sight (LOS) fading channels [49, 50, 55]. Thus, the random variable H is *Rice* distributed:

$$f_H(h) = \frac{2h(k+1)}{\Omega} \exp\left(-k - \frac{(k+1)h^2}{\Omega}\right) I_0\left(2h\sqrt{\frac{k(k+1)}{\Omega}}\right), \quad h \geq 0, \tag{26}$$

where I_0 denotes the modified Bessel function of the first kind of order zero. The *Rician*-k-factor is the ratio between the power in the direct path and the power in the scattered paths. For $k = 0$ the *Rice* distribution is equal to the *Rayleigh* distribution. For $k \to \infty$ the *Rician* fading channel converges to the AWGN channel. Different UWB measurement campaigns show a good fit with the distribution of the path gains with a *Rice* distribution [20, 26, 43]. Using (18), the probability density function of the random SNR is:

$$f_{\Gamma}(\gamma) = \frac{k+1}{\overline{\gamma}} \exp\left(-k - \frac{(k+1)\gamma}{\overline{\gamma}}\right) I_0 \left(2\sqrt{\frac{k(k+1)\gamma}{\overline{\gamma}}}\right), \quad \gamma \geq 0. \tag{27}$$

Combining (20) and (27) we get an integral of the form:

$$\Phi_D(\overline{\gamma}, a_m, \rho_m, k) = \frac{k+1}{\overline{\gamma}} \int_0^{\infty} \mathcal{Q}_D\left(a_m u, \sqrt{\rho_m}\right) \exp\left(-K - \frac{(K+1)u^2}{\overline{\gamma}}\right)$$
$$\cdot I_0 \left(2\sqrt{\frac{K(K+1)u^2}{\overline{\gamma}}}\right) u\, du. \tag{28}$$

There exists only a closed form solution for $D = 1$ [39, eq. 45]. In this case, we get:

$$\Phi_1(\overline{\gamma}, a_m, \rho_m, k) = \mathcal{Q}_1\left(\sqrt{\frac{2ka_m^2\gamma}{(k+1) + a_m^2\gamma}}, \sqrt{\frac{(k+1)\rho_m}{(k+1) + a_m^2\gamma}}\right). \tag{29}$$

Thus, the average SEP in a *Rician* fading channel is with a degree of freedom of $2D = 2$:

$$\overline{P}_{e,ric}(\overline{\gamma}, \mathbf{a}, \boldsymbol{\rho}, k, M) = \frac{1}{M}\left[M - 1 - \mathcal{Q}_1\left(0, \sqrt{\rho_1}\right)\right.$$
$$\left. + \sum_{m=1}^{M-1} \Phi_1\left(\overline{\gamma}, a_m, \rho_m, k\right) - \sum_{m=1}^{M-2} \Phi_1\left(\overline{\gamma}, a_m, \rho_{m+1}, k\right)\right]. \tag{30}$$

A closed form solution for the integral in (28) is not known for $D > 1$. In this case, the integral has to be calculated numerically.

3.3.4. Nakagami-m fading

The probability density function of the *Nakagami-m* distribution of the random path gains is related to the χ^2 distribution:

$$f_H(h) = \frac{2h^{2m-1}}{\Gamma(m)} \left(\frac{m}{\Omega}\right)^m \exp\left(-\frac{m}{\Omega}h^2\right), \quad h \geq 0, \tag{31}$$

where m denotes the *Nakagami-m* fading parameter with $m \in [1/2, \infty)$ and Γ denotes the Gamma function. The *Nakagami-m* distribution includes as special cases the one-sided normal distribution ($m = 1/2$) and the *Rayleigh*-distribution ($m = 1$). For $m \to \infty$ the *Nakagami-m* fading channel converges to the AWGN channel. Different UWB measurement campaigns show a good fit to the *Nakagami-m* distribution [4, 19]. *Nakagami-m* distribution is also used in the IEEE channel model 802.15.4a to model the path gains [31]. The probability density function of the random SNR Γ is with (18) and (31):

$$f_\Gamma(\gamma) = \frac{\gamma^{m-1}}{\Gamma(m)} \left(\frac{m}{\overline{\gamma}}\right)^m \exp\left(-\frac{m}{\overline{\gamma}}\gamma\right), \quad \gamma \geq 0. \tag{32}$$

In this case we have to solve the following integral by substituting $\gamma = u^2/2$:

$$\int_0^\infty u^{2m-1} \exp\left(-\frac{mu^2}{a\overline{\gamma}}\right) \mathcal{Q}_D\left(au, \sqrt{\rho_m}\right) du. \tag{33}$$

(33) can be solved recursively [13]. The average symbol error rate in a fading channel with *Nakagami-m* distributed fading gains is:

$$
\begin{aligned}
\overline{P}_{e,nak}(\overline{\gamma}, m, \mathbf{a}, \boldsymbol{\rho}, M, D) = \frac{1}{M} \Bigg[& M - 1 - \mathcal{Q}_D\left(0, \sqrt{\rho_1}\right) \\
& + \sum_{v=1}^{M-1} \left(A\left(\overline{\gamma}, m, a_v, \rho_v\right) + \beta_v^m B\left(\overline{\gamma}, m, a_v, \rho_v\right)\right) \\
& - \sum_{v=1}^{M-2} \left(A\left(\overline{\gamma}, m, a_v, \rho_{v-1}\right) + \beta_v^m B\left(\overline{\gamma}, m, a_v, \rho_{v+1}\right)\right) \Bigg],
\end{aligned} \tag{34}
$$

where

$$A\left(\overline{\gamma}, m, a, \rho\right) = \exp\left(-\frac{\beta\rho}{2}\right) \left[\beta^{m-1} L_{m-1}\left(-\frac{(1-\beta)\rho}{2}\right) + (1-\beta) \sum_{i=0}^{m-2} \beta^i L_i\left(-\frac{(1-\beta)\rho}{2}\right)\right]$$

and

$$B\left(\overline{\gamma}, m, a, \rho\right) = \exp\left(-\frac{\rho}{2}\right) \sum_{n=1}^{D-1} \frac{\rho_w^n}{2^n n!} {}_1F_1\left(m; n+1; \frac{(1-\beta)\rho}{2}\right) \quad \text{with } \beta_v = \frac{2m}{2m + a_v \overline{\gamma}}.$$

L_i is the *Laguerre* polynomial of degree i [18, eq. 8.970] and ${}_1F_1$ is the confluent hypergeometric function [18, eq. 9.210.1].

Figure 12 shows the bit error probability in a flat fading channel with *Rayleigh* and *Rice* distributed channel gains. Figure 13 shows the bit error probability in a flat fading channel with *Nakagami-m* distributed channel gains.

3.4. Diversity reception

Now we analyse the SEP of an energy detection receiver with diversity reception. The goal is to increase the SNR to improve its performance. Because of the architecture of the receiver, detecting only the energy of the received signal, the possibilities to improve its performance are limited and many combining techniques like maximum ratio combining (MRC) or equal gain combining (EGC) are not feasible. Thus we concentrate on square law combining (SLC) and square law selection (SLS) [29].

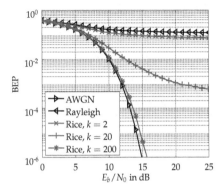

Figure 12. BEP in a flat fading channel with *Rayleigh* and *Rice* distributed channel gains (OOK, $D = 2$)

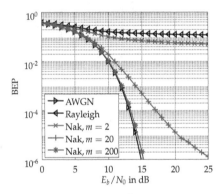

Figure 13. BEP in a flat fading channel with *Nakagami-m* distributed channel gains (OOK, $D = 2$)

The channel model used here is based on flat fading with independent and correlated fading gains H_l for all l diversity paths. The instantaneous SNR at the energy detector l is:

$$\gamma_l = h_l^2 E_S / N_{0,l}$$

and the average SNR at the lth detector is:

$$\overline{\gamma}_l = \Omega_l E_S / N_{0,l}$$

with $\Omega_l = \mathbb{E}(H_l^2)$. Figure 14 shows the model of multichannel receiver.

3.4.1. Square law combining

At the SLC receiver we have a new SNR Y_{SLC} at the receiver output based on the sum of the SNR Y_l at the l detectors:

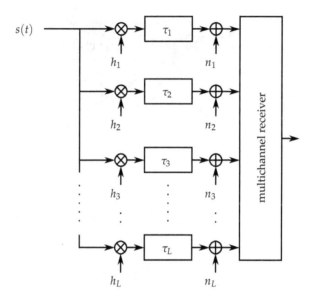

Figure 14. Channel model for multichannel receiver

$$Y_{\text{SLC}} = \sum_{l=1}^{L} Y_l.$$

The new random variable Y_{SLC} has a central χ^2 distribution for $a_0 = 0$ and a noncentral χ^2 distribution for $a_m > 0$ with the degree of freedom of $2LD$. The conditional distribution function for SLC is:

$$F_{Y|A}(y|a_m) = \mathcal{Q}_{LD}(a_m\sqrt{\gamma_{\text{SLC}}}, \sqrt{y}) \tag{35}$$

with the non centrality parameter:

$$\mu_{\text{SLC}} = 2\gamma_{\text{SLC}} = 2\sum_{l=1}^{L} \gamma_l.$$

Thus the SEP in L parallel AWGN channels with different SNR γ_l using SLC is:

$$P_e(\gamma_{\text{SLC}}, \mathbf{a}, \tilde{\rho}, M, D, L) = \frac{1}{M}\Bigg[M - 1 + \sum_{m=1}^{M-1} \mathcal{Q}_{LD}\left(\sqrt{a_m^2\gamma_{\text{SLC}}}, \sqrt{\tilde{\rho}_m}\right)$$
$$- \sum_{m=0}^{M-2} \mathcal{Q}_{LD}\left(\sqrt{a_m^2\gamma_{\text{SLC}}}, \sqrt{\tilde{\rho}_{m+1}}\right)\Bigg],$$

where $\tilde{\rho}$ are the optimal interval thresholds (section 3.2.1.1), based on the new random variable Y_{SLC}.

3.4.1.1. Independent and identically distributed Rayleigh distributed channel gains

Now we derive the SEP for SLC with independent and identical distributed (i.i.d.) Rayleigh distributed channel gains H_l. The SNR Γ_l is exponential distributed for all l. The probability density function $f_{\Gamma_{SLC}}$ of $\Gamma_{SLC} = \sum_{l=1}^{L} \Gamma_l$ with $\overline{\gamma} = \mathbb{E}(\Gamma_l)$ is [27, eq. 10.61]:

$$f_{\Gamma_{SLC}}(\gamma_{SLC}) = \frac{\gamma_{SLC}^{L-1}}{(L-1)!\overline{\gamma}^{L}} \exp\left(\frac{\gamma_{SLC}}{\overline{\gamma}}\right). \tag{36}$$

Comparing (36) with the probability density function of the SNR with *Nakagami-m* distributed channel gains we get the average SEP in I.I.D. *Rayleigh* fading channels:

$$\overline{P}_{e,SLC}(\overline{\gamma}, \mathbf{a}, \tilde{\rho}, M, D, L) = \overline{P}_{e,nak}(L\overline{\gamma}, L, \mathbf{a}, \tilde{\rho}, M, LD),$$

substituting $\overline{\gamma}$ by $L\overline{\gamma}$, m by L, ρ by $\tilde{\rho}$ and $2D$ by $2LD$. The intervall thresholds $\tilde{\rho}$ are based on $L\overline{\gamma}$.

3.4.1.2. Correlated Rayleigh distributed channel gains

In the next step, we assume correlated *Rayleigh* distributed channel gains H_l. In this case the probability density function of the SNR Γ is a sum of weighted exponential distributions [27, eq. 10.60]:

$$f_{\Gamma_{SLC}}(\gamma_{SLC}) = c_1 \sum_{l=1}^{L} c_{2,l} \exp\left(\frac{\gamma_{SLC}}{\overline{\gamma}}\right) \tag{37}$$

with

$$c_1 = \frac{1}{\prod_{l=1}^{L} \lambda_l}, \quad c_{2,l} = \frac{1}{\prod_{l \neq k}^{L}(1/\lambda_k - 1/\lambda_l)},$$

where λ_l are the eigenvalues of the $L \times L$ covariance matrix $\boldsymbol{\Sigma}$ of the normalized received signal. Using the function Y (24) the SEP in a fading channel with correlated *Rayleigh* distributed fading gains can be written as:

$$\overline{P}_{e,SLC}(\overline{\gamma}, \mathbf{a}, \tilde{\rho}, M, D, L) = \frac{1}{M}\left[M - 1 - \mathcal{Q}_{LD}\left(0, \sqrt{\rho_1}\right) \right.$$

$$+ c_1 \sum_{m=1}^{M-1} \sum_{l=1}^{L} \lambda_l c_{2,l} Y_{LD}\left(\lambda_l, a_m, \tilde{\rho}_m\right)$$

$$\left. - c_1 \sum_{m=1}^{M-2} \sum_{l=1}^{L} \lambda_l c_{2,l} Y_{LD}\left(\lambda_l, a_m, \tilde{\rho}_{m+1}\right) \right].$$

3.4.2. *Square law selection*

The receiver based on SLS chooses the detector with the highest received energy Y_{max}:

$$Y_{max} = \max(Y_1, Y_2, \dots, Y_L). \tag{38}$$

Thus, this receiver collects only a fraction of the total received energy. Using (9) and (38), the SEP with SLS in the AWGN channel is:

$$P_e(\gamma, \mathbf{a}, \boldsymbol{\rho}, D, L) = 1 - P_c(\gamma, \mathbf{a}, \boldsymbol{\rho}, D)$$

$$= 1 - \sum_{m=0}^{M-1} \mathbb{P}(\rho_m < Y_{\max} \le \rho_{m+1} | A = a_m) \mathbb{P}(A = a_m). \qquad (39)$$

In order to calculate (39), we calculate the probability that $Y_{\max}$ is in the interval $(\rho_m, \rho_{m+1}]$:

$$\mathbb{P}(\rho_m < Y_{\max} \le \rho_{m+1} | A = a_m)$$

$$= \mathbb{P}\left(\bigcap_{l=1}^{L} \{Y_l < \rho_{m+1} | A = a_m\}\right) - \mathbb{P}\left(\bigcap_{l=1}^{L} \{Y_l < \rho_m | A = a_m\}\right). \qquad (40)$$

If all received energies Y_l are independent, (40) reduces to:

$$\mathbb{P}\left(\bigcap_{l=1}^{L} \{Y_l < \rho_m | A = a_m\}\right) = \prod_{l=1}^{L} \mathbb{P}(Y_l < \rho_m | A = a_m). \qquad (41)$$

Using the distribution function (14), the probability can be written as:

$$\mathbb{P}(Y_l < \rho_m | A = a_m) = 1 - Q_D(a_m \sqrt{\gamma}, \sqrt{\rho_m}). \qquad (42)$$

Combining (40), (41) and (42) and assume $\mathbb{P}(A = a_m) = 1/M$, (39) can be written as:

$$P_e(\gamma, \mathbf{a}, \boldsymbol{\rho}, D, L) = 1 - \frac{1}{M} \sum_{m=0}^{M-1} \left[\prod_{l=1}^{L} \left(1 - Q_D\left(a_m \sqrt{\gamma_l}, \sqrt{\rho_{m+1,l}}\right)\right) \right.$$

$$\left. - \prod_{l=1}^{L} \left(1 - Q_D\left(a_m \sqrt{\gamma_l}, \sqrt{\rho_{m,l}}\right)\right) \right], \qquad (43)$$

where $\rho_{m,l}$ denotes the optimal interval thresholds which are based on the SNR γ_l. If all received energies Y_l are independent and identical distributed, (41) reduces to:

$$\prod_{l=1}^{L} \mathbb{P}(Y_l < \rho_m | A = a_m) = (1 - Q_D(a_m \sqrt{\gamma}, \sqrt{\rho_m}))^L. \qquad (44)$$

In this case, (43) reduces to:

$$P_e(\gamma, \mathbf{a}, \rho, D, L) = 1 - \frac{1}{M} \sum_{m=0}^{M-1} \left[(1 - Q_D(a_m \sqrt{\gamma}, \sqrt{\rho_{m+1}}))^L \right.$$

$$\left. - (1 - Q_D(a_m \sqrt{\gamma}, \sqrt{\rho_m}))^L \right].$$

If the SNR is independent and identically *Rayleigh* distributed with $\overline{\gamma} = \mathbb{E}(\Gamma_l)$ for all l, the SEP in a fading channel with i.i.d. *Rayleigh* distributed channel gains can be written using the function Y (24):

$$P_e(\overline{\gamma}, \mathbf{a}, \boldsymbol{\rho}, D, L) = \frac{1}{M} \left[M + 1 + \left(1 - Q_D \left(a_0 \sqrt{\overline{\gamma}}, \sqrt{\rho_1} \right) \right)^L \right.$$
$$+ \sum_{m=1}^{M-2} \left(1 - Y_D(\overline{\gamma}, a_m, \rho_{m+1}) \right)^L$$
$$\left. - \sum_{m=1}^{M-1} \left(1 - Y_D(\overline{\gamma}, a_m, \rho_m) \right)^L \right].$$

Figure 15 shows the gain that can be achieved using SLC. Figure 16 shows the gain that can be achieved using SLS. The SLC based receiver can collect more energy, but also more noise than the SLS based receiver. The SLS based receiver chooses only the dominant path.

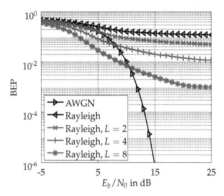

Figure 15. BEP in the *Rayleigh* channel with SLC

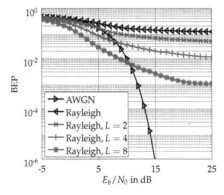

Figure 16. BEP in the *Rayleigh* channel with SLS

3.5. Frequency selective fading channel

Now we focus on the frequency selectivity of a channel and the effect on the SEP for an energy detection receiver. The received signal $\boldsymbol{R}$ is in the baseband equivalent model:

$$R = \underbrace{xH}_{S} + W,$$

where $\boldsymbol{H} = (H_1, \ldots, H_L)^T$ denotes the random channel impulse response, circular symmetric complex white Gaussian noise $\boldsymbol{W} \sim \mathcal{CN}_L\left(0, \sigma_n^2 \mathbf{I}\right)$ and $\boldsymbol{x}$ denotes the convolution matrix, containing shifted versions of the transmitted signal:

$$\mathbf{x} = \begin{pmatrix} s_1 & 0 & \ldots & 0 \\ s_2 & s_1 & \ldots & 0 \\ \vdots & \vdots & \ddots & \vdots \\ s_L & s_{L-1} & \ldots & s_1 \end{pmatrix}.$$

The effect of intersymbol interference has not been taken into account by choosing a gap between symbols larger than the channel delay spread. Using the quadratic form of $\boldsymbol{S}$ the SNR Γ at the receiver can be written as:

$$\Gamma = \frac{E_R}{N_0} = \frac{S^H S}{4 N_0 B}.$$

3.5.1. NLOS fading channel

In an environment with no line of sight between transmitter and receiver, the expectation of the random samples of the channel impulse response $\boldsymbol{H}$ is zero:

$$H \sim \mathcal{CN}_D\left(0, \Sigma\right) \quad \text{with } \Sigma(i,j) = h_i h_j e^{-\frac{|i-j|}{\beta}},$$

where $\beta > 0$ describes the correlation of the elements in the covariance matrix Σ. If the samples of the channel impulse response are uncorrelated, Σ is a diagonal matrix. The expectation the noise free received signal is also zero:

$$S = xH \sim \mathcal{CN}_D\left(0, \Sigma_s\right) \quad \text{with } \Sigma_s = x\Sigma x^H.$$

Let $\Sigma_s^{\frac{1}{2}}$ denote a matrix that fulfils $\Sigma_s^{\frac{1}{2}}\Sigma_s^{\frac{1}{2}} = \Sigma_s$, then a whitened vector $\mathbf{S}' \sim \mathcal{CN}_L(0, \mathbf{I})$ is defined as

$$\Sigma_s^{-\frac{1}{2}}\mathbf{S} = \mathbf{S}'. \tag{45}$$

Note that the existence of $\Sigma_s^{\frac{1}{2}}$ is guaranteed for any positive definite matrix Σ_s. To calculate the distribution of instantaneous SNR Γ, we need to analyse the distribution of $\mathbf{S}^H\mathbf{S}$ first. Using (45) yields

$$\mathbf{S}^H\mathbf{S} = \mathbf{S}'^H\Sigma_s\mathbf{S}'. \tag{46}$$

Using eigenvalue decomposition and special properties of the central χ^2-distribution, a closed form expression for the PDF of $\mathbf{r}^H\mathbf{r}$ can be found [21]. Performing an eigenvalue decomposition of Σ_S in (46) leads to

$$\mathbf{S}^H \mathbf{S} = \mathbf{S}'^H \mathbf{U}^H \mathbf{\Lambda} \mathbf{U} \mathbf{S}'$$

with $\mathbf{\Lambda}$ being a diagonal matrix containing the eigenvalues $\lambda_1, \ldots, \lambda_L$ of Σ_s and the rows of unitary matrix $\mathbf{U}$ being the corresponding eigenvectors. Substituting $\mathbf{G} = \mathbf{U}\mathbf{S}'$, the random variable V is:

$$V = \mathbf{S}^H \mathbf{S} = \mathbf{G}^H \mathbf{\Lambda} \mathbf{G} = \sum_{l=1}^{L} \lambda_l G_l G_l^* \tag{47}$$

with G_l representing the elements of $\mathbf{G}$ and G_l^* denoting the complex conjugate of G_l. Note that $\mathbf{G}$ and $\mathbf{H}$ are equally distributed because the rows of $\mathbf{U}$ are orthonormal among each other. Therefore

$$\mathbf{G} \sim \mathcal{CN}_L (0, \Sigma_s).$$

The random variable $G_l' = G_l G_l^* = |G_l|^2$ is central χ^2 distributed with two degrees of freedom. That is a special case of the central χ^2 distribution which is equivalent to an exponential distribution. As a linear combination of independent and identically exponential distributed variates, V is general *Gamma* or general *Erlang* distributed with the PDF given as [21, eq. 19.147]:

$$f_V(v) = \sum_{j=1}^{D} \left(\prod_{k \neq j} \left(\lambda_j - \lambda_k \right)^{-1} \right) \lambda_j^{D-2} e^{-\frac{v}{\lambda_j}}. \tag{48}$$

The average SNR $\bar{\gamma}$ is based on the expectation of the random variable V:

$$\bar{\gamma} = \frac{\mathbb{E}(V)}{4N_0 B}. \tag{49}$$

Because all random variables G_i are independent $\mathbb{E}(V)$ can be written as:

$$\mathbb{E}(V) = \sum_{i=1}^{D} \lambda_i \mathbb{E}(|G_i|^2) = 2 \sum_{i=1}^{D} \lambda_i \quad \text{with } \mathbb{E}(|G_i|^2) = 2.$$

Introducing a change of variable in (48) yields:

$$f_{\Gamma}(\gamma) = 2N_0 B \, f_V (2N_0 B \gamma). \tag{50}$$

Using the relative SNR $\gamma' = \frac{\gamma}{\bar{\gamma}}$, (50) can be expressed as:

$$f_{\Gamma'}(\gamma') = 2N_0 B \bar{\gamma} \, f_V (2N_0 B \bar{\gamma} \gamma'). \tag{51}$$

The SEP in a fading channel can be calculated using (20). Combining (20) and (51) yields to the following integral:

$$2N_0 B \sum_{j=1}^{D} \left(\prod_{k \neq j} \left(\lambda_j - \lambda_k \right)^{-1} \right) \lambda_j^{D-2} \int_0^{\infty} Q_D \left(\sqrt{a_m^2 \bar{\gamma} \gamma'}, \sqrt{\rho_m} \right) e^{-\frac{2N_0 B}{b \lambda_j} \bar{\gamma} \gamma'} \, d\gamma'.$$

In the case of partial CSI, the integral can be solved using [40, eq. 12] with adequate substitutions:

$$\Theta_D(\bar{\gamma}, N_0, \boldsymbol{\lambda}, B, a_m, \rho_m) = \frac{1}{\bar{\gamma}} \sum_{j=1}^{D} \left(\prod_{k \neq j} (\lambda_j - \lambda_k)^{-1} \right) \lambda_j^{D-2} e^{-\frac{\rho_m}{2}} \left\{ \left(1 + \frac{2N_0 B}{a_m^2 \lambda_j} \right)^{D-1} \right.$$
$$\left. \cdot \left[\exp \left(\frac{\rho_m}{2} \frac{1}{1 + \frac{2N_0 B}{a_m^2 \lambda_j}} \right) - \sum_{n=0}^{D-2} \frac{1}{n!} \left(\frac{\rho_m}{2} \frac{1}{1 + \frac{2N_0 B}{a_m^2 \lambda_j}} \right)^n \right] + \sum_{n=0}^{D-2} \frac{1}{n!} \left(\frac{\rho_m}{2} \right)^n \right\}. \quad (52)$$

Using the function Θ_D (52), the average SEP in a frequency selective fading channel can be expressed in closed form [3]:

$$\bar{P}_e(\bar{\gamma}, N_0, B, \boldsymbol{a}, \boldsymbol{\rho}, \boldsymbol{\lambda}, M, D) = \frac{1}{M} \left[M - 1 + \mathcal{Q}_D \left(0, \sqrt{\rho_1} \right) \right.$$
$$+ \sum_{m=1}^{M-1} \Theta_D(\bar{\gamma}, N_0, \boldsymbol{\lambda}, B, a_m, \rho_m)$$
$$\left. - \sum_{m=1}^{M-2} \Theta_D(\bar{\gamma}, N_0, \boldsymbol{\lambda}, B, a_m, \rho_{m+1}) \right].$$

This equation is only valid if two constraints are met:

1. The receiver may only use $\bar{\rho}$ to determine the decision threshold (partial CSI) and
2. all eigenvalues λ_j are pairwise disjunct.

If there exist two or more identical eigenvalues, one might rearrange (47) and decrease D used in the subsequent calculations to meet the second constraint. Figure 17 shows the BEP in the frequency selective fading channel with i.i.d. and correlated channel gains.

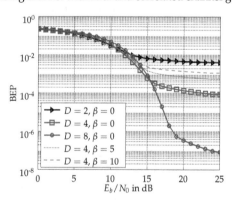

Figure 17. BEP for OOK in the frequency selective fading channel

3.5.2. LOS fading channel

In a channel model with LOS, **H** is expected to have a non-zero mean vector **M**. Therefore V is equal in distribution to a linear combination of noncentral χ^2 distributed variates. Unfortunately, there is no such convenient correspondence for that distribution as the general *Erlang* distribution for V in section 3.5.1.

Even though the PDF of a linear combination of noncentral χ^2 distributed variates is given in [48], there is no known closed-form expression for the integral of the product of this PDF and the MARCUM-Q-function in (20).

The average SEP may still be computed using a discrete approximation of the PDF of V (histogram). This histogram can be generated with arbitrary precision by sampling the PDF given in [48] or by performing a Monte-Carlo-simulation based on the distribution of **H**. For the Monte-Carlo-simulations the samples of the LOS channel impulse response **H** follow a normal distribution

$$\mathbf{H} \sim \mathcal{CN}_L\left(\mathbf{M}, \mathbf{\Sigma_h}\right)$$

with the mean $\mathbf{M} = (M_1, \ldots, M_D)$ with $M_i = \sqrt{\frac{\kappa}{\kappa+1}}\ \forall\ i$ and the covariance matrix $\mathbf{\Sigma_h} = \frac{1}{\kappa+1}\mathbf{I}$.

4. Interference investigations for non-coherent multiband UWB

As the unlicensed MIR-UWB system has no exclusive frequency range there is an increased interference potential from present and from future radio systems operating in the same frequency range. Hence, the performance of the MIR-UWB system can be reduced and a reliable communication cannot be guaranteed at any time. For this reason it is imperative to realise an efficient and low complex interference mitigation.

Section 4 is structured as follows: In section 4.1 an analysis of the interference robustness of an energy detection receiver is presented. This allows the identification of suitable MIR-UWB system parameters which can be configured preferably robust against interferences before initial operation of the MIR-UWB system. The following section 4.2 deals with coexistence-based approaches which are focused on an efficient and adaptive interference mitigation with low complexity. As the mitigation of narrowband interference (NBI) is a crucial issue of the MIR-UWB section 4.3 analyses the non-linear Teager-Kaiser (TK) operation. Thereby, the potential to mitigate NBI and to integrate the TK operation into the existing MIR UWB system is shown.

4.1. Interference robustness of energy detection

A basic issue of the MIR-UWB's energy detection receiver is its high sensitivity with respect to interferences passing the analogue front-end. A significant reduction of the instantaneous Signal-to-Interference-and Noise Ratio (SINR) can occur so that a reliable communication is not guaranteed.

For this reason it is required to investigate the interference robustness of an OOK and BPPM specific energy detection [7, 9]. The analysis bases on an analytical investigation of the interference robustness of an energy detector within an arbitrary but fixed subband. Thereby, dependencies between system- and interference specific parameters can be identified which promise an increase of interference robustness. The analysis bases on one hand on [53]. Therein, the performance of a BPPM specific energy detector is analysed in presence of out-of-band interference. On the other hand it relies on [60] in which the performance of a BPPM specific correlation receiver is investigated in presence of interference.

4.1.1. Signal model

Binary data transmission within an MIR-UWB subband of bandwidth B is considered. Thereby, based on OOK/BPPM the rectangular pulse

$$p(t) = \begin{cases} \sqrt{\frac{2}{T_p}} \cos\left(2\pi f_c t\right) & ,0 < t < T_p \\ 0 & ,\text{else} \end{cases} \tag{53}$$

with carrier frequency f_c and pulse duration T_p is emitted with energy 1 ($f_c \gg 1/T_p$). The resulting signal to be transmitted conducts to

$$s_O(t) = \sqrt{E_p^O} \sum_{n=-\infty}^{\infty} b_n p_i \left(t - nT_b\right) \tag{54}$$

for OOK and

$$s_P(t) = \sqrt{E_p^P} \sum_{n=-\infty}^{\infty} p_i \left(t - nT_b - b_n \frac{T_b}{2}\right) \tag{55}$$

for BPPM. The uniformly distributed data bit $b_n \in \{0,1\}$ is characterised by bit energy E_b as well as by bit duration $T_b = \frac{T_p}{d_s}, d_s > 0$ with duty cycle $d_s \leq \frac{1}{2}$. Finally, $E_p^i, i \in \{O,P\}$ stands for the modulation specific pulse energy which equals $E_p^O = 2E_p^P = 2E_b$.

Assuming perfect synchronisation between the transmitter and the receiver, the signal $s_i(t), i \in \{O,P\}$ is superposed with zero mean white Gaussian noise $n(t)$ of two-sided spectral density $\frac{N_0}{2}$ and interference $j(t)$ leading to (no fading)

$$y_i(t) = s_i(t) + n(t) + j(t). \tag{56}$$

Interference is described as band-limited wide-sense stationary, time-continuous zero mean Gaussian process $J(t)$ characterised by the autocorrelation function ($\tau = t_1 - t_2$)

$$R_J(\tau) = P_J \frac{\sin\left(\pi B_J \tau\right)}{\pi B_J \tau} \cos\left(2\pi f_J \tau\right). \tag{57}$$

It depends on the mean interference power P_J determined by the ratio of the interferer's bit energy $E_{b,J}$ and bit duration $T_{b,J} = qT_b, q > 0$. Further parameters are the interference center frequency f_J as well as its bandwidth B_J.[1] The resulting interferer's signal duration $T_{p,J} \approx \frac{1}{B_J} \leq T_{b,J}$ leads to an interference duty cycle of $d_J = \frac{T_{p,J}}{T_{b,J}} = \frac{d_s T_{p,J}}{qT_p}$.

---

[1] Eq. (57) holds if the interference source is completely inside the MIR-UWB subband. In case B_J overlaps completely or only partially with the subband f_J, B_J and P_J have to be properly modified. However, as it can be ascribed to (57) the following investigations focus solely on an interference source being completely inside the subband.

The received signal of (56) is first bandpass filtered and afterwards subjected to non-coherent energy detection with integration time T_p. At its input stage the available SINR is given by

$$\text{SINR}_E = 10 \log_{10} \frac{E_b}{T_b \left(P_J + P_N \right)}, \tag{58}$$

which is identical for OOK/BPPM. In (58) P_N stands for the mean noise power of the passband noise signal, which is modelled as band-limited wide-sense stationary, time-continuous zero mean Gaussian process $N(t)$: $(\tau = t_1 - t_2)$

$$R_N(\tau) = P_N \frac{\sin(\pi B \tau)}{\pi B \tau} \cos(2\pi f_c \tau). \tag{59}$$

At the output of energy detection the decision variable differs for OOK/BPPM.

Energy detection of OOK
For OOK, the asymmetric decision variable

$$x^O = \int_0^{T_p} y^2(t)\, dt = x_s^O + \Delta x^O \tag{60}$$

occurs. The resulting energy value x^O consists of a deterministic signal-only part

$$x_s^O = \begin{cases} 0 & , b_n = 0 \\ 2E_b & , b_n = 1 \end{cases} \tag{61}$$

of mean E_b and second order moment $2E_b^2$. The component $\Delta x^O = x_{sjn}^O + x_{jn}^O$ contains the mixed signal-noise and signal-interference term

$$x_{sjn}^O = \begin{cases} 0 & , b_n = 0 \\ 2\sqrt{\frac{2E_b^O}{T_p}} \int_0^{T_p} \cos(2\pi f_c t)\,(n(t) + j(t))\, dt dt & , b_n = 1 \end{cases} \tag{62}$$

as well as the contribution

$$x_{jn}^O = \int_0^{T_p} (n(t) + j(t))^2\, dt \quad , b_n = 0, 1 \tag{63}$$

due to noise and interference-only.

Energy detection of BPPM
In contrast to OOK the BPPM specific decision variable at the output of energy detection is symmetric:

$$x^P = \int_0^{T_p} y_P^2(t)\, dt - \int_{\frac{T_b}{2}}^{\frac{T_b}{2} + T_p} y_P^2(t)\, dt = x_s^P + \Delta x^P. \tag{64}$$

The decision variable x^P compares energy values within two observation intervals of duration T_p. [2] It is composed of a signal-only contribution

$$x_s^P = \begin{cases} E_b & , b_n = 0 \\ -E_b & , b_n = 1, \end{cases} \tag{65}$$

which is characterised by mean zero and second order moment E_b^2. The additional term $\Delta x^P = x_{sjn}^P + x_{jn}^P$ is on one hand composed of a mixed signal-noise and signal-interference component

$$x_{sjn}^P = \begin{cases} a \int\limits_0^{T_p} \cos\left(2\pi f_c t\right) \left(n\left(t\right) + j\left(t\right)\right) dt & , b_n = 0 \\ -a \int\limits_{\frac{T_p}{2}}^{\frac{T_p}{2}+T_p} \cos\left(2\pi f_c \left(t - \frac{T_p}{2}\right)\right) \left(n\left(t\right) + j\left(t\right)\right) dt & , b_n = 1 \end{cases} \tag{66}$$

with $a = 2\sqrt{2E_p^P/T_p}$. On the other hand it consists of the noise and interference-only part

$$x_{jn}^P = \int\limits_0^{T_p} \left(n\left(t\right) + j\left(t\right)\right)^2 dt - \int\limits_{\frac{T_p}{2}}^{\frac{T_p}{2}+T_p} \left(n\left(t\right) + j\left(t\right)\right)^2 dt. \tag{67}$$

4.1.2. Statistical analysis of interference robustness

To make statements on the interference robustness of an OOK and BPPM specific energy detection a proper quality criterion has to be introduced. A possible measure is the processing gain (PG) of the energy detector. It refers the available SINR at its output to the SINR_E at its input. For OOK this can be described as

$$\text{PG}^O = 10\log_{10}\left(\frac{2E_b^2}{0,5Q_1^O + Q_2^O}\right) - 10\log_{10}\left(\text{SINR}_E\right), \tag{68}$$

which differs from the PG of the BPPM based energy detection receiver expressed as

$$\text{PG}^P = 10\log_{10}\left(\frac{E_b^2}{Q_1^P + Q_2^P}\right) - 10\log_{10}\left(\text{SINR}_E\right). \tag{69}$$

In (68) and (69) $Q_1^i, i \in \{O,P\}$ stands for the second order moment of the mixed signal-noise and signal-interference component $x_{sjn}^i, i \in \{O,P\}$. In contrast, $Q_2^i, i \in \{O,P\}$ describes the second order moment of the noise and interference-only part $x_{jn}^i, i \in \{O,P\}$.

Based on PG, separate statements on the detection performance can be made for each modulation scheme, i.e., a low modulation specific PG indicates an increased error probability

---

[2] It is assumed that the pulse of duration $T_p \leq \frac{T_p}{2}$ occurs at the beginning of an interval of duration $\frac{T_p}{2}$. Hence, the position of a pulse within the interval is perfectly known.

and vice versa. Hence, the smaller Q_1^i and $Q_2^i, i \in \{O, P\}$ the lower the modulation related error detection probability. In the following, Q_1^i and $Q_2^i, i \in \{O, P\}$ are determined for both modulation schemes.

OOK: For OOK the second order moment of the signal-noise and signal-interference part x_{sjn}^O can be formulated as ($\tau = t_1 - t_2$)

$$Q_1^O = \int\limits_{-t_1}^{T_p - t_1} \int\limits_{0}^{T_p} (R_J(\tau) + R_N(\tau)) \cdot p(t_1) \cdot p(t_1 + \tau) \, dt_1 d\tau. \tag{70}$$

A general solution of Q_1^O can be found using Parseval's theorem under the assumptions $2|f_c + f_J| \gg B_J$ and $4f_c \gg B$. This leads to the closed-form expression

$$
\begin{aligned}
Q_1^O = E_p^O P_J \sum_{n=0}^{\infty} (-1)^n (2\pi T_p)^{2n} & \left(\frac{r_{2n+1} \sum_{l=0}^{2n+1} u_{n,l}}{2\pi f_c (2n+1)} \right. \\
& \left. - \frac{2\pi T_p r_{2n+2} \sum_{l=0}^{2n+2} v_{n,l}}{2\pi f_c (2n+2)} + \frac{4 T_p r_{2n+1}}{(2n+2)! (2n+1)} \right) \\
+ \frac{E_p^O P_N}{2\pi f_c} \sum_{n=0}^{\infty} & \frac{(-1)^n (\pi T_p B_T)^{2n}}{(2n+1)} \left(\frac{8\pi T_p f_c}{(2n+2)!} + (1+ \right. \\
& \left. \cos(4\pi f_c T_p)) \sum_{l=0}^{2n+1} w_{n,l} - \sin(4\pi f_c T_p) \sum_{l=0}^{2n+1} z_{n,l} \right),
\end{aligned} \tag{71}
$$

whereas, with $\Delta_{f_{cJ}} = f_c - f_J$, the following notations are used:

$$r_\nu = \frac{1}{B_J} \left(\left(\frac{B_J}{2} + \Delta_{f_{cJ}} \right)^\nu - \left(-\frac{B_J}{2} + \Delta_{f_{cJ}} \right)^\nu \right),$$

$$w_{n,l} = \frac{\sin\left(4\pi f_c T_p + \frac{1}{2}l\pi\right)}{(2n+1-l)! \, (4\pi f_c T_p)^l},$$

$$u_{n,l} = w_{n,l} + \frac{(-1)^l}{(2n+1-l)!} \sum_{k=0}^{l} \frac{\sin\left(4\pi f_c T_p + \frac{1}{2}k\pi\right)}{(l-k)! \, (4\pi f_c T_p)^k},$$

$$z_{n,l} = \frac{\cos\left(4\pi f_c T_p + \frac{1}{2}l\pi\right)}{(2n+1-l)! \, (4\pi f_c T_p)^l},$$

$$v_{n,l} = \frac{z_{n,l}}{(2n+2-l)} - \frac{(-1)^l}{(2n+2-l)!} \sum_{k=0}^{l} \frac{\cos\left(4\pi f_c T_p + \frac{1}{2}k\pi\right)}{(l-k)! \, (4\pi f_c T_p)^k}.$$

Eq. (71) shows that Q_1^O is depending from the system parameters E_p^O, T_p, f_c, B as well as from the interference parameters P_J, B_J, f_J. In addition, concerning the special case $B_J \to 0$, e.g., a cosine tone, r_ν has to be replaced with $r_\nu^m = \lim\limits_{B_J \to 0} r_\nu = \nu \Delta_{f_{cJ}}^{\nu-1}$. Note that this result is consistent to [9] if $P_N = 0$. In this case (71) simplifies to

$$
Q_1^O = \frac{E_p^O P_J}{T_p \pi^2 \left(f_c^2 - f_J^2 \right)^2} \left[f_c^2 + 3f_J^2 + \left(f_J^2 - f_c^2 \right) \cos \left(4\pi f_c T_p \right) \right.
$$
$$
- 2f_J \left(f_J + f_c \right) \cos \left(2\pi \left(f_c - f_J \right) T_p \right)
$$
$$
\left. - 2f_J \left(f_J - f_c \right) \cos \left(2\pi \left(f_c + f_J \right) T_p \right) \right].
\tag{72}
$$

The second order moment of the noise and interference-only part x_{jn}^O can be generally described as

$$
Q_2^O = \int_{-t_1}^{T_p - t_1} \int_0^{T_p} \left[\left(P_N + P_J \right)^2 + 2 \left(R_N \left(\tau \right) + R_J \left(\tau \right) \right)^2 \right] dt_1 d\tau.
\tag{73}
$$

Thereby, using the theorem of Price [44], (73) can be written in terms of the noise and interference related autocorrelation functions. With Parseval and the assumptions $2f_c \gg B$, $2f_J \gg B_J$ and $|f_c + f_J| \gg \left(B_J \text{ or } \left(B - B_J \right) \right)$ (73) results in

$$
Q_2^O = 2T_p^2 \left[P_J^2 + P_J P_N + P_N^2 + \sum_{k=1}^{\infty} \frac{(-1)^k \left(2\pi T_p \right)^{2k} \left(P_J^2 B_J^{2k} + P_N^2 B_T^{2k} \right)}{(2k+1)! \, (2k+1) \, (k+1)} \right.
$$
$$
\left. + \sum_{k=2}^{\infty} \frac{(-1)^k \left(2\pi T_p \right)^{2k-2} \left(P_J^2 B_J^{2k-2} + P_N^2 B_T^{2k-2} \right)}{(k) \, (2k)!} \right]
$$
$$
+ \frac{2P_J P_N}{\pi B \left(f_p - f_m \right)} \sum_{k=0}^{\infty} \frac{(-1)^k \left(2\pi \right)^{2k+1} T_p^{2k+2}}{(2k+2)!} \cdot \left(\frac{f_1 \left(k \right)}{2k+2} + \frac{f_2 \left(k \right)}{2k+1} \right).
\tag{74}
$$

Thereby, with $f_p = \frac{B}{2} + \frac{B_J}{2}$ and $f_m = \frac{B}{2} - \frac{B_J}{2}$, $f_1 \left(k \right)$ and $f_2 \left(k \right)$ are defined as:

$$
f_1 \left(k \right) = \left(-f_m - \Delta_{f_{cJ}} \right)^{2k+2} - \left(-f_p - \Delta_{f_{cJ}} \right)^{2k+2}
$$
$$
+ \left(-f_m + \Delta_{f_{cJ}} \right)^{2k+2} - \left(-f_p + \Delta_{f_{cJ}} \right)^{2k+2}
$$

and

$$
f_2 \left(k \right) = \left(f_p + \Delta_{f_{cJ}} \right) \left[\left(-f_m - \Delta_{f_{cJ}} \right)^{2k+1} - \left(-f_p - \Delta_{f_{cJ}} \right)^{2k+1} \right]
$$
$$
+ \left(f_p - \Delta_{f_{cJ}} \right) \left[\left(-f_m + \Delta_{f_{cJ}} \right)^{2k+1} - \left(-f_p + \Delta_{f_{cJ}} \right)^{2k+1} \right]
$$
$$
+ \left(f_p - f_m \right) \left[\left(f_m - \Delta_{f_{cJ}} \right)^{2k+1} - \left(-f_m - \Delta_{f_{cJ}} \right)^{2k+1} \right].
$$

Q_2^O is influenced by the system parameters T_p, f_c, B as well as by the interference parameters P_J, B_J, f_J. However, in contrast to the second order moment Q_1^O it cannot be reduced via E_p^O. Eq. (74) simplifies for $B_J \rightarrow 0$ due to $P_J^2 B_J^{2k} = P_J^2 B_J^{2k-2} = 0$, $\frac{f_1(k)}{f_p - f_m} = (2k+2) \left[\left(-\frac{B}{2} - \Delta f_{c,J} \right)^{2k+1} + \left(-\frac{B}{2} + \Delta f_{c,J} \right)^{2k+1} \right]$ as well as $\frac{f_2(k)}{f_p - f_m} = (2k+1) \left(\frac{B}{2} + \Delta f_{c,J} \right) \left(-\frac{B}{2} - \Delta f_{c,J} \right)^{2k}$. Assuming $P_N = 0$ (74) equals the result of [9]:

$$Q_2^O = 2T_p^2 P_J^2 + \frac{P_J^2}{8\pi^2 f_J^2} \left[1 - \cos\left(4\pi f_J T_p \right) \right] . \tag{75}$$

BPPM: Considering BPPM the second order moment of the signal-noise and signal-interference part x_{sjn}^P is: $Q_1^P = \frac{1}{2} Q_1^O$. Q_1^P differs from Q_1^O solely in a factor of two which can be ascribed to the reduced modulation specific pulse energy. In contrast to Q_1^P there is a significant difference concerning the second order moment of the noise and interference-only part x_{jn}^P. With the theorem of Price this can be generally described in terms of the noise and interference specific autocorrelation functions: ($\tau = t_1 - t_2$)

$$Q_2^P = 4 \int_{-t_1}^{T_p - t_1} \int_0^{T_p} \left[R_J^2(\tau) + R_N^2(\tau) + R_J(\tau) R_N(\tau) \right] dt_1 d\tau$$

$$- 4 \int_{-t_1}^{T_p - t_1} \int_{\frac{T_b}{2}}^{T_p + \frac{T_b}{2}} \left[R_J^2(\tau) + R_N^2(\tau) + R_J(\tau) R_N(\tau) \right] dt_1 d\tau. \tag{76}$$

Therefore, using the theorem of Parseval for $2f_J \gg B_J$, the closed-form result

$$Q_2^P = 2 \sum_{k=1}^{\infty} \frac{(-1)^k (2\pi)^{2k} \left(P_J^2 B_J^{2k} + P_N^2 B_T^{2k} \right) g_{2k+2}}{(2k+1)! \, (2k+1) \, (k+1)}$$

$$+ \sum_{k=2}^{\infty} \frac{(-1)^k 2^{2k} (\pi)^{2k-2} \left(P_J^2 B_J^{2k-2} + P_N^2 B_T^{2k-2} \right) g_{2k}}{(2k)! \, (2k)}$$

$$+ \frac{2 P_J P_N}{\pi B_T \left(f_p - f_m \right)} \sum_{k=0}^{\infty} \frac{(-1)^k (2\pi)^{2k+1} g_{2k+2}}{(2k+2)!} \cdot \left(\frac{f_1(k)}{2k+2} + \frac{f_2(k)}{2k+1} \right) \tag{77}$$

with

$$g_v = 2T_p^v - \left(T_p - \frac{T_b}{2} \right)^v + 2 \left(\frac{T_b}{2} \right)^v - \left(T_p + \frac{T_b}{2} \right)^v$$

can be found. Q_2^P is influenced by the system parameters T_p, T_b, f_c, B as well as by the interference parameters P_J, B_J, f_J. Similar to Q_2^O it cannot be reduced via E_p^P. Eq. (77) reveals that for low data rates ($T_b \rightarrow \infty$) $g_v \approx 2T_p^v$ resulting in a negligible influence of Q_2^P. In contrast, the larger the data rate the higher its impact, e.g., for $T_b = 2T_p$ g_v conducts to $g_v = 4T_p^v - (2T_p)^v$. In addition, for $B_J \rightarrow 0$ and $P_N = 0$, (77) allows the same simplifications as for

Q_2^O. In this case (77) conducts to

$$Q_2^P = \frac{P_J^2}{8\pi^2 f_J^2} \left[2 - 2\cos\left(2\pi f_J T_b\right) + \cos\left(2\pi f_J \left(T_b - 2T_p\right)\right) \right.$$
$$\left. - 2\cos\left(4\pi f_J T_p\right) + \cos\left(2\pi f_J \left(T_b + 2T_p\right)\right) \right]. \tag{78}$$

4.1.3. Results

Based on the previous analysis the interference robustness of an OOK/BPPM based energy detection receiver can be identified. Thereby, assuming regulation of ECC [14] an MIR-UWB system with four subbands of equal bandwidth $B = 625\,\mathrm{MHz}$ is taken into account. Without loss of generality, the analysis focuses solely on the first subband located at $f_c = 6.3125\,\mathrm{GHz}$. However, an extension to other subbands or other MIR-UWB system configurations, which are possibly based on other frequency masks is easily possible. Further common system parameters used in the following are the pulse duration $T_p = 3.2\,\mathrm{ns}$, a mean transmit power normalized to one, the modulation specific pulse energy $E_p^i, i \in \{O, P\}$ as well as a constant $\mathrm{SNR_E} = 10\,\mathrm{dB}$ at the input of the energy detector. Fixed interference parameter is the interference specific bit duration $T_{b,J} = 16 T_b = 102.4\,\mathrm{ns}$.

In Fig. 18 (a) the PG is plotted vs. the $\mathrm{SINR_E}$ for the duty cycle $d_s = \frac{1}{2}$. An interference with the two bandwidths $B_{J,1} = 20\,\mathrm{MHz}$ or $B_{J,2} = 400\,\mathrm{MHz}$ located at $f_J = f_c + 50\,\mathrm{MHz}$ is considered leading to the fixed duty cycles $d_{J,1} = \frac{1}{B_{J,1} T_{b,J}} = 0.4883$ or $d_{J,2} = \frac{1}{B_{J,2} T_{b,J}} = 0.0244$. For OOK/BPPM, the PG increases with higher $\mathrm{SINR_E}$ up to the interference-free PG at $\mathrm{SNR_E} = 10\,\mathrm{dB}$. Furthermore, it can be observed that the OOK/BPPM based PG varies with the interference bandwidth. For OOK, the PG increases with a larger interference bandwidth because of the minor impact of the mixed signal-interference as well as the interference-only component involved in the energy detection. A PG of energy detection can be achieved from an $\mathrm{SINR_E} = -3.5\,\mathrm{dB}$ ($B_{J,1} = 20\,\mathrm{MHz}$) and from an $\mathrm{SINR_E} = -5.5\,\mathrm{dB}$ ($B_{J,2} = 400\,\mathrm{MHz}$), respectively. For strong interference no PG results as the energy detector's decision variable is significantly corrupted. In contrast, for BPPM a PG can be achieved for small interference bandwidths, e.g., $B_{J,1} = 20\,\mathrm{MHz}$, over nearly the complete $\mathrm{SINR_E}$ range. For $B_{J,2} = 400\,\mathrm{MHz}$ a PG occurs from $\mathrm{SINR_E} = -2\,\mathrm{dB}$. The reason for this behaviour lies in a different amount of energy resulting from the mixed signal-interference and interference-only term within the two observation periods of duration T_p. Finally, the consideration of OOK/BPPM with respect to their relative PG shows that for strong NBI BPPM is more robust whereas OOK is more robust for mean and low interference.

The detector efficiency in terms of PG can be increased by increasing T_b, e.g., with multiples of T_p. This is illustrated in Fig. 18 (b) for $d_s = \frac{1}{4}$. The enlargement of T_b via d_s results in an increase of E_b for fixed signal power. Thereby, the interference related second order moments Q_1^i and $Q_2^i, i \in \{O, P\}$ will be reduced as the interferer's energy is only collected during integration time T_p within T_b. Larger T_b can be implemented into an MIR-UWB transmitter with minor complexity. However, the trade-off to increase the detector's interference robustness is a reduction of data rate.

Fig. 19 (a) shows the PG of an OOK and BPPM specific energy detection vs. f_J for $d_s = \frac{1}{2}$, which varies from $f_c - \frac{B}{2}$ to $f_c + \frac{B}{2}$ for fixed $\mathrm{SINR_E} = 0\,\mathrm{dB}$, $\mathrm{SNR_E} = 10\,\mathrm{dB}$, $T_{b,J} = 102.4\,\mathrm{ns}$ as

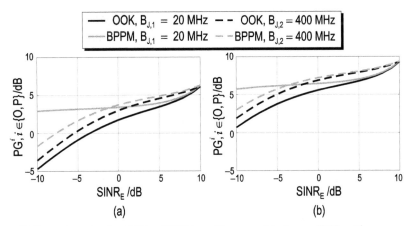

Figure 18. Processing gain of an OOK and BPPM specific energy detection vs. SINR_E with $B_{J,1} = 20\,\text{MHz}$ and $B_{J,2} = 400\,\text{MHz}$, $f_J = f_c + 50\,\text{MHz}$, $\text{SNR}_E = 10\,\text{dB}$, $T_{b,J} = 102.4\,\text{ns}$, $d_J\,(B_{J,1}) = 0.4883$ and $d_J\,(B_{J,2}) = 0.0244$ for $T_b = 2T_p$ (a) and $T_b = 4T_p$ (b).

well as for $B_{J,1} = 20\,\text{MHz}$ and $B_{J,2} = 400\,\text{MHz}$. As long as the interference is completely inside the subband $T_{p,J}$, d_J and hence P_J are fix. In particular, the modulation specific PG at $f_J = f_c + 50\,\text{MHz}$ coincides with the one of Fig. 18. In addition, both modulation schemes show an increase of PG the more the interference is located at the subband's boundary ($f_J = f_c \pm \frac{1}{T_p}$). This can be on one hand ascribed to the subband pulse's sinc spectrum which is zero at the subband's boundary. On the other hand, the more f_J is located at the subband's boundary the minor the interference bandwidth falling into the subband. In case interference overlaps with the subband's boundary, the effective interference parameters B_J, f_J and d_J changes resulting in a reduction of the actual interference power P_J. Fig. 19 (b) shows again that a significant increase of the modulation specific PG can be achieved for $d_s = \frac{1}{4}$. However, without proper pulse shaping the increase depends strongly on the interferer's position inside the subband.

4.2. Coexistence-based approaches

The MIR-UWB system has no exclusive frequency range within the available transmission bandwidth. For this reason there is an increased interference potential from a possible large number of radio systems operating in the same frequency domain. As shown in [12] the impact of interference can result in a significant decrease of the bit error rate (BER) performance of the MIR-UWB system. To maintain system performance, the following section gives a short overview of various coexistence-based approaches with respect to an efficient and easy-to-realise interference mitigation.

Coexistence-based approaches aim at the reliable on-line mitigation of interferences which occur in the environment of the MIR-UWB system. Thus, a best possible trade-off between a maximum data rate and a minimum BER can be obtained for arbitrary interference situations. In this context, an essential requirement is the integration of coexistence-based approaches into the existing MIR-UWB system with only minor complexity increase. Thus, the MIR-UWB system configuration should not be changed in presence of interference on one hand; on

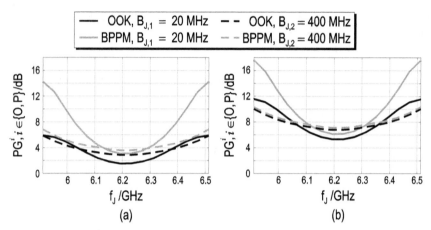

Figure 19. Processing gain of an OOK and BPPM specific energy detection vs. f_J for $SINR_E = 0\,dB$, $SNR = 10\,dB$, $B_{J,1} = 20\,MHz$, $B_{J,2} = 400\,MHz$ and $T_{b,J} = 102.4\,ns$ for $T_b = 2T_p$ (a) as well as $T_b = 4T_p$ (b).

the other hand it should be possible to realise coexistence without complex estimations of interference specific parameters such as the interference power, the interference bandwidth, the interference carrier frequency or the number of instantaneous available interferences.

4.2.1. Static coexistence approach

From a complexity point of view a static coexistence approach should be used. Thereby interfered subbands are deactivated by means of the system specific band plan before any data transmission occurs. Best trade-off between system performance and effort for interference handling will be achieved. However, this approach does not consider dynamic interference situations and hence does not contribute to efficient spectrum usage. Therefrom, the necessity for further efficient and low complex, but more flexible alternatives arises.

4.2.2. Detect and Avoid (DAA)

In [6, 34] an easy-to-realise DAA approach is presented allowing a reliable detection of temporary NBI after system initialisation or within data transmission.

For this purpose, the regularly transmitted preamble is adjusted to simultaneous subband specific signal and noise energy estimation used by the DAA approach. Thereby, a static interference-free working point (WP) is defined, ensuring a determinated BER in each subband. After initialisation, the estimated SNRs are compared with the WP, leading to subband deactivation, if the SNR is lower than the WP. Otherwise, the subband is (re)enabled for data transmission.

During data transmission, the initially estimated signal and noise energy values of enabled subbands are recursively updated in dependence of the actual subband specific bit decision to adapt the initial decision threshold. In addition, this process allows the possibility of fast and reliable detection of suddenly weak or strong interfered subbands. For this reason, a characteristic weighting factor k is used to manipulate the instantaneous energy value's influence on the recursive estimation. Simulation results [6] show a robust interference detection with only a marginal BER performance loss for $k = 10$.

4.2.3. Image-based thresholding

In [10], a simple cluster-based coexistence approach (coexistence approach 1) is analysed resulting in a decision threshold ϵ_{th} with respect to interfered or not interfered noise energy values. Based on the image-based thresholding method of [41] occupied time-frequency slots can be automatically detected and recorded within an extended time-frequency band plan used for initialisation or data transmission.

Assuming knowledge of the interferers' periodicity and perfect synchronisation, m_0 binary zeros are sent over each subband $i \in \{1, \dots, N_{sub}\}$. The measured energy values are written as an energy matrix $\underline{X} = \begin{bmatrix} x_{j,i} \end{bmatrix}, j = 1, \dots, m_0, i = 1, \dots, N_{sub}$ and assigned to their nearest quantized energy levels $E_{min} = \epsilon_1 < \epsilon_2 < \dots < \epsilon_U = E_{max}$. Thereby, E_{min} and E_{max} stands for the minimum and the maximum occurring energy value in $\underline{X}$. The allocation leads to an energy distribution, which is described as

$$p_i = \frac{n_i}{m_0 N_{sub}}, i \in 1, \dots, U. \tag{79}$$

To obtain a separation between interfered and not interfered energy values, two energy classes $C_0 = \{\epsilon_1, \dots, \epsilon_u\}$ and $C_1 = \{\epsilon_{u+1}, \dots, \epsilon_U\}$ can be defined, which include interfered and not interfered energy values (Fig. 20 (a))

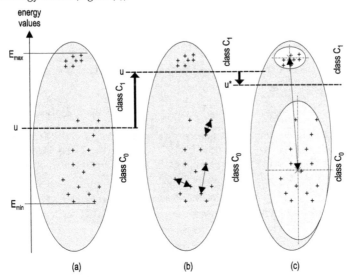

Figure 20. Energy classes, threshold determination. Sub-figure (a) shows the average of all values, sub-figure (b) optimizes according to eq. (81) and sub-figure (c) according to eq. (84).

In order to classify energy values, index u has to be determined using two optimisation criteria.

The first optimisation criterion consists of the minimisation of the combined empirical energy class variance

$$s_w^2(u) = s_{C_0}^2(u) P_{C_0}(u) + s_{C_1}^2(u) P_{C_1}(u). \tag{80}$$

This step aims at an adjustment of the initially set index u, whereas a correct allocation of energy outliers to the corresponding other class is achieved (Fig. 20 (b)). It is based on a weighted sum of the classes' occurrence probabilities

$$P_{C_0}(u) = \sum_{i=1}^{u} p_i, \qquad P_{C_1}(u) = 1 - P_{C_0}(u), \tag{81}$$

as well as on the empirical energy classes' variances

$$s_{C_l}^2(u) = \begin{cases} \sum_{i=1}^{u} \left(\epsilon_i - \bar{x}_{C_l}(u)\right)^2 \frac{p_i}{P_{C_l}(u)}, & l = 0 \\ \sum_{i=u+1}^{U} \left(\epsilon_i - \bar{x}_{C_l}(u)\right)^2 \frac{p_i}{P_{C_l}(u)}, & l = 1 \end{cases} \tag{82}$$

with

$$\bar{x}_{C_l}(u) = \begin{cases} \sum_{i=1}^{u} \frac{\epsilon_i p_i}{P_{C_l}(u)}, & l = 0 \\ \sum_{i=u+1}^{U} \frac{\epsilon_i p_i}{P_{C_l}(u)}, & l = 1 \end{cases} \tag{83}$$

describing the empirical energy class mean levels.

The second optimisation criterion considers the maximisation of the empirical variance between both energy classes

$$s_b^2(u) = \left(\bar{x}_{C_0}(u) - \bar{x}_{tot}\right)^2 P_{C_0}(u) + \left(\bar{x}_{C_1}(u) - \bar{x}_{tot}\right)^2 P_{C_1}(u) \tag{84}$$

standing for the weighted variance of the energy class means $\bar{x}_{C_i}, i = 0, 1$ themselves around the total mean $\bar{x}_{tot} = \bar{x}_{C_0} + \bar{x}_{C_1}$ of the time-frequency pattern (Fig. 20 (c)). Hence, a separation of both classes with respect to the mean value of the total time-frequency pattern is obtained, leading to a more accurate adaptation of index u.

As both optimisation criteria have opposing effects with respect to the best index $u^\star$, they are combined into one characteristic optimisation criterion which is defined as [41]:

$$u^\star = \arg \max_{u=1,\dots,U} \frac{s_b^2(u)}{s_w^2(u)}. \tag{85}$$

This leads to an adjustment of the initially arbitrary index u. Thereby, a correct allocation of energy outliers to the corresponding energy classes is achieved. Evaluated index $u^\star$ leads to interference detection threshold $\epsilon_{th} = \epsilon_{u^\star}$. Afterwards a binary decision has to be done for all received energy values $x_{j,i}, j = 1, \ldots, m_0, i = 1, \ldots, N_{sub}$, which are logged in band plan $\underline{L}$. In this context, we define an energy cell as interfered if $x_{j,i} \geq \epsilon_{th}$.

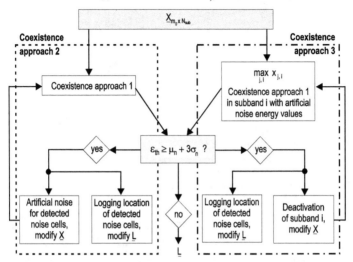

Figure 21. Schematic procedure of global iterative method (coexistence approach 2) and local hierarchical method (coexistence approach 3).

Since the main drawback of the global coexistence approach is its reduced efficiency in case of simultaneously operating broad- or narrowband interference, it should be improved by using iterative coexistence methods [12]. In a global iterative method (coexistence approach 2, left side of Fig. 21) the plausibility of the resulting interference detection threshold is verified via the $3\sigma_n$ standard deviation termination criterion

$$\epsilon_{th} \overset{?}{\geq} \mu_n + 3\sigma_n \tag{86}$$

for which Gaussian distribution as well as an interference-free noise source $n(t)$ with mean μ_n and standard deviation σ_n is assumed at the receiver side. If ϵ_{th} exceeds the confidence interval, deactivation flags are logged within the band plan.

Simultaneously, all labelled possibly corrupted noise energy values are replaced with artificially generated noise energy values resulting from the noise source mentioned above. This procedure is iteratively repeated until the termination criterion is achieved. In this case, the final iterative band plan is delivered and used for initialisation and data transmission.

In contrast, a local hierarchical iterative method (coexistence approach 3, right side of Fig. 21) handles every subband individually to achieve a local based interference detection threshold. The subband with maximum occurring energy value is identified and delivered to receiver side interference detection in conjunction with a sufficient number of additional artificial noise energy values of the above mentioned noise source. For the detection of the coexistence approach's termination, the $3\sigma_n$ standard deviation is used again. If the interference threshold lies above the confidence interval, detected noise cells of the considered subband are labelled

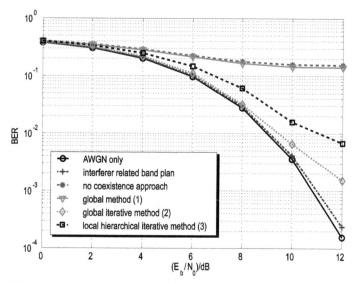

Figure 22. BER vs. E_b/N_0 for AWGN and fixed SIRs of -10 dB (IEEE 802.11a WLAN [47]), 0 dB and 5 dB (IEEE 802.15.3a MB OFDM UWB 1 and 2 [2]) regarding an interferer related band plan, no coexistence approach and the coexistence approaches 1 to 3 [12].

within the band plan. In parallel, the corresponding subband is deactivated and the procedure is iteratively repeated. If the coexistence approach terminates, the binary band plan is allocated to the transmitter for initialisation and data transmission.

Performance analysis [12] demonstrates the high capability of both iterative coexistence methods in presence of multiple interferers having the same or different interference powers (Fig. 22).

4.3. Narrowband interference mitigation

The MIR-UWB system is characterised by a particular high vulnerability to NBI as all interferences inside the passband of the analogue front-end are considered by the energy detection receiver. Its performance can decrease significantly resulting in an increase of the error probability. Hence, a crucial issue concerns the efficient mitigation of NBI [58].

This section analyses the potential of the non-linear Teager-Kaiser (TK) operation [22] to mitigate NBI without the knowledge of the interference related carrier frequency. Based on the definition of the TK operation and some of its most important properties the mitigation potential of a TK based energy detector is analysed. In this context a modified TK operation is introduced and compared with the traditional TK operation [8]. The analysis of the traditional and the modified TK operation considers first one narrowband signal in the baseband. Finally, the analysis is extended to the bandpass domain for one and multiple NBI [5]. Thereby, the potential to integrate the TK operation into the existing MIR-UWB system is discussed for one NBI. Based on the proposal of [42] it is shown that the integration can be realised with only minor complexity increase.

4.3.1. Teager-Kaiser operation

The continuous TK operation is a non-linear differential operator of order two defined as

$$\Psi\left(x\left(t\right)\right) = \dot{x}^2\left(t\right) - x\left(t\right)\ddot{x}\left(t\right). \tag{87}$$

To illustrate the effectiveness of the TK operation the harmonic oscillation $x\left(t\right) = A\cos\left(\omega_0 t + \phi_0\right)$ is considered. Using (87) the signal at the output of the TK operation can be described as

$$\Psi\left(x\left(t\right)\right) = A^2\omega_0^2\left(\sin^2\left(\omega_0 t + \phi_0\right) + \cos^2\left(\omega_0 t + \phi_0\right)\right) = A^2\omega_0^2.$$

Hence, for the special case of a simple harmonic oscillation the TK operation leads to a frequency shift to DC. Fig. 23 highlights the spectrum of a harmonic oscillation (a) as well as the resulting spectrum at the output of TK operation (b).

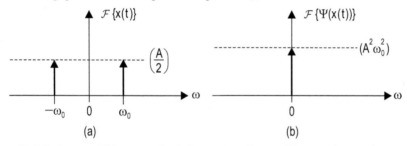

Figure 23. Effectiveness of TK operation for the harmonic oscillation $x\left(t\right) = A\cos\left(\omega_0 t + \phi_0\right)$.

An interpretation of this behaviour is given in [22]. In case of a harmonic oscillation the output of the TK operation describes the required energy to generate the oscillation. Considering (88) the energy depends not only by the amplitude A but also by the oscillation frequency ω_0. Thus, for a constant A the required energy to generate an exemplary 10 Hz signal is lower than the one of an 1000 Hz signal.

When using the continuous TK operation of (87) basic operator specific rules have to be generally considered. A detailed description can be found in [23]. In context with interference mitigation the most important property can be the behaviour of the TK operation in presence of several overlapping signals. In case of K_S overlapping signals $x_1\left(t\right), x_2\left(t\right), \ldots, x_{K_S}\left(t\right)$ the resulting signal at the output of TK operation conducts to

$$\Psi\left(\sum_{i=1}^{K_S} x_i\left(t\right)\right) = \sum_{i=1}^{K_S}\Psi\left(x_i\left(t\right)\right) + \sum_{j=1}^{K_S-1}\sum_{i=j+1}^{K_S}\Psi_c\left(x_j\left(t\right),x_i\left(t\right)\right). \tag{88}$$

It consists on one hand of K_S summands describing the TK operation of the signals $x_1\left(t\right)$, $x_2\left(t\right), \ldots, x_{K_S}\left(t\right)$. Furthermore, due to the non-linearity of the TK operation the additional cross component

$$\begin{aligned}\Psi_c\left(x_j\left(t\right),x_i\left(t\right)\right) &= \Psi_{c_j}\left(x_j\left(t\right),x_i\left(t\right)\right) + \Psi_{c_i}\left(x_i\left(t\right),x_j\left(t\right)\right) \\ &= 2\dot{x}_j\left(t\right)\dot{x}_i\left(t\right) - x_j\left(t\right)\ddot{x}_i\left(t\right) - x_i\left(t\right)\ddot{x}_j\left(t\right) \end{aligned} \tag{89}$$

occurs. This component considers the mutual influence of the two signals $x_j(t)$ and $x_i(t)$. It is composed of the generally non-symmetric signal parts $\Psi_{c_1}(x_1(t), x_2(t)) = \dot{x}_1(t)\dot{x}_2(t) - x_1(t)\ddot{x}_2(t)$ and $\Psi_{c_2}(x_2(t), x_1(t)) = \dot{x}_2(t)\dot{x}_1(t) - x_2(t)\ddot{x}_1(t)$. From (89) two special cases can be immediately concluded. Firstly, the cross component $\Psi_c(x(t), x(t)) = 2\Psi(x(t))$ if $x(t)$ is overlapped with itself. Secondly, $\Psi_c(a, x(t)) = -a\ddot{x}(t)$. The cross component of a constant a and a signal $x(t)$ can be expressed as the product of a with the second derivative of $x(t)$. In particular, the cross component disappears completely, if $a = 0$.

The TK operation of (87) can be modified with the weighting factor $k \neq 0$ as follows [8]:

$$\Psi_k(x(t)) = k\dot{x}(t)^2 - x(t)\ddot{x}(t). \tag{90}$$

This definition contains the traditional TK operation if $k = 1$. Thereby, the Fourier transform of $\Psi_k(x(t))$ of a signal $x(t)$ is generally given as

$$\mathcal{F}\{\Psi_k(x(t))\} = 4\pi^2\left(X(f) * f^2 X(f) - kf X(f) * f X(f)\right). \tag{91}$$

Considering again the special case of a harmonic oscillation $x(t) = A\cos(\omega_0 t)$ the Fourier transform conducts to

$$\mathcal{F}\{\Psi_k(x(t))\} = \frac{1}{2}A^2(2\pi f_0)^2\left((k+1)\delta(f) + \frac{1}{2}(1-k)(\delta(f + 2f_0) + \delta(f - 2f_0))\right). \tag{92}$$

The effectiveness of the modified TK operation is highlighted in Fig. 24.

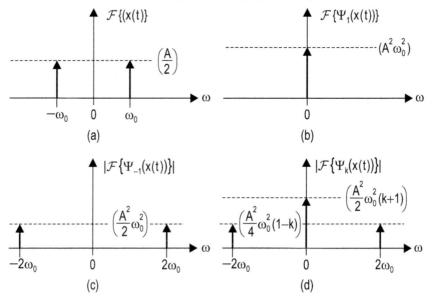

Figure 24. Effectiveness of the traditional and the modified TK operation for $x(t) = A\cos(\omega_0 t)$.

In contrast to the traditional TK operation (Fig. 24 (b)) additional spectral parts at $2\omega_0$ can be observed for $k \neq 1$ (Fig. 24 (d)). For the special case $k = -1$ the complete energy is even shifted to this frequency (Fig. 24 (c)). Hence, it can be concluded that a modified TK operation operates not only as DC frequency shifter.

Due to the modification of the TK operation the property of overlapping signals changes. If two signals $x_1(t)$ and $x_2(t)$ occur at the input of the TK operation its output conducts to [8]:

$$\Psi_k(x_1(t) + x_2(t)) = \Psi_k(x_1(t)) + \Psi_k(x_2(t)) + \Psi_k^c(x_1(t), x_2(t)), \qquad (93)$$

where

$$\Psi_k^c(x_1(t), x_2(t)) = 2k\dot{x}_1(t)\dot{x}_2(t) - x_1(t)\ddot{x}_2(t) - x_2(t)\ddot{x}_1(t) \qquad (94)$$

stands for the modified cross component of $x_1(t)$ and $x_2(t)$.

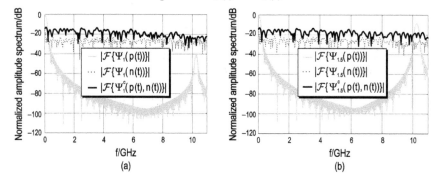

Figure 25. Normalized amplitude spectrum at the output of the traditional TK operation for $k = 1$ (a) and $k = 1.5$ (b), $E_b/N_0 = 12\,\mathrm{dB}$.

Fig. 25 shows the difference of the traditional (a) and the modified TK operation (b) for OOK in case of a binary one. Thereby, Fig. 25 (a) highlights the normalized amplitude spectrum of the occurring components for the interference-free AWGN case for $k = 1$ and $E_b/N_0 = 12\,\mathrm{dB}$. The considered pulse is a cosine-shaped pulse with pulse duration $T = 12.8\,\mathrm{ns}$ and carrier frequency $f_c = 5.13\,\mathrm{GHz}$ [8]. Obviously, the TK operation mixes a large signal part of the pulse to around DC. However, a minor contribution around $2f_c$ can also be observed. As this contribution is much lower than the spectral contributions of $\Psi_1(n(t))$ and $\Psi_1^c(p(t), n(t))$ a possible detection may be limited to the frequency range around DC.

In contrast, Fig. 25 (b) illustrates that the modification of the TK operation with, e.g. $k = 1.5$, has the potential to improve the detection for AWGN. Also in this case large pulse related spectral contributions around DC occur. Furthermore, a significant increase of $\approx 30\,\mathrm{dB}$ of the pulse's amplitude spectrum $\mathcal{F}\{\Psi_{1.5}(p(t))\}$ occurs around $2f_c$. The resulting amplitude spectrum of the pulse is both around DC and around $2f_c$ larger than the contributions of $\Psi_{1.5}(n(t))$ and $\Psi_{1.5}^c(p(t), n(t))$. Hence, an improved detection performance can be expected for AWGN if frequencies around DC as well as around $2f_c$ are considered.

The reason for this behaviour can be ascribed to the modified TK operation of the pulse. Assuming negligible influences of the first and the second derivative of $p(t)$ at $|t| = \frac{T}{2}$ the output of the TK operation on $p(t)$ can be described as [8]:

$$
\begin{aligned}
\Psi_k\left(p\left(t\right)\right) = \frac{2}{3T} \Big[&0.5\left(k+1\right)\left(\omega_T^2 + 3\omega_c^2\right) \\
&- 0.25\left(k-1\right)\left(\omega_T - \omega_c\right)^2 \cos\left(2\left(\omega_c - \omega_T\right)t\right) \\
&+ 0.5\left(\left(k+1\right)\omega_T^2 - 3\left(k-1\right)\omega_c^2\right)\cos\left(2\omega_c t\right) \\
&+ \left(0.5\omega_T^2 + \left(k-1\right)\left(\omega_c\omega_T - \omega_c^2\right)\right)\cos\left(\left(2\omega_c - \omega_T\right)t\right) \\
&+ \left(0.5\omega_T^2 - \left(k-1\right)\left(\omega_c\omega_T + \omega_c^2\right)\right)\cos\left(\left(2\omega_c + \omega_T\right)t\right) \\
&- 0.25\left(k-1\right)\left(\omega_T + \omega_c\right)^2 \cos\left(2\left(\omega_c + \omega_T\right)t\right) \\
&+ 0.5\left(\omega_c^2\left(k+1\right) - \omega_T^2\left(k-1\right)\right)\cos\left(2\omega_T t\right) \\
&+ \left(2\omega_c^2\left(k+1\right) + \omega_T^2\right)\cos\left(\omega_T t\right) \Big]
\end{aligned}
\tag{95}
$$

with $\omega_T = \frac{2\pi}{T}$ and $\omega_c = 2\pi f_c$. Obviously, a modification of the TK operation leads to additional spectral parts at its output. Hence, it is possible to influence the distribution of pulse energy via a simple weighting of the TK operation. It should be noted that the traditional as well as the modified TK operation acts not only as a frequency-to-DC shifter due to additional spectral parts around $2f_c$. For the traditional TK operation ($k = 1$) the two spectral components at $2\left(\omega_c - \omega_T\right)$ and $2\left(\omega_c + \omega_T\right)$ disappear completely resulting in a lower energy concentration around $2f_c$.

4.3.2. Mitigation potential of Teager-Kaiser operation

Assuming negligible noise this section analyses the potential of the traditional and the modified TK operation to mitigate NBI. For this the TK operation's effectiveness is first described in the baseband. Afterwards it is extended to the bandpass domain.

Baseband Domain

To make statements on the effectiveness of the TK operation for a narrowband signal we first consider a narrowband baseband signal of bandwidth $2\omega_N$ which can be modelled as

$$
j\left(t\right) = \sum_{i=1}^{N} A_i \sin\left(\omega_i t + \phi_i\right).
\tag{96}
$$

It consists of N overlapping sinusoids of amplitudes $A_i, i = 1, \ldots, N$, of phases $\phi_i, i = 1, \ldots, N$ as well as of frequencies $\omega_i = 2\pi f_i, i = 1, \ldots, N$. Given to the modified TK operation its output can be described for $\phi_i = 0, i = 1, \ldots, N$ as[3]

$$
\Psi_k\left(j\left(t\right)\right) = C_k\left(t\right) + \sum_{i=1}^{N}\sum_{j>i}^{N} z_i^j \cos\left(\left(\omega_i - \omega_j\right)t\right) + z_j^i \cos\left(\left(\omega_i + \omega_j\right)t\right),
\tag{97}
$$

---

[3] The TK operation is generally characterised by a strong phase dependency. To simplify the descriptions we only consider the special case $\phi_i = 0, i = 1, \ldots, N$ in the following.

with

$$C_k(t) = \frac{1}{2} \sum_{i=1}^{N} A_i^2 \omega_i^2 \left[(k+1) + (k-1) \cos(2\omega_i t) \right],$$

$$z_i^j = k A_i A_j \omega_i \omega_j + \frac{1}{2} A_i A_j \omega_i^2 + \frac{1}{2} A_i A_j \omega_j^2$$

and

$$z_j^i = k A_i A_j \omega_i \omega_j - \frac{1}{2} A_i A_j \omega_i^2 - \frac{1}{2} A_i A_j \omega_j^2.$$

For the traditional TK operation ($k = 1$) the resulting signal is composed of the DC component $C_1 = \sum_{i=1}^{N} A_i^2 \omega_i^2$. In addition, further components around $\omega_i - \omega_j$ and $\omega_i + \omega_j$ occur. Its quantity depends on the two factors $z_i^j = 0,5 A_i A_j \left(\omega_i + \omega_j \right)^2$ and $z_j^i = -0,5 A_i A_j \left(\omega_i - \omega_j \right)^2$. Thereby, a higher energy concentration can be observed for lower frequencies. In contrast, the modified TK operation ($k \neq 1$) shows spectral contributions around DC, around $\omega_i + \omega_j$ and $\omega_i - \omega_j$ as well as around $2\omega_i, i = 1, \ldots, N$. In particular, the contributions at $\omega_i + \omega_j$ and $\omega_i - \omega_j$ are influenced with larger weighting factors z_i^j and z_j^i. Hence, using the modified TK operation a high energy concentration still occurs for low frequencies. However, due to the additional spectral components at $2\omega_i, i = 1, \ldots, N$ the relative difference of energy concentration reduces.

Bandpass Domain

NBI influencing the MIR-UWB system operates in the bandpass domain. Based on the insight for a baseband signal an analytical description of the effectiveness of the traditional and the modified TK operation is done for one or more narrowband bandpass signals. Due to analytical tractability it is assumed that the lth bandpass signal can be described as

$$j_l(t) = 2A_l \sum_{i=1}^{N_l} \sin(\omega_{l,i} t) \cos(\omega_{c_l} t)$$

$$= A_l \underbrace{\sum_{i=1}^{N_l} \sin\left((\omega_{c_l} + \omega_{l,i}) t \right)}_{\alpha_l(t)} - A_l \underbrace{\sum_{i=1}^{N_l} \sin\left((\omega_{c_l} - \omega_{l,i}) t \right)}_{\beta_l(t)}. \tag{98}$$

The bandpass signal $j_l(t)$ of amplitude $2A_l$ consists of N_l sinusoids of frequencies $\omega_{l,1} < \omega_{l,2} < \ldots < \omega_{l,N_l}, \omega_{l,i} = 2\pi f_{l,i}, i = 1, \ldots, N_l$ located around the carrier frequency $\omega_{c_l} = 2\pi f_{c_l}$ with the bandwidth $B_l = 2\omega_{N_l}$.

If K_S bandpass signals $j_l(t), l = 1, \ldots, K_S$ are present at the input of TK operation its output can be described as

$$\Psi_k \left(\sum_{l=1}^{K_S} j_l(t) \right) = \sum_{l=1}^{K_S} \Psi_k(j_l(t)) + \Psi_{k,m}^c(t). \tag{99}$$

It consists of the components $\Psi_k(j_l(t)) = \Psi_k(\alpha_l(t)) + \Psi_k(\beta_l(t)) + \Psi_k^c(\alpha_l(t), \beta_l(t)), l = 1, \ldots, K_S$ which always occur in presence of one bandpass signal. Assuming $\omega_{c_l} \gg B_l$ the lth bandpass signal $j_l(t), l = 1, \ldots, K_S$ results in [8]

$$
\Psi_k(j_l(t)) \approx
$$

$$
A_l^2 \omega_{c_l}^2 \sum_{u=1}^{N_l} \left[(k+1) + 0.5\,(k-1) \cdot \left[\cos\left(2\left(\omega_{c_l} - \omega_{l,u} \right) t \right) + \cos\left(2\left(\omega_{c_l} + \omega_{l,u} \right) t \right) \right] \right]
$$

$$
+ A_l^2 \omega_{c_l}^2 \sum_{u=1}^{N_l} \sum_{v>u}^{N_l} \left[2(k+1) \cos\left(\left(\omega_{l,u} - \omega_{l,v} \right) t \right) + (k-1) \cos\left(\left(2\omega_{c_l} + \omega_{l,u} + \omega_{l,v} \right) t \right) \right.
$$

$$
\left. + (k-1) \cos\left(\left(2\omega_{c_l} - \omega_{l,u} - \omega_{l,v} \right) t \right) \right]
$$

$$
- A_l^2 \omega_{c_l}^2 \sum_{u=1}^{N_l} \sum_{v=1}^{N_l} \left[(k+1) \cos\left(\left(\omega_{l,u} + \omega_{l,v} \right) t \right) - (k-1) \cos\left(\left(2\omega_{c_l} + \omega_{l,u} - \omega_{l,v} \right) t \right) \right] \quad (100)
$$

This result reveals that the traditional TK operation ($k = 1$) acts as a frequency-to-DC shifter for each bandpass signal $j_l(t), l = 1, \ldots, K_S$. In this case the corresponding spectral range goes from DC to the largest occurring bandwidth $B_l, l = 1, \ldots, K_S$ of the K_S bandpass signals. For the modified TK operation additional spectral components around $2\omega_{c,l}, l = 1, \ldots, K_S$ occur. For this reason a mitigation of NBI by the modified TK operation is critical. Finally, for the special case $k = -1$ the complete energy is shifted to $2\omega_{c,l}, l = 1, \ldots, K_S$. This confirms the statement that energy parts can be shifted between frequency ranges with a modified TK operation.

If the output of the traditional or the modified TK operation would only consist of components from $\Psi_k(j_l(t)), l = 1, \ldots, K_S$ a mitigation of the K_S bandpass signal could be possible as $B_T \gg B_l, l = 1, \ldots, K_S$. However, as can be seen in (99), the additional signal component

$$
\Psi_{k,m}^c(t) = \sum_{r=1}^{K_S-1} \sum_{l=r+1}^{K_S} \left[\Psi_k^c\left(\alpha_r(t), \alpha_l(t) \right) + \Psi_k^c\left(\beta_r(t), \beta_l(t) \right) \right] + \sum_{r=1}^{K_S} \sum_{l \neq r} \Psi_k^c\left(\alpha_r(t), \beta_l(t) \right) \quad (101)
$$

occurs in case of at least two bandpass signals. The component $\Psi_{k,m}^c(t)$ describes the cross components between different bandpass signals $j_l(t)$ and $j_r(t), l \neq r$. Thereby, assuming $\omega_{c_l} \gg B_l$ and $\omega_{c_r} \gg B_r$ the two cross components can be described as

$$
\Psi_k^c\left(\alpha_r(t), \alpha_l(t) \right) + \Psi_k^c\left(\beta_r(t), \beta_l(t) \right) \approx
$$

$$
\sum_{u=1}^{N_r} \sum_{v=1}^{N_l} Z_{1,k} \left[\cos\left(\left(\omega_{c_r} - \omega_{c_l} + \omega_{r,u} - \omega_{l,v} \right) t \right) + \cos\left(\left(\omega_{c_r} - \omega_{c_l} - \omega_{r,u} + \omega_{l,v} \right) t \right) \right]
$$

$$
+ Z_{2,k} \left[\cos\left(\left(\omega_{c_r} + \omega_{c_l} + \omega_{r,u} + \omega_{l,v} \right) t \right) + \cos\left(\left(\omega_{c_r} + \omega_{c_l} - \omega_{r,u} - \omega_{l,v} \right) t \right) \right] \quad (102)
$$

with the amplitudes

$$
Z_{1,k} = \frac{A_r A_l}{2} \left(\omega_{c_r}^2 + 2k\omega_{c_r}\omega_{c_l} + \omega_{c_l}^2 \right)
$$

and

$$
Z_{2,k} = -\frac{A_r A_l}{2} \left(\omega_{c_r}^2 - 2k\omega_{c_r}\omega_{c_l} + \omega_{c_l}^2 \right).
$$

Finally, the third cross component $\Psi_k^c\left(\alpha_r\left(t\right),\beta_l\left(t\right)\right)$ is given for $r \neq l$ as

$$\Psi_k^c\left(\alpha_r\left(t\right),\beta_l\left(t\right)\right) = \Psi_{k,r>l}^c\left(\alpha_r\left(t\right),\beta_l\left(t\right)\right) + \Psi_{k,l>r}^c\left(\alpha_r\left(t\right),\beta_l\left(t\right)\right)$$

$$\approx \sum_{u=1}^{N_r}\sum_{v=1}^{N_l} -Z_{1,k}\left[\cos\left(\left(\omega_{c_r}-\omega_{c_l}-\omega_{r,u}-\omega_{l,v}\right)t\right) + \cos\left(\left(\omega_{c_r}-\omega_{c_l}+\omega_{r,u}+\omega_{l,v}\right)t\right)\right]$$

$$-Z_{2,k}\left[\cos\left(\left(\omega_{c_r}+\omega_{c_l}-\omega_{r,u}+\omega_{l,v}\right)t\right) + \cos\left(\left(\omega_{c_r}+\omega_{c_l}+\omega_{r,u}-\omega_{l,v}\right)t\right)\right] \quad (103)$$

Hence, in presence of more than one bandpass signal additional spectral components occur arround $|\omega_{c_r}-\omega_{c_l}|$ and $\omega_{c_r}+\omega_{c_l}$. The frequency parts depend on the carrier frequencies of the bandpass signals. The spectral components are influenced by the weighting factor k. E.g., for the traditional TK operation ($k = 1$) the spectral component around $|\omega_{c_r}-\omega_{c_l}|$ dominates. In contrast for the modified TK operation ($k \neq 1$) additional relevant spectral components can be identified around $\omega_{c_r}+\omega_{c_l}$ making the usage of the frequency at twice the subband's center frequency $2\omega_c$ critical.

To verify the results a subband of bandwidth 625 MHz and carrier frequency 5.2 GHz is considered for $k = 1$. $K_S = 3$ bandpass signals of amplitudes $A_1 = 1$, $A_2 = 1/3$ and $A_3 = 2/3$, of bandwidths $B_1 = 5$ MHz, $B_2 = 10$ MHz and $B_3 = 1$ MHz as well as of carrier frequencies $f_{c_1} = 4.98$ GHz, $f_{c_2} = 5.04$ GHz and $f_{c_3} = 5.28$ GHz are assumed at the input of TK operation. Fig. 26 shows the positive frequency range for the resulting to one

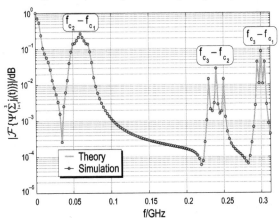

Figure 26. Normalized amplitude spectrum at the output of the traditional TK operation ($k = 1$) in presence of three narrowband bandpass signals.

normalized baseband spectrum at the output of the traditional TK operation. Simulation as well as analytical results show the spectral contributions of the three signals occurring at $f_{c_2} - f_{c_1} = 60$ MHz, $f_{c_3} - f_{c_2} = 240$ MHz and $f_{c_3} - f_{c_1} = 300$ MHz with the bandwidths 30 MHz, 22 MHz and 12 MHz. The spectral components are distributed over the complete bandwidth of the subband. For this reason the consideration of the TK operation with an additional filtering operation is critical. In particular, the mitigation scheme proposed in [42] becomes inefficient to efficiently mitigate all occurring interferences.

4.3.3. Integration of Teager-Kaiser operation

As illustrated in the previous section an integration of TK operation into the MIR-UWB system is possible with only minor complexity increase if at most one NBI occurs in each subband. In this case the approach of [42] can be used. It bases on the interplay of the TK operation with a highpass filtering. As illustrated in Fig. 27 only two additional analogue components have to be integrated into each subband of the existing non-coherent MIR-UWB receiver. Thereby, received subband signals are given to TK operation which acts as a frequency-to-DC shifter. The resulting low-frequency signal is afterwards highpass filtered to mitigate interfered signal components without any a priori information of the interference specific carrier frequency. As the bandwidth of the subband signal is larger than the interference bandwidth energy detection might be possible.

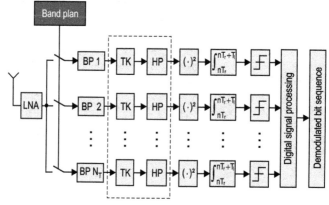

Figure 27. Integration of TK operation into the existing non-coherent MIR-UWB receiver.

In the following the potential of interference mitigation with the TK operation is shown for OOK in case of a binary one. An MIR-UWB subband of carrier frequency 5.13 GHz and effective bandwidth 162.5 MHz is considered for SNR = 11 dB. It is assumed that an IEEE 802.11a WLAN signal [47] of bandwidth 20 MHz and of carrier frequency 5.14 GHz interferes the MIR-UWB subband with an SIR of −5 dB.

Fig. 28 (a) shows the to one normalized amplitude spectrum of all occurring signal components at the output of TK operation. Thereby, the UWB signal spectrum ranges from DC to 162.5 MHz whereas the lower frequency regions have a higher energy concentration. A similar behaviour occurs for the narrowband WLAN signal. Its corresponding amplitude spectrum ranges from DC to 20 MHz whereas energy is strongly distributed around DC. Furthermore, additional spectral cross components between signal, noise and interference occur which can be ascribed to the non-linearity of the TK operation. To mitigate the WLAN signal highpass filtering is done after the TK operation.

Fig. 28 (b) illustrates the to one normalized amplitude spectrum after highpass filtering. The used highpass filter is characterised by the order six, a passband ripple of 0.1 dB, a 50 dB stopband attenuation as well as a 50 MHz wide stopband. Obviously, the narrowband WLAN signal is mitigated after highpass filtering. In contrast the subband signal has an

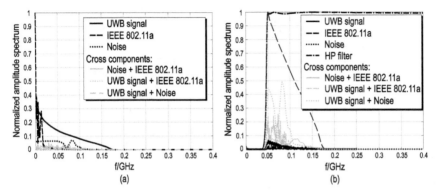

Figure 28. Normalized amplitude spectrum at the output of TK operation (a) and after highpass filtering (b), SNR = 11 dB.

amplitude spectrum which ranges from 50 MHz to 162.5 MHz. As this spectrum dominates the amplitude spectra of the occurring cross components energy detection is possible.

5. Summary

This chapter deals with an easy-to-realise non-coherent MIR-UWB system which is a promising approach for high data rate and energy efficient communication over short distances. Due to its low complexity the MIR-UWB system is an alternative to already existing UWB systems for high data rate applications such as Multiband OFDM UWB.

The MIR-UWB system is based on an energy detection receiver. Thus the first part of this chapter deals with the performance of this component. To understand the energy detection receiver we look at the bit and symbol error probability in different wireless channels.

First we introduce a closed form expression of the SEP for an energy detection receiver with M-PAM in the AWGN channel. Based on this result, we optimise the interval thresholds to minimise the SEP. Optimal interval thresholds guarantee a minimal SEP for M-PAM. In the next step we look into the optimal amplitudes for M-PAM using an energy detection receiver. This approach enables to reduce the SEP for M-PAM with medium to large degrees of freedom.

To understand the characteristics of the energy detection receiver in fading channels we look into different approaches to model the energy at the receiver. It has been shown, that the flat fading channel model can be used to model the energy at a receiver for a receiver bandwidth $B > 100$ MHz. Based on this assumption we introduce closed form expressions for the SEP of the energy detection receiver with M-PAM for different fading statistics such as *Rayleigh*, *Rice* and *Nakagami-m*. We also analyse the SEP of an multichannel receiver using different combining techniques. Square Law Combing and Square Law Selection are possible combining schemes for an energy detection receiver. A closed form solution for SLC and SLS is introduced for the AWGN and for the *Rayleigh* fading channel including i.i.d. and correlated fading gains.

The first part ends with the analysis of the SEP in a frequency selective fading channel. Based on *Rayleigh* distributed fading gains, representing a non line-of-sight channel (NLOS), we introduce a closed form expression for the energy detection receiver with M-PAM. The result also contains the possibility to analyse the effect of correlated fading gains. This is the case in typical UWB wireless channels. If the fading gains are not *Rayleigh* distributed, we present a numerical solution for any fading distributions. The results of the first part enable a precise prediction of an energy detection receiver with M-PAM in many different scenarios.

Since the MIR-UWB system is highly sensitive to interference, the second part of the chapter considers three different aspects regarding an efficient interference mitigation.

The first aspect deals with the analysis of the interference robustness of an OOK and BPPM specific energy detection receiver being the essential component of the non-coherent MIR-UWB receiver. Thereby, taking into account thermal noise a general frame work is presented which can be used to give statements on the detector's interference robustness for an interference with arbitrary bandwidth. Furthermore, possible system parameters can be identified to increase the detector's interference robustness.

The second aspect considers the coexistence capability of the MIR-UWB system. Thereby, various easy-to-realise adaptive coexistence-based approaches are presented. Starting with a static coexistence approach a DAA coexistence approach for temporary NBI is presented being integrated into the system specific initialisation and data phase. The proposed method allows a reliable adaptive mitigation of temporary NBI. A further adaptive coexistence approach bases on image-based thresholding which can be integrated into the initialisation phase of the MIR-UWB system. Based on an exemplary interference scenario the potential to efficiently mitigate multiple interferences of different interference power is shown.

Lastly, the third aspect focuses on the analytical investigation of the potential to mitigate NBI inside an energy detection receiver. Hereby, the TK operation as well as a modified TK operation is analysed. It is shown that for a narrowband baseband signal the output of the TK operation is characterised by a larger energy concentration in the lower frequency range. In contrast, for the modified TK operation further spectral components occur for higher frequencies. A subsequent analysis of one NBI in the bandpass domain shows that the TK operation acts like a frequency-to-DC shifter. This reveals the potential to mitigate a single NBI without the knowledge of the NBI's carrier frequency. In contrast, for the modified TK operation additional spectral components at twice the NBI's carrier frequency occur making interference mitigation critical. In case of multiple NBI further spectral components occur at the TK operation's output which can be ascribed to the mutual interference influences. Due to a possible distribution of the spectral components within the total MIR-UWB subband interference mitigation depends on the interference position inside the MIR-UWB subband.

Acknowledgements

This work was supported within the priority program No. 1202 (UKoLoS) by the German Research Foundation (DFG).

Author details

Rainer Moorfeld and Adolf Finger
Communications Laboratory, Dresden University of Technology, Germany

Hanns-Ulrich Dehner, Holger Jäkel, Martin Braun and Friedrich K. Jondral
Communications Engineering Lab, Karlsruhe Institute of Technology (KIT), Germany

6. References

[1] Abramowitz, M. & Stegun, I. [1964]. *Handbook of mathematical functions with formulas, graphs, and mathematical tables*, Applied mathematics series, Dover Publications.

[2] Batra, A. [2008]. *Multiband OFDM Physical Layer Proposal for IEEE 802.15 Task Group 3a*, IEEE P802.15-03/268r1.

[3] Bober, M., Moorfeld, R. & Jorswieck, E. [2011]. Performance of Energy Detection in NLOS frequency-selective Fading Channels, *Proc. of IEEE International Symposium on Personal, Indoor and Mobile Communications (PIMRC)*.

[4] Cassioli, D., Win, M. Z. & Molisch, A. F. [2002]. The ultra-wide bandwidth indoor channel: from statistical model to simulations, *IEEE Journal on Selected Areas in Communications* 20(6): 1247–1257.

[5] Dehner, H., Jäkel, H., Burgkhardt, D. & Jondral, F. K. [2010]. The Teager-Kaiser Energy Operator in Presence of Multiple Narrowband Interference, *IEEE Communications Letters* Vol. 14(No, 8): 716 – 718.

[6] Dehner, H., Jäkel, H., Burgkhardt, D., Jondral, F. K., Moorfeld, R. & Finger, A. [2010]. Treatment of temporary narrowband interference in noncoherent multiband impulse radio UWB, *Proc. of IEEE Mediterranean Electrotechnical Conference*, pp. 1335 – 1339.

[7] Dehner, H., Jäkel, H. & Jondral, F. K. [2011a]. Narrow- and broadband Interference Robustness for OOK/BPPM based Energy Detection, *Proc. of IEEE International Conference on Communications*.

[8] Dehner, H., Jäkel, H. & Jondral, F. K. [2011b]. On the modified Teager-Kaiser energy operator regarding narrowband interference, *Proc. of IEEE Wireless Telecommunications Symposium*.

[9] Dehner, H., Koch, Y., Jäkel, H., Burgkhardt, D., Jondral, F. K., Moorfeld, R. & Finger, A. [2010]. Narrow-band Interference Robustness for Energy Detection in OOK/PPM, *Proc. of IEEE International Conference on Communications*.

[10] Dehner, H., Linde, M., Moorfeld, R., Jäkel, H., Burgkhardt, D., Jondral, F. K. & Finger, A. [2009]. A low complex and efficient coexistence approach for non-coherent multiband impulse radio UWB, *Proc. of IEEE Sarnoff Symposium*.

[11] Dehner, H., Moorfeld, R., Jäkel, H., Burgkhardt, D., Finger, A. & Jondral, F. K. [2009]. Multi-band Impulse Radio – An Alternative Physical Layer for High Data Rate UWB Communication, *Frequenz, Journal of RF-Engineering and Telecommunications* Vol. 63(No, 9-10): 200–204.

[12] Dehner, H., Romero, A., Jäkel, H., Burgkhardt, D., Moorfeld, R., Jondral, F. K. & Finger, A. [2009]. Iterative coexistence approaches for noncoherent multi-band impulse radio UWB, *Proc. of IEEE International Conference on Ultra-Wideband*, pp. 734 – 738.

[13] Digham, F. F., Alouini, M.-S. & Simon, M. K. [2007]. On the Energy Detection of Unknown Signals Over Fading Channels, *IEEE Transactions on Communications* 55(1): 21–24.

[14] ECC [2006]. ECC decision of 24 march 2006 on the harmonised conditions for devices using ultra-wideband (UWB) technology in bands below 10.6 GHz, *Technical Report* .

[15] Eisenacher, M. [2006]. Optimierung von Ultra-Wideband-Signalen (UWB), *Dissertation, Forschungsberichte aus dem Institut für Nachrichtentechnik der Universität Karlsruhe (TH), Band 16, 2006* .

[16] Fisher, R., Kohno, R., Laughlin, M. & Welborn, M. [2005]. *DS-UWB Physical Layer Submission to 802.15 Task Group 3a*, IEEE P802.15-04/0137r1.

[17] Ghassemzadeh, S. S., Jana, R., Rice, C. W., Turin, W. & Tarokh, V. [2004]. Measurement and modeling of an ultra-wide bandwidth indoor channel, *IEEE Transactions on Communications* 52(10): 1786–1796.

[18] Gradshteyn, I. S. & Ryzhik, I. M. [2007]. *Table of Integrals, Series, and Products*, 7 edn, Academic Press.

[19] Hentilä, L., Taparungssanagorn, A., Viittala, H. & Hämäläinen, M. [2005]. Measurement and modelling of an UWB channel at hospital, *Proc. of IEEE International Conference on Ultra-Wideband (ICU)*.

[20] Hovinen, V., Hämäläinen, M. & Pätsi, T. [2002]. Ultra wideband indoor radio channel models: preliminary results, *IEEE Digest of Papers Ultra Wideband Systems and Technologies*.

[21] Johnson, N. L., Kotz, S. & Balakrishnan, N. [1994]. *Continuous Univariate Distributions*, Vol. 1, 2 edn, Wiley.

[22] Kaiser, J. [1990]. On a simple algorithm to calculate the energy of a signal, *Proc. of IEEE International Conference on Acoustics, Speech, and Signal Processing*, pp. 381 – 384.

[23] Kaiser, J. [1993]. Some Useful Properties of Teager's Energy Operators, *Proc. of IEEE International Conference on Acoustics, Speech, and Signal Processing*, pp. 149 – 152.

[24] Karedal, J., Wyne, S., Almers, P., Tufvesson, F. & Molisch, A. F. [2004]. Statistical analysis of the UWB channel in an industrial environment, *Proc. of IEEE Vehicular Technology Conference (VTC-Fall)*.

[25] Kay, S. M. [1998]. *Fundamentals of Statistical Signal Processing: Detection Theory*, Vol. 2, Prentice Hall.

[26] Kunisch, J. & Pamp, J. [2002]. Measurement results and modeling aspects for the UWB radio channel, *IEEE Digest of Papers Ultra Wideband Systems and Technologies*.

[27] Lee, W. C. [1997]. *Mobile Communications Engineering: Theory and Applications*, McGraw-Hill, Inc.

[28] Li, L., Moorfeld, R. & Finger, A. [2011a]. Bit and power loading for the multiband impulse radio UWB architecture, *Proc. of 8th Workshop Positioning Navigation and Communication (WPNC)*.

[29] Li, L., Moorfeld, R. & Finger, A. [2011b]. Closed form expressions for the symbol error rate for non-coherent UWB impulse radio systems with energy combining, *Proc. of 8th Workshop Positioning Navigation and Communication (WPNC)*.

[30] Mittelbach, M., Moorfeld, R. & Finger, A. [2006]. Performance of a Multiband Impulse Radio UWB Architecture, *Proc. of 3rd International conference on mobile technology, applications and systems*.

[31] Molisch, A. F., Cassioli, D., Chong, C.-C., Emami, S., Fort, A., Kannan, B., Karedal, J., Kunisch, J., Schantz, H. G., Siwiak, K. & Win, M. Z. [2006]. A comprehensive standardized model for ultrawideband propagation channels, *IEEE Transactions on Antennas and Propagation* 54(11): 3151–3166.

[32] Molisch, A. F., Foerster, J. R. & Pendergrass, M. [2003]. Channel models for ultrawideband personal area networks, *IEEE Wireless Communications Magazine* 10(6): 14–21.

[33] Moorfeld, R. & Finger, A. [2009]. Multilevel PAM with optimal amplitudes for non-coherent energy detection, *Proc. of Wireless Communications and Signal Processing (WCSP)*.

[34] Moorfeld, R., Finger, A., Dehner, H., Jäkel, H. & Jondral, F. K. [2008]. A simple and fast detect and avoid algorithm for non-coherent multiband impulse radio UWB, *Proc. of IEEE International Symposium on Spread Spectrum Techniques and Applications*, pp. 587 – 591.

[35] Moorfeld, R., Finger, A., Dehner, H.-U., Jäkel, H. & Jondral, F. K. [2009]. Performance of a high flexible non-coherent multiband impulse radio UWB system, *Proc. of 9th IASTED International Conference on Wireless and Optical Communications (WOC)*.

[36] Moorfeld, R., Finger, A. & Zeisberg, S. [2004]. High data rate UWB performance with reduced implementation complexity, *Proc. of IEEE International Symposium on Spread Spectrum Techniques and Applications (ISSSTA)*.

[37] Moorfeld, R., Lu, Y. & Finger, A. [2012]. Energy detection with optimal symbol constellation for M-PAM in UWB fading channels, *Proc. of IEEE International Conference on Ultra-Wideband (ICUWB)*.

[38] Moorfeld, R., Zeisberg, S., Pezzin, M., Rinaldi, N. & Finger, A. [2004]. Ultra-Wideband Impulse Radio (UWB-IR) Algorithm Implementation Complexity and Performance, *Proc. of IST Mobile and Wireless Communications Summit*.

[39] Nuttall, A. H. [1972]. Some integrals involving the Q function, *Technical report*, Naval Underwater Systems Center (NUSC).

[40] Nuttall, A. H. [1975]. Some integrals involving the Q_M function, *IEEE Transactions on Information Theory* 21(1): 95–96.

[41] Otsu, N. [1979]. A threshold selection method from gray-level histograms, *IEEE Transactions on Systems, Man, and Cybernetics* Vol. 20: 62 – 66.

[42] Ozdemir, O., Sahinoglu, Z. & Zhang, J. [2008]. Narrowband Interference Resilient Receiver Design for Unknown UWB Signal Detection, *Proc. of IEEE International Conference on Communications*, pp. 785 – 789.

[43] Pagani, P. & Pajusco, P. [2006]. Characterization and Modeling of Temporal Variations on an Ultrawideband Radio Link, *IEEE Transactions on Antennas and Propagation* 54(11): 3198–3206.

[44] Papoulis, A. [2002]. *Probability, Random Variables and Stochastic Processes*, 4. edn, McGraw-Hill.

[45] Paquelet, S. & Aubert, L. M. [2004]. An energy adaptive demodulation for high data rates with impulse radio, *Proc. of IEEE Radio and Wireless Conference*, pp. 323–326.

[46] Paquelet, S., Aubert, L. M. & Uguen, B. [2004]. An impulse radio asynchronous transceiver for high data rates, *Proc. of Joint Conference on Ultrawideband Systems and Technologies Ultra Wideband Systems (UWBST & IWUWBS)*.

[47] *Part 11: Wireless LAN Medium Access Control (MAC) and Physical Layer (PHY) Specifications: High-speed Phyiscal Layer in the 5 GHz Band* [1999]. IEEE Std. 802.11a-1999.

[48] Press, S. J. [1966]. Linear combinations of non-central chi-square variates, *The Annals of Mathematical Statistics* 37(2): 480–487.

[49] Proakis, J. G. [2001]. *Digital Communications*, 4th edn, McGraw-Hill.

[50] Rappaport, T. S. [2002]. *Wireless Communications: Principles and Practice*, Prentice Hall.

[51] Saleh, A. A. M. & Valenzuela, R. [1987]. A statistical model for indoor multipath propagation, *IEEE Journal on Selected Areas in Communications* 5(2): 128–137.

[52] Simon, M. K. & Alouini, M.-S. [2006]. *Digital Communication over Fading Channels: A Unified Approach to Performance Analysis*, Wiley.

[53] Steiner, C. & Wittneben, A. [2007]. On the interference robustness of ultra-wideband energy detection receivers, *Proc. of IEEE International Conference on Ultra-Wideband*.

[54] Stoica, L. [2008]. *Non-coherent energy detection transceivers for Ultra Wideband Impulse radio systems*, PhD thesis, Centre for Wireless Communications, Oulu, Finland.

[55] Tse, D. & Viswanath, P. [2005]. *Fundamentals of wireless communication*, Cambridge University Press.

[56] Urkowitz, H. [1967]. Energy detection of unknown deterministic signals, *Proc. of the IEEE* 55(4): 523–531.

[57] Weiß, T. [2004]. OFDM-basiertes Spectrum Pooling, *Dissertation, Forschungsberichte aus dem Institut für Nachrichtentechnik der Universität Karlsruhe (TH), Band 13* .

[58] Witrisal, K., Leus, G., Janssen, G., Pausini, M., Troesch, F., Zasowski, T. & Romme, J. [2009]. Noncoherent ultra-wideband systems, *IEEE Signal Processing Magazine* Vol. 26(No, 4): 48 – 66.

[59] Zhang, H., Udagawa, T., Arita, T. & Nakagawa, M. [2002]. A statistical model for the small-scale multipath fading characteristics of ultra wideband indoor channel, *Proc. of IEEE Digest of Papers Ultra Wideband Systems and Technologies*.

[60] Zhao, L. & Haimovich, A.-M. [2002]. Performance of ultra-wideband communications in the presence of interference, *IEEE Journal on Selected Areas in Communications* Vol. 20(No, 9): 1684 – 1691.

Chapter 2

Chip-to-Chip and On-Chip Communications

Josef A. Nossek, Peter Russer, Tobias Noll, Amine Mezghani,
Michel T. Ivrlač, Matthias Korb, Farooq Mukhtar,
Hristomir Yordanov, Johannes A. Russer

Additional information is available at the end of the chapter

1. Introduction

In high-performance integrated circuits manufactured in CMOS deep sub-micron technology, the speed of global information exchange on the chip has developed into a bottleneck, that limits the effective information processing speed. This is caused by standard on-chip communication based on multi-conductor interconnects, e.g., implemented as parallel interconnect buses. The supported clock frequency of such wired interconnects - at best - remains constant under scaling, but - for global interconnects - reduces by a factor of four, as the structure size is reduced by half. Such multi-conductor interconnects also exhibit some undesirable properties when used for chip-to-chip communication. The much larger distances that have to be bridged, force the clock frequencies for the chip-to-chip interconnects to much lower values than those for on-chip circuitry. In widening up this bottleneck by increasing the number of parallel wires, the separation between the wires has to decrease. This causes increased mutual coupling between neighboring wires, which reduces the supported clock frequency and counters the effect of having more wires in the first place.

The high clock frequencies used in on-chip interconnects and the huge information rate of chip-to-chip communication lets possible solutions belong to the domain of ultra-wideband (UWB) technology. Pursuing suitable solutions, we explore firstly the improvement of the multi-conductor interconnect by signal processing and coding. From information theory, it is known that information can be transmitted through a noisy channel with arbitrary low probability of error as long as the rate is lower than the channel capacity given by the Shannon theorem. Achieving this capacity requires, however, sophisticated digital signal processing and coding. In particular, the DAC (Digital-to-analog converters) and the ADC (analog-to-digital converter) components which are formed by the output or the input of a logic CMOS inverter, respectively, turns to be a limiting factor. In fact, the ADC and DAC components, perform a single-bit conversion between the analog and the digital domain. With such coarse quantization, all state of the art techniques for signal processing fail. We provide information theoretic bounds on the improvements possible by coding the transmission, and propose methods to design suitable codes which allow decoding with low latency.

Thereby, an analytical field-theoretical modeling of multi-conductor interconnects is needed. Moreover, modifications to standard signal processing techniques which make them suitable for medium-low resolution quantization are developed and analyzed, and their performance is studied.

As a promising alternative solution, wireless Ultrawideband (UWB) enables high speed communication at short distances. In fact, it is anticipated that even higher performance is achievable in chip-to-chip and on-chip communication, when multi-conductor interconnects are replaced by wireless ultra-wideband multi-antenna interconnects. Hereby, the signal pulses do not necessarily increasingly disperse as they travel along their way to the receiving end of the interconnect. The propagating nature of the wireless interconnect, the extreme high available bandwidth and the very short distances can offer a much more attractive channel for chip-to-chip and on-chip communications. In addition, applying multiple antennas at the transmitter side as well as the receiver side can drastically improve the data rate and the reliability of UWB systems at the cost of certain computational complexity. This chapter provides theoretical and empirical foundations for the application of ultra-wideband multi-antenna wireless interconnects for chip-to-chip communication. Appropriate structures for integrated ultra-wideband antennas shall be developed, their properties theoretically analyzed and verified against measurements performed on manufactured prototypes. Qualified coding and signal processing techniques, which aim at efficient use of available resources of bandwidth, power, and chip area shall be developed. Since Analog-to-Digital Converters (ADCs) are considered critical components for the UWB, main focus is hereby given to low resolution signal quantization and processing. Therefore, the analysis and the design of UWB systems with low resolution signal quantization (less than 4 bits) is a vital part of this chapter, where optimized receive and transmit strategies are obtained.

On the other hand, detailed cost-models for the digital hardware architecture, which are based on signal flow charts and VLSI implementations of dedicated functional blocks shall be developed, which allow for an informative analysis of elementary trade-offs between computational speed, required chip area, and power consumption. In fact, quantitative optimization in terms of silicon area (manufacturing costs) and even more important in terms of energy dissipation (usage costs) is mandatory already in the standardization and conception phase of digital systems to be highly integrated as System-on-Chips (SoC). This is especially true for digital communication systems where e.g. in the optimization of channel coding traditionally only the transmission power has been considered. In general this leads to highly complex and energy intensive receivers. Actually a proper optimization of such systems requires a joint optimization of the transmitter and receiver cost features, e.g. the minimization of the total energy per transmitted bit. For such a quantitative optimization quite accurate cost models for the components of the transmitter and receiver are required. Instead, if any, only oversimplified cost models are applied today. While quite accurate cost models are available for many communication system components there is a lack of such models for channel decoders like Viterbi, Turbo, and Low-Density-Parity-Check (LDPC) decoders. Out of these, especially the derivation of sufficiently accurate cost models for LDPC decoders is challenging: The realization of the extensive internal exchange of messages between the so-called bit and check nodes in such a decoder results in non-linear dependencies between decoder features and code parameters. For example in high-throughput decoders the data exchange is performed via a complex dedicated interconnect structure. Its realization frequently requires an artificial expansion of silicon area. In the past various decoder architectures have been proposed to reduce the interconnect impact and trading throughput for silicon area and energy. All that together makes the derivation of LDPC decoder cost models a challenging task.

The Chapter is organized as follows. Radio frequency engineering aspects involved in wired and wireless interconnects are investigated first. There is a multitude of requirements for chip-to-chip communication, which an integrated antenna has to fulfill, like large bandwidth, small geometrical profile, and so on. Therefore, a detailed study of the possible solutions for an integrated on-chip antenna is performed. Novel solutions, which make use of the digital circuit's ground plane as a radiating element, are investigated. In the third section, the signal processing and coding aspects involved are carried out based on the obtained channel models, where both multiconductor interconnects and wireless multiantenna interconnects are interpreted as discrete-time, multi-input-multi-output (MIMO) systems. In the last section of this chapter appropriate silicon area, timing, and energy cost models for high-throughput LDPC decoders, which reproduce accurately the non-linear dependencies and being applicable to bit-parallel as well as to bit-serial decoder architectures are presented. These models allow for a quantitative comparison of different decoder architectures revealing the most area and energy efficient architecture for a given code and throughput specification. Additionally, a new highly area and energy efficient architecture based on a bit-serial interconnect is derived. This architecture is the result of a systematic architecture search and proper optimization based on the cost models.

2. Multi-conductor interconnects and on-chip antennas

2.1. Multi-conductor interconnects

With the increase of the on-chip data transfer rate to several 10 Gbit/s the spatio-temporal intersymbol interference (auto-interference) within the multiwired bus systems becomes a limiting factor for the circuit performance. Due to the limited available space for the bus systems shielding between the wires of the bus should be omitted. This allows for larger wire cross sections and thereby to reduce the signal distortion. An appropriate signal coding and signal processing will compensate for the effects of the coupling between the wires.

The wiring inside high speed MOS circuits exhibits sub-micron cross-sectional dimensions and conductor width and conductor thickness are of similar size. Within the signal frequency band the cross sectional dimensions are in the order of the skin effect penetration depth. The signal transmission properties of the bus system is detrained by the capacitances per unit of length and the resistance per unit of length.

The TEM modes of a lossless multiconductor transmission line with equidistant conductors of equal cross section and filled with homogenous isotropic dielectric material used for bus have been discussed in [18]. Figure 1 a shows the cross-sectional drawing of the bus. The quasi-electrostatic parameters of the bus embedded in the substrate between two ground planes have been computed. The bus capacitance per unit of length – matrix, describing capacitances with respect to ground and mutual capacitances, has been derived from the conductor geometry using an analytical technique based on even-odd mode analysis [10, 18]. The analytical technique is based on the inversion of the Schwarz-Christoffel conformal mapping [5, pp. 191–201]. The advantages of the proposed method are its accuracy, the lack of geometrical limitations and the algorithm efficiency. The results for the ground and coupling

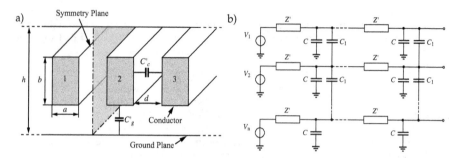

Figure 1. a) A cross section of three-wire digital bus with a coupling and a ground capacitance [10] and b) equivalent lumped element circuit [8].

capacitances per unit length for the multi conductor transmission line, filled in with silicon, are presented in Fig. 2. Since the capacitance depends only on the ratio of the line dimensions, all geometrical data are normalized to the distance h between the ground planes.

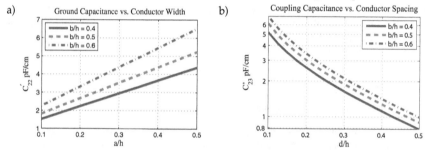

Figure 2. a) Ground capacitance vs. geometry for digital transmission line, filled in with Si, $d/h = 0.125$, b) Coupling capacitance vs. geometry for digital transmission line, filled in with Si, $a/h = 0.25$ [10].

The obtained results have been used to compute the transmission line parameters of the bus [7, 8, 10, 18]. The bus model is based on multiconductor TEM transmission line theory [5, pp. 356–363]. In case of the TEM transmission line the inductance per unit of length matrix follows directly from the capacitance per unit of length matrix and the material [18]. in case of small conductor cross sections the resistance per unit of length becomes such high that the inductance per unit of length matrix can be neglected in comparison with the resistantce per unit of length. In this case the impedance per unit of length matrix becomes diagonal [8]. The ohmic losses in the conductors are modeled by resistance per unit length R'. These parameters determine the lumped element equivalent circuit of the bus shown in Fig. 1 b.

The crosstalk between the conductors of the bus has been investigated in [18]. Figure 3 a shows the response of the crosstalk voltage at the end of the line. The results, obtained by solving the transmission line equations have been compared with numerical Method of Moment (MoM) full-wave simulations. The analytic model exhibits good accuracy up to frequencies beyond 10 GHz. Figure 3 b shows the pulse distortion at the end of the transmission line. Analytical data computed with the transmission line model have been compared with numerical data obtained from SPICE simulation.

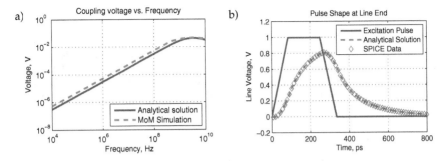

Figure 3. a) Frequency response of the crosstalk voltage at the end of the line, b) pulse response at the end of the line [18].

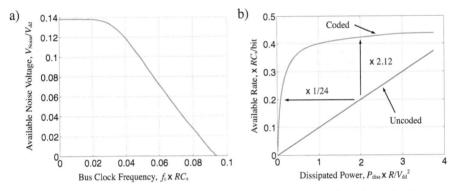

Figure 4. a) Maximum permissible noise-voltage at the receiver of a four-conductor bus as function of the bus clock frequency, b) achievable information rates of coded and uncoded transmission as function of the dissipated power [7].

The space-time intersymbol interference present in on-chip interconnection buses is a limiting factor of the performance of digital integrated circuits. This effect has greater influence as the transfer data rate increases and the circuit dimensions decrease. In order to be able to develop coding techniques for reducing the detrimental effects of intersymbol interference, an efficient and precise method for calculating the impulse response of the interconnect is required [31]. In [7] a quasi-analytical method was applied for computing the impulse response of a digital interconnection bus. The fundamental performance limits of bus systems due to information theory have been analyzed. Figure 3 shows the maximum permissible noise-voltage V_{Noise} at the receiver of a four-conductor bus as function of the bus clock frequency f_c and the achievable information rates of coded and uncoded transmission as function of the dissipated power P_{diss}. The clock frequency for the coded transmission is set to $0.11/(RC_s)$, which is above the cutoff of the uncoded bus and proves to work well with the coded system. Here V_{dd}, R and C_s are the magnitude of the signal voltage at the input, the total resistance, and the total average substrate capacitance, respectively. Here, $C_c = 6C_s$ is assumed, but similar results are obtained for other ratios.

Conclusion

The developed methods allow to compute the impulse response of the multi-conductor bus, and – building on this ground – to compute information theoretic measures, like mutual information. Those measures allow to quantify the possible gains in performance that can be achieved by employing suitable coding schemes to the multi-conductor interconnection bus. The obtained results reveal a huge potential of coded transmission both in terms of increasing the data rate and in decreasing the dissipated power.

2.2. On-chip antennas

An interesting future possibility for handling Gbit/s data streams on chip and from chip to chip will be wireless intra-chip and inter-chip communication. This section describes investigations of integrated on-chip antennas for broad-band intra-chip and inter-chip communications. At frequencies of 60 GHz and beyond antennas can be made sufficiently small to be integrated on monolithic circuits [1, 19]. However, there are still problems when integrating millimeterwave antennas on CMOS circuits. The integration of millimeterwave antennas on silicon requires a high resistivity substrate in order to achieve low losses, whereas for CMOS circuits the substrate resistivity has to be low in order to provide isolation of the circuit elements. Furthermore, chip surface is a cost factor and should not be wasted for antennas.

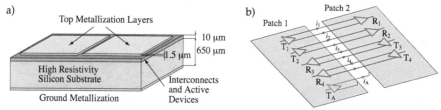

Figure 5. a) Schematic drawing of a chip with an integrated antenna, b) Differential lines, connecting the digital circuits under the separate antenna patches.

An integrated on-chip antenna for chip-to-chip wireless communication, based on the usage of the digital circuits' ground planes as radiating elements was presented in [12–17]. Figure 5 a shows schematically the realization of this principle in silicon technology. The integrated circuit is fabricated on a high resistivity silicon substrate ($\geq$ 1kΩ·cm) with a thickness in the order of of 650 μm. The substrate is backed by a metallic layer. On top of the substrate a low-resistivity layer ($\approx$ 5Ω·cm) of few micrometer thickness is grown. A homogeneous low-resistivity layer of 3 μm to 5 μm thickness is followed by a top with embedded CMOS circuitry and the interconnects. A low resistivity top layer is required for the circuit insulation. The electromagnetic field of the circuits is mainly confined in this top layer. The antenna field is spreading over the whole thickness of the substrate. Due to the high resistivity of the substrate the antenna losses are low. Since only a small fraction of the antenna near-field energy is stored in the low-resistivity layer, the coupling between the antenna near-field and the circuit field is weak. Furthermore, the interference between the CMOS circuits and the antenna field can be reduced when the main part of the circuit is operating in a frequency band distinct from the frequency band used for the wireless transmission.

The utilization of the electronic circuit ground planes as radiating elements for the integrated antennas allows for optimal usage of chip area, as the antennas share the chip area with the circuits. It has to be taken care that the interference between the antenna field and the field propagating in the circuit structures stays within tolerable limits. Consider the structure represented schematically in Fig. 5 b. The structure contains two antenna patches 1 and 2. Both antenna patches serve as the ground planes of circuits. These circuits contain line drivers $T_1 \ldots T_4$ driving over symmetrical interconnection lines the line receivers $R_1 \ldots R_4$. Furthermore there is a driver T_A, the output of which is connected to both patches, however only one conductor bridges the gap between the patches. The currents $i_1 \ldots i_4$ all are flowing back over the symmetric lines. The sum of the currents $i_1 \ldots i_4$ flowing in both directions through the transmission line modes vanishes and is not exciting the antenna. Different from this, the current i_A excites an antenna radiation mode. The circuit for this current is closed via the displacement current in the near-field of the antenna. By exciting the interconnection structures in transmission line modes and the antennas in antenna modes the interference between circuit and antennas can be minimized. We need not to use differential lines between the patches. In general a interconnection structure consisting of N conductors can guide up to $N - 1$ quasi-TEM transmission line modes and one antenna mode.

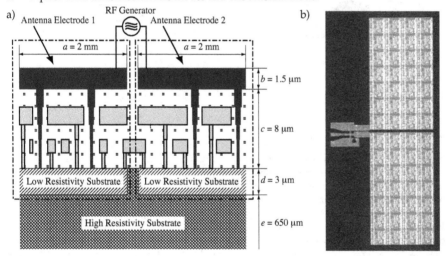

Figure 6. Cross-sectional view of integrated on-chip antenna, using the ground planes as antenna electrodes [11, 14].

Figure 6 a shows a cross-sectional view of the integrated on-chip antenna, using the ground planes in layer b on top of the integrated circuit as the antenna electrodes. Layer c with a total thickenss of 8 μm contains the active and passive circuit elements and the interconnect wiring. The low-resistivity layer d provides isolation of the circuit elements. The thick substrate layer e is of high resistivity. Figure 6 b shows a photograph of the fabricated open-circuit slot antenna with CMOS circuits under the antenna electrode [15]. Figure 7 a shows the simulated current distribution of a two-patch V-band antenna [14, 15]. The current distribution in both patches mainly is concentrated in the neighborhood of the slots. The antenna behaves as an open-circuited slot antenna. The guided wavelength is in the range of a millimeter. The open-circuited slot with a length of about one millimeter is a transmission line resonator with a resonance frequency in the V-band. The standing wave in the slot excites the radiation field.

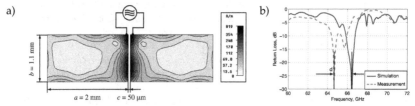

Figure 7. (a) Top view and current distribution of a two-patch dipole antenna, operating at 66 GHz, (b) Measured return loss of the on-chip open slot antenna [14, 15].

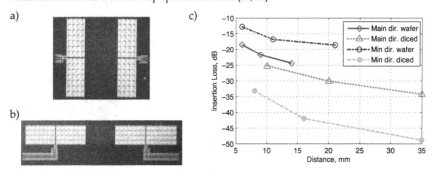

Figure 8. Antenna orientation with a) collinear and b) parallel slots, c) Measured insertion loss of a wireless chip-to-chip link depending [14, 15].

The antennas have been measured on-wafer and diced. The measured return loss of the diced slot antenna from Fig. 7 a is compared with the simulation results in Fig. 7 b. The insertion loss of a transmission link has been measured for the two antenna alignments shown in Figs. 8 a and b, where the antennas were positioned in each other radiation minima and maxima. Figure 8 s c shows the measured insertion loss of a wireless links formed by two antennas. When the antennas are oriented such that their slots are collinear, they are in each other's direction of minimum radiation. When they are oriented such that their slots are parallel, they are in each other's direction of maximum radiation. Both cases were investigated for on-wafer and for diced chips. The chip-to-chip links with both antennas on different chips exhibit higher insertion loss. The lower insertion loss of links between antennas on the same chip is due to the contribution of surface waves. The worst-case transmission link (gain-chip-to-chip link) in the direction of minimum radiation shows an insertion loss of -47dB, which is sufficient for high-rate data links.

Lumped element circuit models can provide a compact description of wireless transmission links [20–22, 25]. Distributed circuits can be modeled also in a broad frequency band with arbitrary accuracy using lumped element network models. A general way to establish network models is based on modal analysis and similar techniques [2–5]. In the case of wireless transmission links, high insertion losses have to be considered. Therefore methods for the synthesis of lossy multiports have to be applied. In [23, 24] a lumped-element two-port antenna model is presented where the antenna near-field is modeled by a reactive two-port and the real resistor R_r terminating the two-port models the energy dissipation in the far-field.

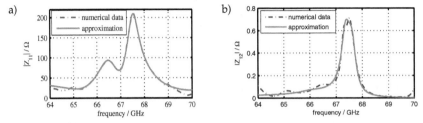

Figure 9. Comparison of the numerical data of (a) $|Z_{11}|$ and (b) $|Z_{12}|$ obtained from the full-wave simulation of the wireless transmission link with the data computed from lumped element model

Figure 9 shows the comparison of the numerical data of the magnitudes of the two-port impedance parameters $|Z_{11}|$ and $|Z_{12}|$ obtained from the full-wave simulation of the wireless transmission link with the data computed from lumped element model. For details of the model see [3]. The numerical full-wave simulations have been performed using CST software. An accurate model of Z_{11} is achieved for the frequency band from 65 GHz to 69 GHz based on four pairs of poles and two single poles at zero and infinity. The frequency range may be extended by increasing the number of poles.

Conclusion

We have investigated methods for an area-efficient design of on-chip integrated antennas, based on the utilization of the same metallization structures both as a CMOS circuit ground plane and as antenna electrodes. An experimental setup has been designed for validating the computed antenna parameters, as well as the interference between the CMOS interconnects and the antenna. Equivalent circuits have been established to model integrated antennas and wireless intra-chip and inter-chip transmission links.

3. Communication theoretical limits, coding and signal processing

Both multiconductor interconnects and wireless multiantenna interconnects can be interpreted as discrete-time, multi-input-multi-output (MIMO) systems. Such systems have been subject to extensive study in the recent past in the field of digital, especially mobile communications. Starting from the analysis of their promising information theoretic capabilities (e.g., [41]), a large amount of signal processing and coding techniques have been developed, that aim at achieving the information theoretic bounds (e.g., [42–44, 46]).

The common approach to handle spatio-temporal interference in MIMO systems, involves either linear or non-linear transmit and receive signal processing, which job is to transform the original MIMO system into a »virtual« MIMO system, where large amounts of spatio-temporal interference have been removed [48]. All state of the art MIMO signal processing techniques have in common that they assume that either the receiver, the transmitter, or both, have access to, or can generate signals with arbitrary precision. This implies, in practice, the existence of ADC and DAC components with a large enough resolution such that the non-linear effects of signal quantization can be neglected. However, in multiconductor, or wireless

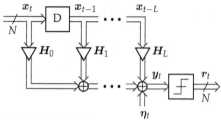

Figure 10. MIMO channel with ISI and single-bit outputs modeling wired as well as the wireless interconnects.

multiantenna interconnects, used for high-speed on-chip or chip-to-chip communication, such an assumption of having available high-resolution ADC and DAC components, cannot be made.

In case of on-chip multiconductor interconnects, the DAC and the ADC components are formed by the output or the input of a logic CMOS inverter, respectively. Hence, the ADC and DAC components, perform a *single-bit* conversion between the analog and the digital domain. With such coarse quantization, all state of the art techniques for MIMO signal processing fail.

In the case of wireless multiantenna interconnects for chip-to-chip communication the situation can be expected to be slightly better. However, because of the huge bandwidth, the requirements on conversion time are extremely high, such that only moderate resolution (4–5 bits) ADC and DAC components are reasonable. As it turns out, such a moderately high resolution is still too low for reliable operation of state of the art MIMO signal processing.

In this section, we treat the ADC and DAC components as an integral part of the MIMO system. We develop signal processing and coding techniques, which utilize the information theoretic gains of MIMO systems with very-low to moderately-low resolution signal quantization. We first provide suitable design principles for low-latency channel-matched codes applied on general frequency selective MIMO channels, which are based on an information theoretic ground.

3.1. Single-bit ADC/DAC: Coding and performance limits

Consider the MIMO channel with inter-symbol interference (ISI) and single-bit output quantization shown in Fig. 10. The channel has a memory of length L and it is governed by the channel law

$$r_t = \mathcal{Q}\{y_t\} = \mathcal{Q}\left\{\sum_{k=0}^{L} H_k x_{t-k} + \eta_t\right\}. \tag{1}$$

Here, $H_k \in \mathbb{C}^{N \times N}$ is the k-th channel matrix. $x_k \in \mathcal{X}$, $\eta_k \in \mathbb{C}^N$, $y_k \in \mathbb{C}^N$ and $r_k \in \mathcal{Y} = \{\alpha + j\beta \,|\, \alpha, \beta \in \{+1, -1\}\}^N$ denote the channel input vector, the noise vector, the unquantized receive vector and the channel output vector, at the k-th time instant, respectively. The single-bit quantization operator $\mathcal{Q}$ returns the sign of the real and imaginary part of each component of the unquantized received signal r_t, i.e.,

$$\mathcal{Q}\{y_t\} = \text{sign}(\text{real}\{y_t\}) + j \cdot \text{sign}(\text{imag}\{y_t\}). \tag{2}$$

The conditional probability of the channel output satisfies

$$\Pr(r_t | x_{-L}^\infty, r_1^{t-1}, r_{t+1}^\infty) = \Pr(r_t | x_{t-L}^t), \ t \geq 0. \tag{3}$$

Here, x_0^∞ and r_1^{t-1} stand for the sequences $[x_{-L}, x_{-L+1}, \ldots]$ and $[r_1, r_2, \ldots, r_{t-1}]$, respectively. The noise is additive white Gaussian with covariance matrix $\mathrm{E}[n_t n_t^{\mathrm{H}}] = \sigma_\eta^2 I_N$. The transmit signal energy is normalized to 1, that is $\|x_t\|_2 = 1$. The signal-to-noise ratio is accordingly defined as

$$\mathrm{SNR} = 1/\sigma_\eta^2. \tag{4}$$

The channel transition probabilities can be calculated via

$$\Pr(r_t | x_{t-L}^t) = \prod_{c \in \{\mathbb{R}, \mathbb{I}\}} \prod_{i=1}^{N} \Phi\left(\frac{[y_t]_{c,i} \cdot [\tilde{y}]_{c,i}}{\sigma_\eta/\sqrt{2}} \right). \tag{5}$$

Here, $[x_t]_{\mathbb{R},i}$ ($[x_t]_{\mathbb{I},i}$) denotes the i-th real (imaginary) component of the input vector x_t, $\tilde{y} = \sum_{k=0}^{L} H_k x_{t-k}$ is the noise-free unquantized receive vector and $\Phi(x) = \frac{1}{\sqrt{2\pi}} \int_{-\infty}^{x} e^{-\frac{t^2}{2}} dt$ is the cumulative normal distribution function. The input symbols are modulated using QPSK (or BPSK with 1-bit DAC). Consequently, we consider ISI channels with the input and output cardinality $|\mathcal{X}| = |\mathcal{R}| = 4^N$.

3.1.1. Code-design

We are looking for codes which maximally increase throughput and that allow fast decoding with good performance. In this way, the code design has to focus on finding low-complexity codes providing good coding gain while having low overhead both with respect to circuit complexity and power dissipation.

Even though linear block and convolution coding schemes are favorable candidates for error correction, they are not able to decrease power consumption and eliminate the residual error floor caused by the crosstalk even in the noiseless case [31]. This is due to their structural properties (linear codes) and the coarse quantization of the channel.

Therefore, the coding schemes which are needed are non-linear and – for having good performance and, at the same time, a low complexity – for instance have memory of order one. In [51–53], an information-theoretic framework was developed as a practical design guideline for novel codes. To this end, the following optimization has been considered in [52]

$$C_{M_s}^{\text{uniform}} = \max_{P_{ij}:(i,j) \in \mathcal{T}} \mathcal{I}(x; y) \quad \text{s.t.} \quad P_{ij} = \Pr(S_k = j | S_{k-1} = i) \in \{0, 2^{-K}\},$$

$$S_k = x_{k-M_s+1}^k \in \mathcal{X}^{M_s},$$

$$M_s \geq L, \tag{6}$$

where K defines the rate $R = K/(2N)$ of the code (for QPSK modulation) and $C_{M_s}^{\text{uniform}}$ is the maximum channel capacity that can be attained with a homogeneous Markov source of order M_s. In general, the capacity of an unconstrained Markov source of order M_s [30] is higher than this *uniform capacity* ($P_{i,j}$ takes 0 or 2^{-K}), i.e., $C_{M_s}^{\text{uniform}} \leq C_{M_s}$.
This coding approach incorporates the following four ideas:

1. In order to avoid the complexity of maximizing an arbitrary Markov source, we restrict the optimization to homogeneous Markov sources.

2. We choose the memory length M_s of the source to be roughly as long as the number of channel taps L but not shorter. The reason is that the information rate of a Markov source of order $M_s = L$ is noticeably larger compared to the i.u.d. capacity, but memory lengths above L yield only a small additional gain in information rate.

3. As we want to avoid the use of distribution shapers, the number of transmit symbols is fixed at 2^K (irrespective of the current state). Thus, the encoder can be realized as a look-up table and we obtain the data encoding rule

$$x_n = \text{ENC}(d_n, [x_{n-1}, \ldots, x_{n-M_s}]),$$

where $d_n \in \{0,1\}^K$ is the data vector of the source. Using QPSK modulation the realizable code rates R are

$$\frac{K}{2N} \in \left\{ \frac{1}{2N}, \frac{2}{2N}, \cdots, 1 \right\}.$$

4. The optimized transition probabilities are uniformly distributed and at the same time they approximate the capacity-achieving input distribution. Hence, the optimized transition probability matrix P_{ij} serves as an inner code that can be concatenated with an outer Turbo-like code, e.g. *low density parity check code* (c.f. section 4), in order to reach information rates (well) above the i.u.d. capacity [53].

Fig. 11 illustrates the channel model together with the encoder and decoder. The optimization

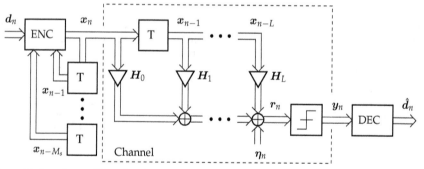

Figure 11. Encoding and Decoding for MIMO Channels with ISI and Single-Bit Output Quantization.

(6) is non-convex but can be solved by an efficient greedy algorithm [52] that delivers an optimized transition probability matrix $P = [P_{i,j}]$ that maximizes the mutual information between the input and the output.

In Figure 12, a coded bus system employing a memory-based code with a code rate of K/N is shown. A fixed bus access time T_{CU}^{cod} is chosen such that two channel temporal taps are significant ($L = 1$). The encoding scheme is time-invariant and has the property that the data vector, $d_n = [d_1[n], \ldots, d_K[n]]^T \in \{0,1\}^{K \times 1}$, is encoded and decoded instantaneously (without latency). The actual code vector x_n depends on the input data vector d_n and the previous transmitted vector x_{n-1}.

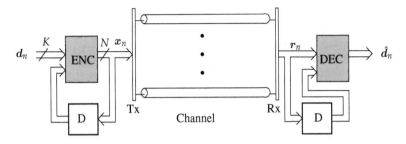

Figure 12. Noisy bus communication with a memory-based Code.

x_{n-1} \ Data	000	001	010	011	100	101	110	111
0000	0000	0001	0011	0111	1000	1100	1110	1111
0001	0000	0001	0111	1000	1001	1100	1101	1111
0010	0000	0010	0011	0100	1010	1011	1110	1111
0011	0000	0010	0011	0100	1010	1011	1110	1111
0100	0000	0001	0100	0101	0111	1100	1101	1111
0101	0000	0001	0100	0101	0111	1100	1101	1111
0110	0000	0011	0100	0110	0111	1100	1110	1111
0111	0000	0001	0100	0110	0111	1100	1110	1111
1000	0000	0001	0011	1000	1001	1011	1110	1111
1001	0000	0001	0011	1000	1001	1011	1100	1111
1010	0000	0010	0011	0100	1010	1011	1110	1111
1011	0000	0010	0011	0100	1010	1011	1110	1111
1100	0000	0001	0100	0101	0111	1100	1101	1111
1101	0000	0001	0100	0101	0111	1100	1101	1111
1110	0000	0010	0011	0110	0111	1000	1110	1111
1111	0000	0001	0011	0111	1000	1100	1110	1111

Table 1. The mapping function of a 3/4 optimized code: $x_n = \mathrm{ENC}(d_n, x_{n-1})$.

At the receiver side, the decoder uses the value of the current and the previous channel outputs to reconstruct an estimate of the data vector $\hat{d}_n$ as

$$\hat{d}_n = g(y_n, y_{n-1}) = \arg\max_d \Pr\left(d_n = d \mid y_n, y_{n-1}\right). \qquad (7)$$

Obviously, the mapping done by this function performs a maximum-likelihood estimation of d_n based on y_n and y_{n-1} [1].

Although this approach seems quite heuristic, its usefulness can be demonstrated by simulation. Table 1 lists the mapping function of a code designed for a bus with $N = 4$ mutually coupled, tapped RC lines as shown in Fig. 1 used at $T_{\mathrm{CU}} = 9RC_s = 1.5RC_c$ symbol time, where R is the serial resistance, C_s is the ground capacitance and C_c is the coupling capacitance (c.f. section 2.1).

---

[1] The reliability of the decoding can be improved if earlier or later outputs are considered using the BCJR algorithm (forward-backward algorithm), at the cost of some latency.

Its performance in terms of symbol error rate (SER) when applied to a noisy bus system, compared to uncoded transmission, is shown in Figure 13. The uncoded transmission reveals an error floor (a residual SER at vanishing noise variance) due to signaling belong the RC-specific time. However, as we see in Figure 13, the optimized code does not see any error floor. Besides, it turns out that the achievable power savings of this code (in terms of energy per transmitted information bit) is 40%, without taking into account the power overhead of the codec circuit. The SER curve of a space-only code, which has been optimized by exhaustive search, is also plotted. Due to its simplicity, this code performs inherently worse than the discussed memory-based code. Although several coding schemes can be found in the literature [29, 32], such a unified framework that jointly address power, rate, and reliability aspects, simultaneously is new.

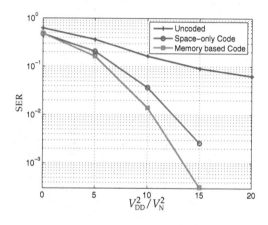

Figure 13. Symbol error rate (SER) as function of V_{DD}^2 / V_N^2 for an interconnect with four lines, employing a memory-based code of rate 3/4, compared to the uncoded case. The performance of a space-only code is also plotted.

We note that, for large buses, it is impractical to encode all bits at once because of the large complexity in the design and the implementation of the codec circuit. Therefore, partial coding can be employed in which the bus is partitioned into sub-buses of smaller width, which are encoded separately. The partitioning requires some additional wires since a shielding wire has to be placed between every two adjacent sub-buses.

3.2. Low-resolution ADC: Linear signal-processing

In the following, we concentrated on receive signal processing and our aim is to study the applicability of standard equalization techniques for our application, where the receiver is equipped with a low to moderate ADC for each antenna or port. A modified version of the standard linear receiver designs is presented in the context of MIMO communication with quantized output, taking into account the presence of the quantizer. An essential aspect of our analysis is that no assumption of uncorrelated white quantization errors is made. The performance of the modified receiver designs as well as the effects of quantization are studied theoretically and experimentally. Thereby, perfect channel state information (CSI) at the receiver is assumed, which can be obtained even with coarse quantization as discussed in Section 3.3.

In [33], the joint optimization of the linear receiver and the quantizer in a MIMO system is addressed. The figure of merit that has been used for the design of the optimum quantizer and receiver is the *mean square error* (MSE). Based on this MSE approach, the communication performance (in terms of channel capacity) of the quantized MIMO channel is studied. Our work [34] generalizes this modified MMSE filter to frequency selective channels. Motivated by the same approach, the authors of [36] optimized the Decision Feedback Equalizer (DFE) for the flat MIMO channel with quantized outputs.

In this and the following Section, we provide a summary of these works. Throughout these sections, $r_{\alpha\beta}$ denotes $E[\alpha\beta^*]$. The operators $(\bullet)^T$, $(\bullet)^H$, $(\bullet)^*$, $\text{tr}[\bullet]$ stand for transpose, Hermitian transpose, complex conjugate, and trace of a matrix, respectively.

3.2.1. System model

Let us now consider a point to point MIMO Gaussian channel, where the transmitter operates M antennas and the receiver employs N antennas. Figure 14 shows the general form of a quantized MIMO system, where $H \in \mathbb{C}^{N \times M}$ is the channel matrix. For simplicity, inter-symbol interference (ISI) is ignored, even though considering it would be straightforward. The vector $x \in \mathbb{C}^M$ comprises the M transmitted symbols with zero-mean and covariance $R_{xx} = E[xx^H]$. The vector η refers to zero-mean complex circularly symmetric Gaussian noise with a covariance matrix $R_{\eta\eta} = E[\eta\eta^H]$, while $y \in \mathbb{C}^N$ is the unquantized channel output:

$$y = Hx + \eta. \tag{8}$$

In our system, the real parts $y_{i,R}$ and the imaginary parts $y_{i,I}$ of the receive signals y_i, $1 \le i \le N$, are each quantized by a b-bit resolution uniform/non-uniform scalar quantizer. Thus, the resulting quantized signals are given by

$$r_{i,l} = Q(y_{i,l}) = y_{i,l} + q_{i,l}, \, l \in \{R, I\}, \, 1 \le i \le N, \tag{9}$$

where $Q(\cdot)$ denotes the quantization operation and $q_{i,l}$ is the resulting quantization error. The matrix $G \in \mathbb{C}^{M \times N}$ represents the receive filter, which delivers the estimate $\hat{x}$

$$\hat{x} = Gr. \tag{10}$$

Our aim is to choose the quantizer and the receive matrix G minimizing the MSE $= E[\|\hat{x} - x\|_2^2]$, taking into account the quantization effect. Since the ADC can drastically affect the performance of the system, it should be also designed carefully.

3.2.2. Quantizer characterization

Each quantization process can be given a distortion factor $\rho_q^{(i,l)}$ to indicate the relative amount of quantization noise generated, which is defined as follows

$$\rho_q^{(i,l)} = \frac{E[q_{i,l}^2]}{r_{y_{i,l}y_{i,l}}}, \tag{11}$$

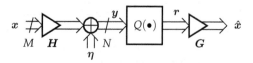

Figure 14. Quantized MIMO System.

where $r_{y_{i,l}y_{i,l}} = \mathrm{E}[y_{i,l}^2]$ is the variance of $y_{i,l}$ and the distortion factor $\rho_q^{(i,l)}$ depends on the number of quantization bits b, the quantizer type (uniform or non-uniform) and the probability density function of $y_{i,l}$. Note that the signal-to-quantization noise ratio (SQNR) has an inverse relationship with regard to the distortion factor. The uniform/non-uniform quantizer design is based on minimizing the *mean square error* (distortion) between the input $y_{i,l}$ and the output $r_{i,l}$ of each quantizer. In other words, the SQNR values are maximized. With this optimal design of the scalar finite resolution quantizer, whether uniform or not, the following equations hold for all $0 \leq i \leq N, l \in \{R, I\}$ [35, 37, 38]

$$\mathrm{E}[r_{i,l}q_{i,l}] = 0 \tag{12}$$

$$\mathrm{E}[y_{i,l}q_{i,l}] = -\rho_q^{(i,l)}r_{y_{i,l}y_{i,l}}. \tag{13}$$

Obviously, (13) follows from (11) and (12). Under multipath propagation conditions and for large number of antennas, the quantizer input signals $y_{i,l}$ will be approximately Gaussian distributed and thus, they undergo nearly the same distortion factor ρ_q, i.e., $\rho_q^{(i,l)} = \rho_q \ \forall i \forall l$. Furthermore, the optimal parameters of the uniform as well as the non-uniform quantizer and the resulting distortion factor ρ_q for Gaussian distributed signal are tabulated in [35] for different resolutions b.

Now, let $q_i = q_{i,R} + jq_{i,I}$ be the complex quantization error. Under the assumption of uncorrelated real and imaginary part of y_i, the following relations are obtained

$$r_{q_iq_i} = \mathrm{E}[q_iq_i^*] = \rho_q r_{y_iy_i}, \text{ and } r_{y_iq_i} = \mathrm{E}[y_iq_i^*] = -\rho_q r_{y_iy_i}. \tag{14}$$

This particular choice of the (non-)uniform scalar quantizer minimizing the distortion between r and y, combined with the receiver developed in the next Section, is also optimal with respect to the total MSE between the transmitted symbol vector x and the estimated symbol vector $\hat{x}$, as we will see later.

3.2.3. Nearly optimal linear receiver

The linear receiver G that minimizes the MSE, $\mathrm{E}[\|\varepsilon\|_2^2] = \mathrm{E}[\|x - \hat{x}\|_2^2] = \mathrm{E}[\|x - Gr\|_2^2]$, can be written as:

$$G = R_{xr}R_{rr}^{-1}, \tag{15}$$

and the resulting MSE equals

$$\mathrm{MSE} = \mathrm{tr}\left(R_{\varepsilon\varepsilon}\right) = \mathrm{tr}\left(R_{xx} - R_{xr}R_{rr}^{-1}R_{xr}^{\mathrm{H}}\right), \tag{16}$$

where R_{xr} equals

$$R_{xr} = \mathrm{E}[xr^{\mathrm{H}}] = \mathrm{E}[x(y+q)^{\mathrm{H}}] = R_{xy} + R_{xq}, \tag{17}$$

and R_{rr} can be expressed as

$$R_{rr} = \mathrm{E}[(y+q)(y+q)^{\mathrm{H}}] = R_{yy} + R_{yq} + R_{yq}^{\mathrm{H}} + R_{qq}. \tag{18}$$

We have to determine the linear filter G as a function of the channel parameters and the quantization distortion factor ρ_q. To this end, we derive all needed covariance matrices by using the fact that the quantization error q_i, conditioned on y_i, is statistically independent from all other random variables of the system. First we calculate $r_{y_i q_j} = \mathrm{E}[y_i q_j^*]$ for $i \neq j$

$$\mathrm{E}[y_i q_j^*] = \mathrm{E}_{y_j}\Big[\mathrm{E}[y_i q_j^* | y_j]\Big]$$

$$= \mathrm{E}_{y_j}\Big[\mathrm{E}[y_i | y_j]\mathrm{E}[q_j^* | y_j]\Big]$$

$$\approx \mathrm{E}_{y_j}\Big[r_{y_i y_j} r_{y_j y_j}^{-1} y_j \mathrm{E}[q_j^* | y_j]\Big] \tag{19}$$

$$= r_{y_i y_j} r_{y_j y_j}^{-1} \mathrm{E}[y_j q_j^*]$$

$$= -\rho_q r_{y_i y_j}. \tag{20}$$

Note that, in (19), we approximate the Bayesian estimator $\mathrm{E}[y_i | y_j]$ with the linear estimator $r_{y_i y_j} r_{y_j y_j}^{-1} y_j$, which holds with equality if the vector y is jointly Gaussian distributed. Eq. (20) follows from (14). Summarizing the results of (14) and (20), we obtain

$$R_{yq} \approx -\rho_q R_{yy}. \tag{21}$$

Similarly, we evaluate $r_{q_i q_j}$ for $i \neq j$ using (21), and with (14) we arrive at

$$R_{qq} \approx \rho_q R_{yy} - (1 - \rho_q)\rho_q \mathrm{nondiag}(R_{yy}), \tag{22}$$

where $\mathrm{nondiag}(A)$ obtained from a matrix A by setting its diagonal elements to zero.
Inserting the expressions (21) and (22) into (18), we obtain

$$R_{rr} \approx (1 - \rho_q)(R_{yy} - \rho_q \mathrm{nondiag}(R_{yy})). \tag{23}$$

Also in a similar way, we get $R_{xq} = \mathrm{E}[xq^{\mathrm{H}}] \approx -\rho_q R_{xy}$, and (17) becomes

$$R_{xr} \approx (1 - \rho_q)R_{xy}. \tag{24}$$

In summary, we get from (23) and (24) the following expression for the Wiener filter from (15) operating on quantized data

$$G_{\mathrm{WFQ}} \approx R_{xy}(R_{yy} - \rho_q \mathrm{nondiag}(R_{yy}))^{-1}, \tag{25}$$

and for the resulting MSE, we obtain using (16)

$$\mathrm{MSE}_{\mathrm{WFQ}} \approx \mathrm{tr}\Big[R_{xx} - (1 - \rho_q)R_{xy}(R_{yy} - \rho_q \mathrm{nondiag}(R_{yy}))^{-1}R_{xy}^{\mathrm{H}}\Big]. \tag{26}$$

We obtain R_{yy} and R_{xy} easily from our system model

$$R_{yy} = R_{\eta\eta} + H R_{xx} H^{\mathrm{H}}, \tag{27}$$

$$R_{xy} = R_{xx} H^{\mathrm{H}}. \tag{28}$$

Let us examine the first derivative of the MSE_{WFQ} in (26) with respect to ρ_q

$$\frac{\partial \text{MSE}_{\text{WFQ}}}{\partial \rho_q} = \text{tr}\left[G_{\text{WFQ}}\text{diag}(R_{yy})G_{\text{WFQ}}^{\text{H}}\right] > 0, \tag{29}$$

where G_{WFQ} is given in (25). Therefore the MSE_{WFQ} is monotonically increasing in ρ_q. Since we choose the quantizer to minimize the distortion factor ρ_q, our receiver and quantizer designs are jointly optimum with respect to the total MSE.

3.2.4. Lower bound on the mutual information and the capacity

In this section, we develop a lower bound on the mutual information rate between the input sequence x and the quantized output sequence r of the system in Figure 14, based on our MSE approach. Generally, the mutual information of this channel can be expressed as [26]

$$I(x, r) = H(x) - H(x|r). \tag{30}$$

Given R_{xx} under a power constraint $\text{tr}(R_{xx}) \leq P_{\text{Tr}}$, we choose x to be Gaussian, which is not necessarily the capacity achieving distribution for our quantized system. Then, we can obtain a lower bound for $I(x, r)$ (in bit/transmission) as

$$I(x, r) = \log_2 |R_{xx}| - h(x|r)$$

$$= \log_2 |R_{xx}| - h(x - \hat{x}|r)$$

$$\geq \log_2 |R_{xx}| - h(x - \hat{x}) \tag{31}$$

$$\geq \log_2 \frac{|R_{xx}|}{|R_{\varepsilon\varepsilon}|}. \tag{32}$$

Since conditioning reduces entropy, we obtain inequality (31). On the other hand, the second term in (31) is upper bounded by the entropy of a Gaussian random variable whose covariance is equal to the error covariance matrix $R_{\varepsilon\varepsilon}$ of the linear MMSE estimate of x. Finally, we get using (26) and (28)

$$I(x, r) \gtrsim -\log_2 \left|I - (1-\rho_q)R_{xy}(R_{yy}-\rho_q\text{nondiag}(R_{yy}))^{-1}H\right|. \tag{33}$$

Considering the case of low SNR values, we get easily with $R_{yy} \approx R_{\eta\eta}$, (33) and (28), the following first order approximation of the mutual information[2][3]

$$I(x, r) \gtrsim (1 - \rho_q)\text{tr}[R_{xx}H^{\text{H}}R_{\eta\eta}^{-1}H] / \log(2). \tag{34}$$

Compared with the mutual information $I(x, y)$ for the unquantized case, also at low SNR [39], the mutual information for the quantized channel degrades only by the factor $(1 - \rho_q)$. For the spacial case $b = 1$, we have $\rho_q|_{b=1} = 1 - \frac{2}{\pi}$ (see [35]) and the degradation of the mutual

---

[2] We assume also that $\rho_q \ll 1$ (or $R_{\eta\eta}$ is diagonal).
[3] Note that $\log |I + \Delta X| \approx \text{tr}(\Delta X)$.

information becomes

$$\lim_{\text{SNR}\to 0} \frac{I(\boldsymbol{x},\boldsymbol{r})}{I(\boldsymbol{x},\boldsymbol{y})}\bigg|_{b=1} \approx \frac{2}{\pi}. \tag{35}$$

In other words, the power penalty due to the 1-bit quantization is approximately equal $\frac{\pi}{2}$ (1.96 dB) at low SNR. This shows that mono-bit ADCs may be used to save system power without an excessive degradation in performance, and confirms the significant potential of the coarsely quantized UWB MIMO channel. Using a different approach, [40] presented a similar result, and showed that the above approximation is asymptotically exact.

3.2.5. Simulation results

The performance of the modified Wiener filter for a 4- and 5-bit quantized output MIMO system (WFQ), in terms of BER averaged over 1000 channel realizations, is shown in Figure 15 for a 10×10 MIMO system (QPSK), compared with the conventional Wiener filter (WF) and Zero-forcing filter (ZF). The symbols and the noise samples are assumed to be uncorrelated, that is, $\boldsymbol{R}_{xx} = \sigma_x^2 \boldsymbol{I}$ and $\boldsymbol{R}_{\eta\eta} = \sigma_\eta^2 \boldsymbol{I}$. Hereby, the SNR (in dB) is defined as

$$\text{SNR} = 10 \cdot \log_{10} \left(\frac{\sigma_x^2}{\sigma_\eta^2} \right). \tag{36}$$

Furthermore, we used a generic channel model, where the entries of $\boldsymbol{H}$ are complex-valued realizations of independent zero-mean Gaussian random variables with unit variance. Clearly, the modified Wiener filter outperforms the conventional Wiener filter at high SNR. This is because the effect of the quantization error is more pronounced at higher SNR values when compared to the additive Gaussian noise variance. Since the conventional Wiener filter converges to the ZF-filter at high SNR values and loses its regularized structure, its performance degrades asymptotically to the performance of the ZF-filter, when operating on quantized data. For comparison, we also plot the BER curves for the WF and ZF filter, for the case when no quantization is applied.

3.3. Channel estimation

Because in general, the MIMO channel cannot be assumed known a-priori, a channel estimation has to be performed. In practice, it is highly desirable that the channel is estimated directly by the communication device – in our case by on-chip digital circuitry. However, this implies that the channel estimator is restricted to use received signal samples of a pilot sequence after single-bit quantization in the extreme case. This motivates investigation of channel estimation with coarse quantization. This problem was first addressed by [27], where a maximum likelihood (ML) channel estimation with quantized observation is presented. In general, the solution cannot be given in closed form, but requires an iterative numerical approach, which hampers the analysis of performance.

In [28], it has been shown that – in contrast to unquantized channel estimation – different orthogonal pilot sequences (with same average total transmit power and same length) yield different performances. Especially, orthogonality in the time-domain (time-multiplexed pilots) can be preferable to orthogonality in space. With orthogonal pilots

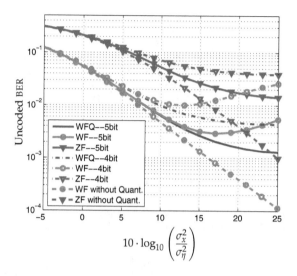

$$10 \cdot \log_{10}\left(\frac{\sigma_x^2}{\sigma_\eta^2}\right)$$

Figure 15. The WFQ vs. the conventional WF and ZF receivers, QPSK modulation with $M = 10$, $N = 10$, 4- ($\rho_q = 0.01154$) and 5- ($\rho_q = 0.00349$) bit uniform quantizer.

that are multiplexed in time, the problem can be reduced from the MIMO to the SIMO (single-input-multiple-output) case, because each line of the multiconductor interconnect is excited separately for time-multiplexed pilots. Finally, the problem can be reduced to the SISO (single-input-single-output) case, when the channel estimation is performed separately in parallel at each receiving end of the multiconductor interconnect. For this case, in [28], a closed-form solution can be found for the maximum likelihood channel estimation problem, which makes performance analysis possible in an analytical fashion.

In [50], a more general setting for parameter estimation based on quantized observations was studied, which covers many processing tasks, e.g. channel estimation, synchronization, delay estimation, Direction Of Arrival (DOA) estimation, etc. An Expectation Maximization (EM) based algorithm is proposed to solve the Maximum a Posteriori Probability (MAP) estimation problem. Besides, the Cramér-Rao Bound (CRB) has been derived to analyze the estimation performance and its behavior with respect to the signal-to-noise ratio (SNR). The presented results treat both cases: pilot aided and non-pilot aided estimation. The paper extensively dealt with the extreme case of single bit quantization (comparator) which simplifies the sampling hardware considerably. It also focused on MIMO channel estimation and delay estimation as application area of the presented approach. Among others, a 2×2 channel estimation using 1-bit ADC is considered, which shows that reliable estimation may still be possible even when the quantization is very coarse, with any desired accuracy, provided the pilot sequence is long enough. Since in on-chip and chip-to-chip communications, the channel almost does not change in time, it is possible to use very long pilot sequences, and run the channel estimation only once, or once in a while.

4. Efficient digital hardware architecture

Sole optimization of transmitting power in the standardization and conception phase of communication channels results in highly complex and energy-intensive receivers with a complex channel decoder as one of its key components. Neglecting the energy dissipation of the integrated decoder in this early phase results in suboptimal and, thus, costly communication systems in terms of manufacturing and usage costs. In the previous part of this chapter approaches to reduce the ADC complexity and, thus, the complexity of the subsequent digital components by using single-bit or medium-low resolution quantizations have been discussed. A quantitative comparison of these new approaches to standard receivers requires accurate cost models of the digital components. Quite accurate cost models are available for most of the communication system components except for channel decoders. While such cost models can be easily derived for Viterbi, Reed-Solomon, and Turbo Decoders, an estimation of the silicon area and the energy dissipation of LDPC decoders is challenging due to the high internal communication effort between the basic components.

Although LDPC codes have already been introduced by Gallager in 1962 [55], up to now they are known to achieve the best decoding performance [56] and are adopted in various communication standards (e.g. [57],[58],[59]) and other applications such as hard-disk drives [60]. They belong to the class of block codes and, thus, can be defined by a parity-check matrix H with m rows and n columns or by the corresponding Tanner Graph. Both are shown in Figure 16 for a very simplified LDPC code. Each row of the parity-check matrix represents one parity check wherein a '1'-entry in column i and row j indicates, that the received symbol i takes part in parity check number j. In the Tanner Graph such a parity check is represented by one so called check node and each column by one bit node. Furthermore, the number of one entries per row d_C (column d_V) defines the number of connected bit (check) nodes per check (bit) node.

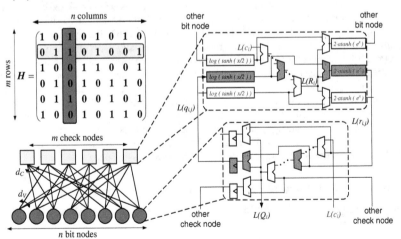

Figure 16. Parity-check matrix, Tanner-Graph and non-linear recursive decoder loop

Figure 16 also illustrates one of the $n \cdot d_V$ non-linear recursive loops of the resulting decoder. In each iteration the extrinsic information $L(q_{i,j})$ on the received symbol i is sent to check node j. Here, new A-posteriori information $L(r_{i,j})$ is derived. The sign of $L(r_{i,j})$ is chosen in such a way, that the confidence in the received symbol i indicated by the magnitude of $L(q_{i,j})$ increases. For the sake of clarity only the magnitude calculation is illustrated in Figure 16. Considering the original Sum-Product decoding algorithm [55] the check node consists of transcendent functions and a multi-operand adder with subsequent subtractor stages. Here, the basic idea is, whenever all participating symbols in that parity check feature a high confidence in their current estimation, the magnitude of the A-posteriori information is high. The A-posteriori information $L(r_{i,j})$ is then sent back to the bit node. Here, all information of symbol i, namely the d_V A-posteriori information and the received information $L(c_i)$, are combined using a multi-operand adder resulting in a new estimation $L(Q_i)$ of symbol i. To avoid decoding-performance-demoting cycles, in the next decoding iteration only the extrinsic information $L(q_{i,j}) = L(Q_i) - L(r_{i,j})$ is used instead of $L(Q_i)$. For more information on the decoding algorithm and possible fix-point realizations refer to [61].

A metric for the code's and, thus, the decoder's complexity is the number of '1'-entries in the matrix $n \cdot d_V = m \cdot d_C$. Each '1'-entry can be assigned to a part of the bit- and check-node logic as highlighted in gray in Figure 16. Thereby, each '1'-entry leads to four two-operand adders/subtractors, a block for the calculation of $log\left(tanh\left(^x/_2\right)\right)$, a block for the calculation of $2 \cdot atanh\left(e^x\right)$ and a register stage at the output of the bit node. Additionally, $n \cdot d_V$ is a measure for the communication between the nodes as $2 \cdot w \cdot n \cdot d_V$ bits are exchanged between the bit and the check nodes in each decoding iteration with w being the word length of the exchanged messages.

In high-throughput applications with a time-invariant parity-check matrix all bit and check nodes are typically instantiated in parallel as in the first integrated LDPC decoder [66]. Here, typically the m check nodes are realized in the center of the decoder floorplan surrounded by the n bit-node instances. The communication between the nodes is then realized by $2 \cdot w \cdot n \cdot d_V$ dedicated interconnect lines. In [66] the logic area, which is the accumulated silicon area of all logic gates, is approximately $25\,mm^2$. However, the total of 26,624 interconnect lines can not be realized on this area. The silicon area needs to be artificially expanded until a successful routing of all interconnect lines could be established. The resulting global interconnect has a length of $80\,m$ on a macro size of $52.5\,mm^2$. Thus, only 50% of the active silicon area is utilized in the final decoder. The impact of the complex global interconnect complicates the derivation of accurate area, timing and energy cost models which might be the reason why no cost models are available in literature so far. However, such models are necessary to avoid costly wrong decisions in early design phases, for example when choosing a certain LDPC code in the system-conception phase. Also in later design phases those models are indispensable, for example for a quantitative exploration of the architecture design space.

4.1. Accurate area, timing, and energy cost models

In general the silicon area of a high-throughput LDPC decoders can be estimated using

$$A_{DEC\_P} = max(A_L, A_R), \tag{37}$$

with A_L being the logic area and A_R the required area to realize the global interconnect. To reduce the logic area typically the approximative Min-Sum algorithm [70] is used which estimates the magnitude of $L(r_{i,j})$ using the minimal and second minimal magnitude of $L(q_{i,j})$ (e.g. [62],[63],[64],[65]). The derivation of A_L for this decoding algorithm as the accumulated silicon area of all logic gates has been presented in [67]. The resulting total logic area can be estimated using

$$A_L = l_L^2 = 1000 \cdot n \cdot d_V \cdot (11.5 \cdot w + 2 \cdot ld\,(d_V)) \cdot \lambda^2. \tag{38}$$

This equation reveals a linear dependency between the code complexity $n \cdot d_V$ and the accumulated gate area.

The major challenge in deriving an accurate routing-area model is the adaptability to different LDPC codes. It is possible to divide the problem into two parts: an estimation of the available and the required Manhattan length. Considering a certain logic area, the available Manhattan length is a measure for the routing resources above the decoder's node logic. Considering that the node layouts require M_L of the total M metal layers in the CMOS stack for the local interconnect, $M_R = M - M_L$ metal layers are available for the realization of the global bit- and check-node communication. The required routing area A_R can then be determined by equalizing the available and the required Manhattan lengths. This means, that the available Manhattan length allows the realization of the required Manhattan length. Thereby, the available Manhattan length can be derived as

$$l_{AVAIL} = \frac{u}{p} \cdot \left(l_L^2 \cdot M_R + min\left(l_{DEC}^2 - l_L^2, 0\right) \cdot M\right), \tag{39}$$

with the routing pitch p, an utilization factor u for each metal layer, and a decoder macro side length of l_{DEC}. Considering that no artificial increase of the decoder is required ($l_{DEC} \leq l_L$) the second term is zero and the available routing resources are on top of the node logic. If the decoder needs to be expanded, the whole metal stack is available between the node instances for the realization of the global interconnect. Therefore, this part is weighted with M.

The estimation of the required Manhattan length is more challenging as it depends on code characteristics as for example the number of interconnect lines and the average length of one interconnect line. An upper bound estimation of the required Manhattan length could be derived by using the maximum possible length l_{MAX} of one bit- and check-node connection which is shown in Figure 17(a). In a typical placement with the bit nodes surrounding the check-node array the longest possible connection runs from one corner of the decoder macro to the opposing corner of the check-node array. An analysis of the logic model [67] shows that the check nodes occupy about 60% of the complete decoder macro leading to a maximum Manhattan length of

$$l_{MAX} = 2 \cdot \left(l_{CNA} + \frac{l_{DEC} - l_{CNA}}{2}\right) = l_{DEC} + l_{CNA} = 1.77 \cdot l_{DEC}. \tag{40}$$

When looking at the wire-length histogram for an exemplary code (see Figure 17(b)) the average Manhattan length is significantly smaller than the maximum length leading to an overestimation of the required Manhattan length and, thus, of the required routing area. An analysis of various LDPC codes showed, that the shape of the wire-length histogram is always similar. Especially, the ratio between the average and the maximum Manhattan length was found to be almost constant as can be seen from Table 2. For the derivation of the average Manhattan lengths all placements have been optimized using a custom simulated annealing process [62]. While code no. 11 is the code adopted in [57], the other codes are taken from

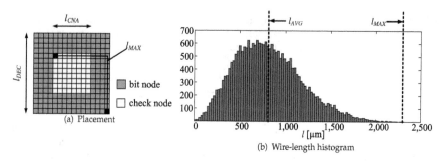

Figure 17. Bit- and check-node architecture

Code nr.	n	m	d_V	d_C	$n{\cdot}d_V$	l_{AVG}/l_{MAX}	ρ_{AVR}/ρ_{MAX} vertical	ρ_{AVR}/ρ_{MAX} horizontal
1	96	48	3	6	288	0.33	0.58	0.59
2	408	204	3	6	1224	0.31	0.56	0.55
3	408	204	3	6	1224	0.30	0.55	0.56
4	408	204	3	6	1224	0.31	0.54	0.50
5	816	408	3	6	2448	0.31	0.52	0.57
6	816	408	5	10	4080	0.34	0.52	0.58
7	816	408	5	10	4080	0.34	0.52	0.57
8	816	408	5	10	4080	0.34	0.50	0.55
9	999	111	3	27	2997	0.37	0.36	0.33
10	1008	504	3	6	3024	0.32	0.54	0.46
11	2048	384	6	32	12288	0.37	0.42	0.40
12	4000	2000	3	6	12000	0.34	0.57	0.54
13	4000	2000	4	8	16000	0.35	0.55	0.55
14	8000	4000	3	6	24000	0.35	0.55	0.45

Table 2. Interconnect properties of various LDPC codes

[68]. For a wide range of LDPC codes with code complexities $n \cdot d_V$ between 300 and 24,000 the ratio varies only between 0.30 and 0.37. Approximating the ratio of the average to the maximum Manhattan length with 0.35 and using (40), the required Manhattan length can be estimated based on the decoder side length as

$$l_{REQ} = 1.2 \cdot n \cdot d_V \cdot w \cdot l_{DEC}. \tag{41}$$

Additionally, an estimation of the achievable utilization is possible based on the comparison of the average routing density ρ_{AVG} and the maximum routing density ρ_{MAX}. The ratios of these values for vertical and horizontal interconnect lines are also given in Table 2. Although there are exceptions (e.g. code no. 9) the utilization $u = \rho_{AVG}/\rho_{MAX}$ is almost constant and will be chosen to $u = 0.5$ in the following.

Considering that the decoder area needs to be expanded and assuming a uniform stretch (39) and (41) still hold. Then, the minimal required decoder area to realize the global interconnect A_R can be calculated by equating (39) and (41) and solving for l_R as

$$A_R = l_R^2 = \left(1.2 \cdot \frac{n \cdot d_V \cdot w}{M} \cdot p + \sqrt{\left(1.2 \cdot \frac{n \cdot d_V \cdot w}{M} \cdot p\right)^2 - l_L^2 \cdot \frac{M_R - M}{M}} \right)^2 . \tag{42}$$

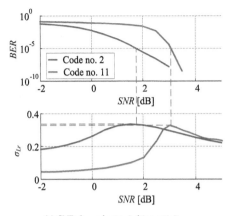

Code nr.	SNR [dB]	$\sigma_{L(q)}$	$\sigma_{L(r)}$
1	2.6	0.14	0.30
2	1.7	0.17	0.34
3	1.7	0.17	0.33
4	2.9	0.14	0.30
5	1.5	0.18	0.34
6	0.5	0.16	0.33
7	0.5	0.16	0.33
8	0.5	0.16	0.33
9	9.8	0.41	0.29
10	1.4	0.19	0.34
11	3.0	0.20	0.33
12	1.2	0.19	0.34
13	0.1	0.20	0.35
14	1.2	0.19	0.34

(a) SNR-dependent switching activity   (b) Table of switching activities ($BER = 10^{-5}$)

Figure 18. Switching activity

In contrast to the logic area, the routing area shows a quadratic dependence on code complexity. By comparing (38) and (42) it can be shown that the bit-parallel decoder is routing dominated as soon as

$$n \cdot d_V \geq 500 \cdot \frac{M_R^2}{w}. \tag{43}$$

The required artificial increase of the silicon area also impacts the other two decoder features: the energy per iteration E_{IT} and the iteration period, which is the required time for one decoding iteration and the inverse of the block throughput [69]. Here, only the interconnect fraction of the decoder energy will be discussed in detail. For more information on the derivation of the iteration period and the total decoder energy refer to [67]. The dynamic energy dissipation of the global interconnect can be estimated using

$$E_{INT} = \frac{1}{2} \cdot \left(\sigma_{Lq}\left(BER\right) + \sigma_{Lr}\left(BER\right) \right) \cdot \alpha \cdot C' \cdot \frac{l_{REQ}}{2} \cdot V_{DD}^2, \tag{44}$$

with V_{DD} being the supply voltage, C' the capacitive load per unit length of a minimum-spaced interconnect line and α a fitting factor to cover the fact, that on average the global interconnect lines are not minimum spaced [67]. Furthermore, the switching activities on the interconnect lines from bit to check nodes (σ_{Lq}) and vice versa (σ_{Lr}) need to be considered. In Figure 18(a) the BER and the switching activity σ_{Lr} for two codes from Table 2 and different signal-to-noise ratios are illustrated. The switching activity highly depends on the considered SNR and is especially high in the so-called waterfall region when the BER starts to get significantly smaller. Furthermore, the two codes strongly differ when it comes to comparing the switching activities for a given SNR (e.g. 1dB). But, considering a specific BER (indicated by the dashed lines) an almost equal switching activity for the two codes can be observed (approx. 0.33 for a BER of 10^{-5}). The comparison of the switching activities σ_{Lq} and σ_{Lr} for all codes listed in Figure 18(b) shows, that this behavior is common for almost all other codes, as well. Therefore, a quite accurate estimation of the decoder energy based on the code parameters n and d_V is possible without knowledge of the actual LDPC code.

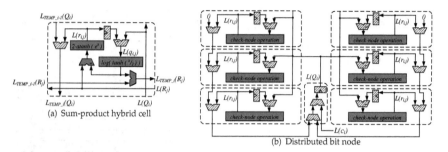

(a) Sum-product hybrid cell

(b) Distributed bit node

Figure 19. Hybrid-cell decoder architecture

4.1.1. Hybrid-cell decoder architecture

The main routing problem of the bit- and check-node architecture arises from the high routing density at the border of the check-node array as it can be seen in the interconnect-density chart in Figure 20(a) for an exemplary code [57]. To overcome this drawback it is possible to break up the bit- and check-node clustering of the logic and rearrange it. The new idea is based on the observation, that each '1'-entry in the parity-check matrix can be assigned to certain parts of the decoder loop. Then, the decoder consists of $n \cdot d_V$ small, equal basic components. A combination of the logic for one '1'-entry (see grey blocks in Figure 16) leads to the block diagram of one hybrid cell, as it is shown in Figure 19(a). This hybrid cell gets the accumulated information $L_{TEMP\_i-1}(Q_i)$ of the received A-priori information $L(c_i)$ and of all A-posteriori information of the previous hybrid cells and adds the A-posteriori information $L(r_{i,j})$ of check node j. The resulting information $L_{TEMP\_i}(Q_i)$ is forwarded to the next hybrid cell. The last hybrid cell in that column calculates $L(Q_i)$ and sends this value back to all participating hybrid cells. A similar structure is used in the check-node part of the hybrid-cell where the calculation of $L(R_i)$ is distributed over d_C hybrid cells. Although, here, the hybrid-cell approach considers a Sum-Product algorithm, it is also applicable to a Min-Sum based decoder. Therefore, the Φ function and the multi-operand adder have to be replaced with basic compare-and-swap cells.

In contrast to the bit- and check-node architecture, in which the $(d_V + 1)$-operand adder in the bit node and the d_C-operand adder in the check node would be realized using a tree topology, the hybrid-cell architecture is based on an adder chain topology. However, it is possible to introduce tree-stages for the bit-node operation as illustrated in Figure 19(b). The $L(r_{i,j})$ values are accumulated in two branches and the intermediate results are added to the channel information $L(c_i)$ in an additional IO cell. A similar topology is possible for the check-node operation. The global interconnect of the hybrid-cell architecture has been realized in a 90-nm CMOS technology using five metal layers for the same code as used for the bit- and check-node architecture in Figure 20(a). In a first step, the placement of the nodes has been optimized using a custom simulated annealing process. Thereby, a placement scheme as depicted in Figure 20(b) with the hybrid cells surrounded by the io cells has been assumed. The advantage of the hybrid-cell architecture becomes obvious when comparing the two interconnect densities. The routing density of the hybrid-cell architecture is distributed more uniformly especially without high density peaks at the border of the bit- and check-node array. Thus, the average routing density of the hybrid-cell interconnect is higher than of the bit- and check-node architecture, promising a smaller silicon area.

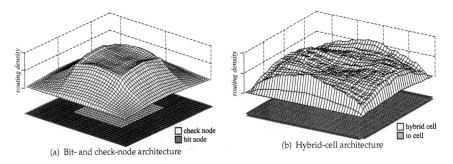

(a) Bit- and check-node architecture

☐ check node
■ bit node

(b) Hybrid-cell architecture

☐ hybrid cell
☐ io cell

Figure 20. Routing density

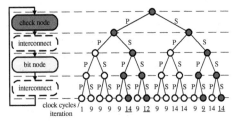

Figure 21. Bit- and check-node architecture design space.

4.1.2. Hardware-efficient partially bit-serial decoder architecture

Another promising approach to reduce the decoder's silicon area is the introduction of a bit-serial interconnect as proposed in [65]. The number of interconnect lines can be reduced by a factor of w resulting in a significant reduction in decoder area because of the quadratic dependency in (42). While the realized minimum search in the check node requires a most-significant-bit-first data flow in the check node the multi-operand adder in the bit node has to be realized using a least-significant-bit-first data flow. Therefore, the order of the bits needs to be flipped twice per iteration resulting in a high number of clock cycles. Although the clock frequency of the decoder is higher due to the bit-serial node logic, the high number of clock cycles per iteration limit the achievable decoder throughput and block latency. However, it is possible to introduce a bit-serial data flow in a more fine-grained way. A systematic architecture analysis is possible by breaking the decoder loop into four parts as shown in Figure 21, namely the bit and check node and the communication between the nodes in both directions. Now, possible architectures can be distinguished by assuming either a bit-serial or a bit-parallel approach in each of the four parts. Obviously, also a digit-serial approach is possible as discussed in [69]. Considering only a bit-serial or bit-parallel data flow, in total 16 different architectures are possible. As a first order metric of the decoder throughput, the number of clock cycles per iteration considering a message word length of $w = 6$ is given. To avoid extensive routing-induced extensions of the silicon area, especially the highlighted architectures with a bit-serial communication in both directions should be taken into account. When comparing the number of clock cycles per iteration for these four architectures, the

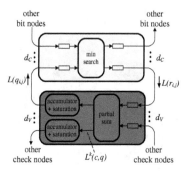

Figure 22. Partially bit-serial architecture

architecture with a bit-parallel bit node allows for the smallest number of clock cycles per iteration and, thus, promises the highest decoder throughput. As the bit-parallel realization of the bit node would result in a large silicon area and a long critical path, further optimizations on arithmetic level have to be done. Here, it is possible to gain from the bit-serial input data stream by realizing the multi-operand adder in the bit node bit-serially using an MSB-first data flow. Within each clock cycle a partial sum $L^k(c, q)$ for the received bit-weight is generated which is accumulated subsequently to derive the new estimation $L(Q_i)$ as is shown in the decoder loop in Figure 22. The long ripple path in the accumulator unit running over the complete word length can be reduced using a carry-select principle. For further details of the realization on arithmetic and circuit level refer to [62].

4.1.3. Quantitative architecture comparison

The cost models have been adapted to the new architecture concepts to allow for a quantitative evaluation of the architecture design space. Figure 23(a) illustrates the resulting silicon area A and iteration period T_{IT} of the fully bit-parallel, fully bit-serial, hybrid-cell and partially bit-serial decoder architecture for three different code complexities $n \cdot d_V = 5,000$, $10,000$ and $15,000$. For all code complexities the new architecture concepts are Pareto optimal as they allow for a trade-off between silicon area and iteration period in comparison to the bit-parallel and bit-serial architectures. Considering small code complexities the decoder architectures with a bit-parallel interconnect show the smallest area-time (AT) product and, therefore, are most AT-efficient. Considering a specified decoder throughput, the hybrid-cell architecture is promising whenever the timing constraints cannot be met by using bit-serial approaches, as it reduces the silicon area significantly in comparison to the bit-parallel bit- and check-node architecture. The new partially bit-serial architecture features the smallest area-time product for all code complexities larger than $9,000$. In comparison to the bit-serial architecture a significantly smaller iteration period with only a slightly increased area is achieved. The architectures with a bit-parallel interconnect are located further and further away from the curve representing the smallest achievable area-time product. Here, the timing advantage of the bit-parallel architectures vanishes for large code complexities. Figure 23(b) depicts the energy per decoding iteration E_{IT} of the four decoder architectures for different code complexities. The advantage with respect to energy of the decoder architectures with a

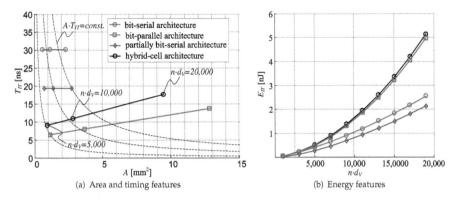

(a) Area and timing features   (b) Energy features

Figure 23. Comparison of area, timing and energy features ($d_V = 6, d_C = 32, M_R = 4, w = 6, \lambda = 40nm$)

bit-serial interconnect becomes apparent. Considering code complexities larger than $10,000$, the energy per iteration of the bit-parallel decoder becomes more than twice as high as the partially bit-serial architecture. The latter allows for the smallest energy per iteration in the complete code complexity range. This emphasizes the efficiency of the new partially bit-serial architecture which allows for the smallest area-time product in a wide range of code complexities and the smallest decoding energy, simultaneously. This work has been supported by the German Research Foundation (DFG) under the priority program UKoLoS (SPP1202).

5. Conclusion

This chapter presented results, accomplished within the frame of the DFG priority program »Ultrabreitband Funktechniken für Kommunikation, Lokalisierung und Sensorik«. Focus was put primarily on the analysis and optimization of on-chip and chip-to-chip multi-conductor/multi-antenna interconnects. While we could show that special techniques of physical optimization, coding and signal processing can improve interconnect performance to a remarkable degree, it is expected that even higher performance is achievable in chip-to-chip communication, when multi-conductor interconnects are replaced by wireless ultra-wideband multi-antenna interconnects. Hereby, the signal pulses do not necessarily increasingly disperse as they travel along their way to the receiving end of the interconnect. The propagating nature of the wireless interconnect can make for a much more attractive channel for chip-to-chip communications. The primary goal has been the development of both theoretical and empirical foundations for the application of ultra-wideband multi-antenna wireless interconnects for chip-to-chip communication. Suitable structures for integrated ultra-wideband antennas have been developed, their properties theoretically analyzed and verified against measurements performed on manufactured prototypes. Qualified coding and signal processing techniques, which aim at efficient use of available resources of bandwidth, power, and chip area has been proposed. In addition, attention was given to the implementation of iterative decoding structure for LDPC codes. Detailed cost-models, which are based on signal flow charts and VLSI implementations of dedicated functional blocks have be developed, which allow for an informative analysis of elementary trade-offs between throughput, required chip area, and power consumption. This work has been supported by the German Research Foundation (DFG) under the priority program UKoLoS (SPP1202).

Author details

Nossek Josef A., Mezghani Amine and Michel T. Ivrlač
Institute for Circuit Theory and Signal Processing, Technische Universität München, Germany

Russer Peter, Mukhtar Farooq, Russer Johannes A. and Yordanov Hristomir
Institute for Nanoelectronics, Technische Universität München, Germany

Noll Tobias and Korb Matthias
Chair of Electrical Engineering and Computer Systems, RWTH Aachen University, Germany

6. References

[1] Russer P (1998) Si and SiGe millimeter-wave integrated circuits, Si and SiGe based monolithic integrated antennas for electromagnetic sensors and for wireless communications. IEEE Transactions on Microwave Theory and Techniques 1998: 590-603.

[2] Johannes A. R, Kuznetsov Y, Russer P, Russer J.A, Kuznetsov K, Russer P (2010) Discrete-time network and state equation methods applied to computational electro magnetics. Mikrotalasma revija: 2-14.

[3] Johannes A.R, Baev A, Kuznetsov Y, Mukhtar F, Yordanov H, Russer P (2011) Combined lumped element network and transmission line model for wireless transmission links. German Microwave Conference (GeMIC): 1-4.

[4] Felsen L.B, Mongiardo M, Russer P (2009) Electromagnetic Field Computation by Network Methods. Heidelberg: Springer-Verlag.

[5] Russer P (2006) Electromagnetics, Microwave Circuit and Antenna Design for Communications Engineering. Boston: Artech House.

[6] Russer J.A, Mukhtar F, Wane S, Bajon D, Russer P (2012) Broad-Band Modeling of Bond Wire Antenna Structures. German Microwave Conference (GeMIC): 1-4.

[7] Yordanov H, Ivrlač M.T, Mezghani A, Nossek J.A, Russer P (2008) Computation of the impulse response and coding gain of a digital interconnection bus. 23rd Annual Review of Progress in Applied Computational Electromagnetics: 494-499.

[8] Yordanov H, Russer P (2008) Computing the Transmission Line Parameters of an On-chip Multiconductor Digital Bus. Time Domain Methods in Electrodynamics: 69-78.

[9] Yordanov H, Ivrlač M.T, Nossek J.A, Russer P (2007) Field modelling of a multiconductor digital bus. Microwave integrated circuit conference, 2007: 579-582.

[10] Yordanov H, Russer P (2008) Computation of the Electrostatic Parameters of a Multiconductor Digital Bus. International Conference on Electromagnetics in Advanced Applications 2007: 856-859.

[11] Yordanov H, Russer P (2010) Antennas embedded in CMOS integrated circuits. Proceedings of the 10th Topical Meeting on Silicon Monolithic Integrated Circuits in RF Systems: Electronics and Energetics 2010: 53-56.

[12] Yordanov H, Russer P (2009) Wireless Inter-Chip and Intra-Chip Communication. Proceedings of the 39th European Microwave Conference: 145-148.

[13] Yordanov H, Russer P (2009) On-Chip Integrated Antennas for Wireless Interconnects. Semiconductor Conference Dresden (SCD) 2009.

[14] Yordanov H, Russer P (2010) Integrated On-Chip Antennas Using CMOS Ground Planes. Proceedings of the 10th Topical Meeting on Silicon Monolithic Integrated Circuits in RF Systems: 53-56.

[15] Yordanov H, Russer P (2010) Area-Efficient Integrated Antennas for Inter-Chip Communication. Proceedings of the 10th Topical Meeting on Silicon Monolithic Integrated Circuits in RF Systems: 401-404.

[16] Yordanov H, Russer P (2008) Chip-to-chip interconnects using integrated antennas. Proceedings of the 38th European Microwave Conference, EuMC 2008: 777-780.

[17] Yordanov H, Russer P (2008) Integrated on-chip antennas for chip-to-chip communication. IEEE Antennas and Propagation Society International Symposium Digest 2008: 1-4.

[18] Yordanov H, Ivrlač M.T, Nossek J.A, Russer P (2007) Field modelling of a multiconductor digital bus. Microwave Conference, 2007. European. 37th European: 1377-1380.

[19] Russer P, Fichtner N, Lugli P, Porod W, Russer A.J, Yordanov H (2010) Nanoelectronics Based Monolithic Integrated Antennas for Electromagnetic Sensors and for Wireless Communications. IEEE Microwave Magazine: 58-71.

[20] Mukhtar F, Yordanov H, Russer P (2010) Network Model of On-Chip Antennas. URSI Conference 2010.

[21] Mukhtar F, Yordanov H, Russer P (2011) Network model of on-chip antennas. Advances in Radio Science: 237-239.

[22] Mukhtar F, Kuznetsov Y, Russer P (2011) Network Modelling with Brune's Synthesis. Advances in Radio Science: 91-94.

[23] Russer J.A, Dončov N, Mukhtar F, Stošić B, Asenov T, Milovanović B, Russer P (2011) Equivalent lumped element network synthesis for distributed passive microwave circuits. Mikrotalasna revija: 23-28.

[24] Russer J.A, Gorbunova A, Mukhtar F, Yordanov H, Baev A, Kuznetsov Y, Russer P (2011) Equivalent circuit models for linear reciprocal lossy distributed microwave two-ports. Microwave Symposium Digest: 1-4.

[25] Russer J.A, Dončov N, Mukhtar F, Stošić B, Milovanović B, Russer P (2011) Compact equivalent network synthesis for double-symmetric four-ports. Telecommunication in Modern Satellite Cable and Broadcasting Services (TELSIKS), 2011 10th International Conference on: 383-386.

[26] Cover T.M, Thomas J.A (1991) Elements of Information Theory. New York: John Wiley and Son.

[27] Lok T.M, Wei V.K.W (1998) Channel Estimation with Quantized Observations. IEEE International Symposium on Information Theory 1998: 333.

[28] Ivrlač M.T, Nossek J.A (2007) On MIMO Channel Estimation with Single-Bit Signal-Quantization. ITG Smart Antenna Workshop 2007.

[29] Sotiriadis P.P, Wang A, Chandrakasan A (2000) Transition Pattern Coding: An Approach to Reduce Energy in Interconnect. ESSCIRC 2000: 320-323.

[30] Kavčić A (2001) On the capacity of markov sources over noisy channels. IEEE Global Communications Conference 2001: 2997-3001.

[31] Ivrlac M.T, Nossek J.A (2006) Chalanges in Coding for Quantized MIMO Systems. IEEE International Symposium on Information Theory 2006: 2114-2118.

[32] Lyuh G, Kim T (2002) Low Power Bus Encoding with Crosstalk Delay Elimination. 15th Annual IEEE International ASIC/SOC Conference 2002: 389-393.

[33] Mezghani A, Khoufi M.S, Nossek J.A (2007) A Modified MMSE Receiver for Quantized MIMO. ITG/IEEE WSA 2007.

[34] Mezghani A, Nossek J.A (2007) Wiener Filtering for Frequency Selective Channels with Quantized Outputs. IEEE SSD 2007.

[35] Max J, Quantizing for Minimum Distortion (1960) IEEE Trans. Inf. Theory 1960: 7-12.

[36] Mezghani A, Khoufi M.S, Nossek J.A (2008) Spatial MIMO Decision Feedback equalizer Operating on Quantized Data. IEEE ICASSP 2008.

[37] Jayant N.S, Noll P (1984) Digital Coding of Waveforms. New Jersey: Prentice-Hall 1984.

[38] Proakis J.G (1995) Digital Communications. New York: McGraw Hill 1995.

[39] Verdú S (2002) Spectral Efficiency in the Wideband Regime - IEEE Trans. Inform. Theory 2002: 1319-1343.

[40] Mezghani A, Nossek J.A (2007) On Ultra-Wideband MIMO Systems with 1-bit Quantized Outputs: Performance Analysis and Input Optimization - IEEE International Symposium on Information Theory 2007.

[41] Teletar E (1999) Capacity of Multi-Antenna Gaussian Channels. European Transactions on Telecommunications 1999: 585-595.

[42] Schubert M, Boche H (2005) Iterative multiuser uplink and downlink beamforming under SINR constraints. IEEE Transactions on Signal Processing: 2324-2334.

[43] Mezghani A, Joham M, Huger R, Utschick W (2006) Transceiver Design for Multi-user MIMO Systems. Proc. ITG Workshop on Smart Antennas 2006.

[44] Joham M, Kusume K, Gzara M.H, Utschick W, Nossek J.A (2002) Transmit Wiener Filter for the Downlink of TDD DS-CDMA Systems. ISSSTA 2002: 9-13.

[45] Ivrlač M.T, Utschick W, Nossek J.A (2002) On Time-Switched Space-Time Transmit Diversity. Proceedings of the 56th IEEE Vehicular Technology Conference 2002: 710-714.

[46] Ivrlač M.T, Choi R.L.U, Murch R, Nossek J.A (2003) Effective use of Long-Term Transmit Channel State Information in Multi-user MIMO Communication Systems. Proceedings of the IEEE Vehicular Technology Conference 2003.

[47] Widrow B, Kollár I (2008) Quantization Noise. Cambridge University Press.

[48] Joham M, Utschick W, Nossek J.A (2003) Linear Transmit Processing in MIMO Communication Systems. IEEE Transactions on Signal Processing 2003.

[49] Mezghani A, Khoufi M.S, Nossek J.A (2008) Maximum Likelihood Detection For Quantized MIMO Systems. In Proc. ITG/IEEE WSA 2008.

[50] Mezghani A, Antreich F, Nossek J.A (2010) Multiple Parameter Estimation With Quantized Channel Output. In Proc. ITG/IEEE WSA 2010.

[51] Lang T, Mezghani A, Nossek J.A (2010) Channel adaptive coding for coarsely quantized MIMO systems. In Proc. ITG/IEEE WSA 2010.

[52] Lang T, Mezghani A, Nossek J.A (2010) Channel-matched trellis codes for finite-state intersymbol-interference channels. In Proc. SPAWC 2010.

[53] Mezghani A, Ivrlac M, Nossek J.A (2008) On Ultra-Wideband MIMO Systems with 1-bit Quantized Outputs: Performance Analysis and Input Optimization. IEEE International Symposium on Information Theory and its Applications 2008.

[54] Gray R.M (1990) Quantization noise spectra. IEEE Trans. Inform. Theory 1990: 1220-1244.

[55] Gallager R. (1962) Low-density parity-check codes. Information Theory, IRE Transactions on, vol. 8, pp. 21-28.

[56] Sae-Young C., Forney, Jr. G. D. , et al. (2001) On the design of low-density parity-check codes within 0.0045 dB of the Shannon limit. Communications Letters, IEEE, vol. 5, pp. 58-60.

[57] IEEE (2006) IEEE Standard for Information Technology-Telecommunications and Information Exchange Between Systems-Local and Metropolitan Area Networks-Specific Requirements Part 3: Carrier Sense Multiple Access With Collision Detection (CSMA/CD) Access Method and Physical Layer Specifications. IEEE Std 802.3an-2006 (Amendment to IEEE Std 802.3-2005), pp. 1-167.

[58] IEEE (2005) 802.16e. Air Interface for Fixed and Mobile Broadband Wireless Access Systems. IEEE P802.16e/D12 Draft, oct 2005.

[59] IEEE (2006) 802.11n. Wireless LAN Medium Access Control and Physical Layer specifications: Enhancements for Higher Throughput. IEEE P802.16n/D1.0, Mar 2006.

[60] Galbraith R. L., Oenning T., et al. (2010) Architecture and Implementation of a First-Generation Iterative Detection Read Channel. Magnetics, IEEE Transactions on, vol. 46, pp. 837-843.

[61] Korb M., Noll T. G. (2012) A Quantitative Analysis of Fixed-Point LDPC-Decoder Implementations using Hardware-Accelerated HDL Emulations. to appear in Embedded Computer Systems (SAMOS) International Conference.

[62] Korb M. Noll T. G. (2011) Area- and energy-efficient high-throughput LDPC decoders with low block latency. ESSCIRC, Proceedings of the, 2011, pp. 75-78.

[63] Mohsenin T., Baas B. M. (2006) Split-Row: A Reduced Complexity, High Throughput LDPC Decoder Architecture. Computer Design, International Conference on, 2006, pp. 320-325.

[64] Zhang Z., Anantharam V., et al. (2009) A 47 Gb/s LDPC Decoder With Improved Low Error Rate Performance. VLSI Circuits, Symposium on, 2009, pp. 286-287.

[65] Darabiha A., Carusone A. C., et al. (2007) A 3.3-Gbps Bit-Serial Block-Interlaced Min-Sum LDPC Decoder in 0.13-ţm CMOS. Custom Integrated Circuits Conference, 2007. CICC '07. IEEE, 2007, pp. 459-462.

[66] Blanksby A. J., Howland C. J. (2002) A 690-mW 1-Gb/s 1024-b, Rate-1/2 Low-Density Parity-Check Code Decoder. Solid-State Circuits, IEEE Journal of, vol. 37, pp. 404-412.

[67] Korb M. Noll T. G. (2010) LDPC decoder area, timing, and energy models for early quantitative hardware cost estimates. System on Chip (SoC), 2010 International Symposium on, 2010, pp. 169-172.

[68] MacKay D. http://www.inference.phy.cam.ac.uk/mackay/codes/data.html

[69] Korb M., Noll. T. G. (2009) Area and Latency Optimized High-Throughput Min-Sum Based LDPC Decoder Architectures. ESSCIRC, Proceedings of the, 2009, pp. 408-411.

[70] Fossorier M. P. C., Mihaljevic M., et al. (1999) Reduced complexity iterative decoding of low-density parity check codes based on belief propagation. Communications, IEEE Transactions on, vol. 47, pp. 673-680, 1999.

Chapter 3

Non-Coherent UWB Communications

Nuan Song, Mike Wolf and Martin Haardt

Additional information is available at the end of the chapter

1. Introduction

The use of ultra wide band (UWB) signals can offer many advantages for communications. It can provide a very robust performance even under harsh multipath and interference conditions, the capability of precision ranging and a reduced power consumption. Since the power spectral density is very low, it is possible to overlay UWB networks with already existing non-UWB emissions.

Early UWB concepts for communications have almost exclusively relied on impulse radio, where the whole available bandwidth, i.e., up to 7.5 GHz, is covered at once by means of very short pulses which are generated with a low duty cycle. Meanwhile, a bandwidth of 7.5 GHz is only available in the US [1, 3]. In Europe, the spectrum which is available with the same transmit power spectral density of -41.3 dBm/MHz ranges only from 6.0 to 8.5 GHz [4], if no detect and avoid techniques are applied[1]. A potential UWB system has therefore to be able to 'live' with a mean transmit power of less than -7.3 dBm.

This is a small value, but fortunately UWB systems may exploit the signal energy very efficiently because firstly, even at data rates in the Gbps range, it is not required to use bandwidth efficient (but energy inefficient) modulation schemes like a 1024-QAM. Secondly, UWB transmission benefits from a good fading resistance.

For a measured indoor channel [7], Fig. 1 shows that even a bandwidth of 'only' 500 MHz ensures a very good fading resistance: If the receiver is moved over all x-y-positions in the non-LOS case, the smallest power value at the receive antenna lies less than 3 dB below the mean power, averaged over all positions. Thus the fading margin could be chosen in the order of 3-4 dB — even for indoor channels which exhibit the largest coherence bandwidth.

The second energy efficiency argument claimed above is underpinned in Fig. 2. It shows the channel capacity depending on the bandwidth, where additive white Gaussian noise (AWGN) is assumed. A value of 83 MHz just corresponds to the total bandwidth available in the 2.4 GHz ISM band, which is chosen for comparison. At 1 Gbps, a 2.5 GHz bandwidth promises an advantage of 25 dB with respect to the required receive-power. Furthermore, even binary modulation (on the inphase and quadrature components) promises high data rates.

---

[1] With detect and avoid techniques, -41.3 dBm/MHz is also permitted between 4.2 and 4.8 GHz.

Unfortunately, a very large signal bandwidth is also associated with some serious problems. These problems are related to the transceiver components itself (availability of broadband antennas, amplifiers etc.), and to the technical effort which is required for synchronization, channel estimation and interference rejection. Since UWB networks operate in frequency bands already assigned to other RF-systems, the probability that narrowband interference occurs at all increases with the bandwidth, too.

Furthermore, by increasing the bandwidth, more and more multipath arrivals with different path gains and delays are resolvable at the receiver, which makes it more difficult to collect the multipath energy coherently — although the power at the receive antenna does not suffer from the fading effect. Fig. 1 shows an example that one may lose 10 dB and more, if an UWB-receiver uses only the strongest signal echo. Thus, especially in non-LOS scenarios, a coherent RAKE receiver requires a very large number of RAKE fingers and a precise channel knowledge to efficiently capture the multipath energy. Such a coherent RAKE receiver will be very complex and costly, such that the hardware itself may consume a lot of power. This fact is the major motivation for systems using non-coherent detection, which are discussed in this chapter.

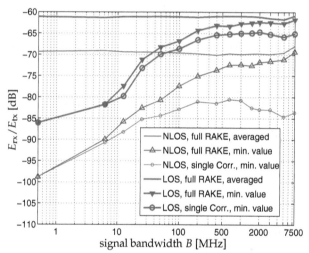

Figure 1. Received energy E_{rx} normalized by the transmitted symbol's energy E_{tx} in dB, versus different signal bandwidths. The thickness of the curves indicates LOS or non-LOS regimes. The curves without markers show E_{rx}/E_{tx} averaged across all x-y-positions within an rectangular area of 30 cm × 40 cm (1 cm grid, data from [7]), if an ideal full RAKE-receiver is used. The curves marked with triangles show the minimum value of E_{rx}/E_{tx} which occurs within these positions, again assuming a full RAKE. Thus the small-scale fading effect becomes visible. The curves with circles depict the normalized receive energy for a receiver which exploits only the strongest propagation path, i.e., a single correlator is applied. Transmitters-receiver separation is about 3 m, the carrier frequency is always set to 6.85 GHz.

Non-coherent UWB transmission is an attractive approach especially if simple and robust implementations with a small power consumption are required. Main application fields are low data rate sensor or personal area networks, which require low cost devices and a long battery life time. It should be noted that the current IEEE802.15.4a UWB-PHY for low data rate communications enables non-coherent detection, too [18]. The main advantage of a

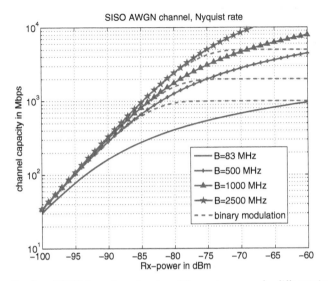

Figure 2. Capacity of an AWGN channel as a function of the receive-power for different signal bandwidths. The 2.4 GHz ISM band offers a bandwidth of 83 MHz.

non-coherent receiver is clearly the dramatically reduced effort which is required for channel estimation, synchronization, and multipath diversity combining. This advantage is, however, bought by a serious drawback: non-coherent detectors are more susceptible to narrowband interference (NBI), multi-user interference (MUI), and inter-symbol interference (ISI).

Non-coherent detection can either rely on envelope detection or on differential detection. In the simplest case, path-diversity combining is carried out via an analog integration device. However, the change from analog to digital combining stimulates new perspectives. Since a digital code matched filter can be applied prior to the non-coherent part of the receiver, the capability to distinguish users (or networks) by means of code division multiple access is improved. We show that digital receiver implementations with user specific filtering have also an enhanced interference rejection capability and energy efficiency. Moreover, we present well suited solutions for the analog-to-digital conversion, the spread-spectrum code sequences, and the modulation format.

2. Non-coherent detection in multipath AWGN

Although non-coherent detection is not restricted to low data rates — even orthogonal frequency division multiplex (OFDM) can be combined with non-coherent modulation and detection [19] — we focus our attention on low data rate single carrier transmission.

Non-coherent detection can either be based on envelope detection or on differential detection. In the simplest case, path-diversity combining can be achieved by means of a single, analog integrate and dump filter, see Fig. 3 and Fig. 4. The integration effectively provides a binary weighting of the multipath arrivals: all components inside the integration window of size T_{int} are weighted with "1", while all the others are weighted with "0". Regardless of whether envelope or differential detection is chosen, we assume that the receiver uses a quadrature

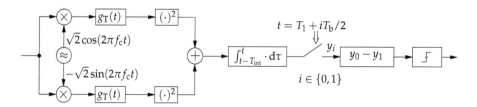

Figure 3. Envelope detection of a 2-PPM signal using a quadrature down-conversion stage. The analog integrate and dump filter (integrator) is required to capture the multipath energy.

down-conversion stage, since an ECC-conform UWB-signal is a 'truth' band-pass signal with a maximum bandwidth of 2.5 GHz and a lower cut-on frequency of 6 GHz. It is also assumed that each the inphase and the quadrature branch contain a low-pass filter, whose impulse response $g_T(t)$ is matched to the transmitted pulse $\psi_1(t)$. Without multipath, this ensures that the energy of the received signal is focused at the sampling times. For example, according to the IEEE 802.15.4a UWB PHY standard a receiver needs to perform a matched de-chirp operation, if the optional "chirp on UWB" pulse is used.

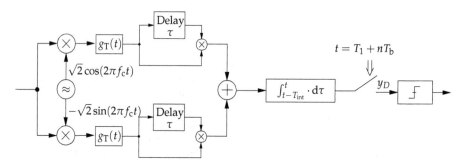

Figure 4. Differential detection for DPSK ($\tau = T_b$) or transmitted-reference PSK ($\tau < T_b$).

In the following, we consider binary pulse-position modulation (2-PPM) and differential phase-shift keying (DPSK) as straightforward modulation schemes to be combined with either envelope detection or with differential detection.

If E_b denotes the mean energy per bit, a 2-PPM signal (single carrier with fixed carrier frequency) given in the complex baseband can be written as

$$s(t) = \sqrt{E_b} \sum_{n=-\infty}^{\infty} (1 - b_n)\psi_1(t - nT_b) + b_n\psi_1(t - nT_b - T_b/2), \qquad (1)$$

where $\psi_1(t)$ needs to be orthogonal to $\psi_1(t - T_b/2)$. T_b is the bit interval, which is split into two subintervals each of length $T_b/2$. Depending on the binary information b_n, $b_n \in \{0, 1\}$, to be transmitted, the waveform $\sqrt{E_b}\psi_1(t)$ is generated either at the time nT_b or $T_b/2$ seconds later. For single carrier transmission with a fixed carrier frequency, the unit energy basis function $\psi_1(t)$ needs to exhibit a bandwidth of at least 500 MHz, i.e., it is a spread spectrum waveform. For example, according to the IEEE 802.15.4a UWB PHY, $\psi_1(t)$ consists of a single pulse with a duration of 2 ns (or less) or a burst of up to 128 such pulses with a scrambled polarity.

A (single carrier) DPSK signal given in the complex baseband can be written as

$$s(t) = \sqrt{E_b} \sum_{n=-\infty}^{\infty} (2\tilde{b}_n - 1)\psi_1(t - nT_b), \qquad (2)$$

where $\tilde{b}_n = b_n \oplus \tilde{b}_{n-1}$, $\tilde{b}_n \in \{0, 1\}$, denotes differentially encoded bits.

2.1. Performance estimation for single-window combining

By increasing the transmission bandwidth B, more and more multipath arrivals are resolvable at the receiver. For example, with $B = 500$ MHz and an assumed excess delay of 50 ns, about 25 individual paths are resolvable in the time domain. In contrast to a channel matched filter (or its RAKE receiver equivalent), which combines all these arrivals coherently and with an appropriate weighting, the integrate and dump filter shown in Fig. 3 and Fig. 4 clearly allows only a non-coherent combining. This suboptimal combining leads to a performance loss, which increases with the product $B \cdot T_{int}$, where it is assumed that the whole signal energy is contained within the integration interval T_{int}. If it is further assumed that $B \cdot T_{int}$ is an integer $N \geq 1$, the BER p_b for 2-PPM (energy detection) can be estimated as [5]

$$p_b = \frac{1}{2^N} \exp\left(-\frac{E_b}{2N_0}\right) \sum_{i=0}^{N-1} \frac{1}{2^i} \mathcal{L}_i^{N-1}\left(-\frac{E_b}{2N_0}\right) \qquad (3)$$

$$\approx \frac{1}{2}\text{erfc}\left(\frac{E_b/N_0}{2\sqrt{E_b/N_0 + N}}\right),$$

where $\mathcal{L}_i^{N-1}$ is a generalized Laguerre polynomial. The second expression corresponds to a Gaussian approximation which can be used for $N > 15$. Eqn. (3) is only valid if the integration interval T_{int} contains the whole bit energy E_b, and if no ISI occurs, which is the case if the channel excess delay is small compared to T_b. The compact solution (3) has its origin in the fact that the samples y_0 and y_1 (see Fig. 3) are χ^2-distributed with $2N$ degrees of freedom [10].

For DPSK and differential detection, the statistical description is very similar. Thus, it is only required to substitute N_0 by $N_0/2$ in (3), i.e., the E_b/N_0-performance differs by exactly 3 dB in favor of DPSK.

Fig. 5 shows the penalty with respect to the required E_b/N_0, if we switch from — rather hypothetical — coherent channel matched filter detection to non-coherent detection with equal gain single-window combining (SinW-C). A value of 1.2 dB at $N = 1$ just corresponds to the E_b/N_0-penalty in AWGN ($p_b = 10^{-3}$).

The results must be interpreted very carefully, since it is assumed that the whole bit energy is concentrated within the interval T_{int}. In reality, this is surely not the case and an appropriate T_{int} must be found. If T_{int} is increased, more and more noise is integrated which leads to the loss shown in Fig. 5, but more signal energy may be collected as well.

From the energy efficiency point of view, any non-coherent combining should take place with respect to the multipath energy only. If chirp or direct sequence spread spectrum (DSSS) signals are used, where the signal energy is spread over time even at the transmitter, it is preferable to equip the receiver with a matched filter $g_T(t)$ which focuses the energy of the chirp or DSSS-waveform before the non-coherent processing takes place, as it was assumed in Fig. 3 and Fig. 4.

Fig. 6 shows the E_b/N_0-performance of 2-PPM in the case of a non-LOS indoor channel, where E_b is interpreted as the bit energy available at the receiver's antenna output. For the results shown, we have used measured UWB channels (5m×5m×2.6m office) including the antennas. The measurements were carried out by the IMST GmbH [6, 7]. B_6 is chosen to be 500 MHz.

The results show that the BER strongly depends on T_{int}, whereas the position of the integration window is always chosen optimally. For the reference indoor channel considered here [10], the optimal value of T_{int} is about 20 ns. At $T_{int} = 20$ ns, SinW-C loses additionally 1.5 dB compared to an optimal non-coherent detector, cf. Section 2.3.

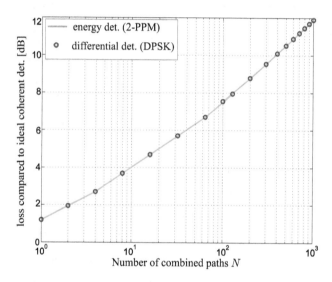

Figure 5. Power penalty due to non-coherent combining as a function of the time bandwidth product N ($p_b = 10^{-3}$). Coherent matched filter detection acts as the reference.

2.2. Weighted sub-window combining

In the case of SinW-C, at least two parameters need to be adjusted, the window size T_{int} and its position T_1 (synchronization). Since the BER over T_{int} performance of SinW-C may also exhibit several local minima, the practical determination of appropriate T_{int} and T_1 values may be more difficult than it seems. These problems are reduced, if weighted sub-window combining (WSubW-C) is used, where the whole integration window is divided into a number of N_{sub} sub-windows of size T_{sub}. From the E_b/N_0 performance point of view, it is preferable to choose the N_{sub} weighting coefficients according to the sub-window energies. In [10] we have shown that WSubW-C with $T_{sub} = 4$ ns (which corresponds to a sampling rate of 250 MHz) outperforms SinW-C (with optimum synchronization) by about 0.5 dB, if indoor channels are considered.

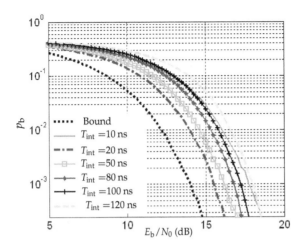

Figure 6. BER performance of non-coherent 2-PPM detection for SinW-C and a 6 dB signal bandwidth of 500 MHz. A measured indoor non-LOS channel is used to obtain the results.

2.3. Performance limit

If the sampling rate of the WSubW-C is equal to the signal bandwidth B, i.e., $T_{sub} = 1/B$, each resolvable multipath component[2] can be weighted according to its energy. In [10] we have shown that this approach ensures the benchmark performance, if non-coherent combining is used. The advantage compared to perfectly synchronized SinW-C is about 1.5 - 2 dB for indoor channels, cf. Fig. 6 (black, dotted curve).

3. Analog receiver implementations

3.1. Feasibility of analog differential detection

ISI will degrade the BER performance of DPSK systems, if the symbol interval is considerably smaller than the channel excess delay. However, it is crucial to realize delays on the order of 50 ns or more in the analog domain, if the ultra-wideband nature of the signals is taken into account. Fig. 7 shows the normalized group delay of a Bessel-Thomson all-pass filter with a maximum flat group delay $t_g(f)$. If a 5 % group delay error is chosen to define the cut-off frequency, it is clear that a 5th order filter can provide a usable frequency range of $\approx 1/t_g(0)$, i.e., for a desired cut-off frequency of $f_g = B_6/2 = 250$ MHz, the delay is only 4 ns. Even a huge and completely unrealistic filter order of 20 could only provide a delay of $5.6 \cdot 4 = 22$ ns, if $f_g = 250$ MHz. It should be noted that two of these analog delay lines have to be implemented, if a quadrature down-conversion stage as shown in Fig. 4 is used.

A basic motivation of impulse radio based on transmitted reference (TR) signaling is that shorter delays can be used. This is possible, since the autocorrelation does not take place with the previous modulated symbol but rather with an additional reference pulse. Our results show that the performance of TR-signaling varies extremely from channel realization

---

[2] At a total transmission bandwidth B, multipath components can be resolved down to $1/B$ in the time domain.

to channel realization, since the autocorrelation process is disturbed by intra-symbol interference. Additionally, if a reference pulse is periodically inserted prior to each modulated pulse, a 3 dB loss occurs. Delay hopping techniques or reference symbol averaging may reduce this 3 dB loss, but require even more (and longer) delay lines.

It is more than unlikely that analog implementations of differential receivers will have a chance to be applied in low cost products. For the multi-user case with analog multipath combining (next section), we have therefore focused on energy detection combined with time-hopping (TH) impulse radio.

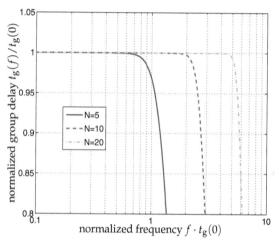

Figure 7. Normalized group delay of a Bessel-Thomson all-pass filter of order N.

3.2. Multiple access for analog multipath combining

As mentioned above, the transmitted pulse $\psi_1(t)$ may be also a chirp or direct-sequence spread spectrum waveform. To realize the filter $g_T(t)$ in the analog domain, SAW-filters (SAW: surface acoustic wave) could be a solution, but only if $\psi_1(t)$ is a fixed, user independent waveform. Therefore, as long as the non-coherent signal processing (or a part of it) takes place in the analog domain, we assume a user independent $\psi_1(t)$.

To still enable multiple access (MA) communications, we assume that each symbol to be transmitted is represented by several short pulses which are generated at distinct times according to a user specific TH-code. The decoding will be carried out digitally, i.e., after the non-coherent signal processing took place, cf. Fig. 8.

Compared to direct sequence MA codes which contain a large number of (non-zero) chips, the sparseness of TH codes facilitates the receiver processing, reduces the complexity and keeps the additional loss due to the non-coherent combining of the code elements within limits. A TH code is determined by two parameters: the number of pulse repetitions N_s, which is equivalent to the code weight, and the number of hopping positions[3] N_h. In [12], we have presented a semi-analytical method to assess the multiple access performance. It

---

[3] One symbol interval can be divided into an integer number N_s of equally spaced intervals (named frames), where each frame contains one pulse. Within each frame, N_h positions are possible.

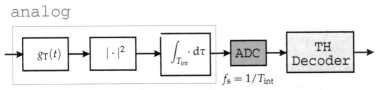

Figure 8. Block diagram of a TH code division multiple access receiver with analog multipath combining. Differential detection is not considered since analog wide band delays are difficult to realize.

is based on the statistics of the total code collisions (or "hits", determined by N_s and N_h) as well as the first- and second-order moments of the multipath channel's energy within the integration window. For non-coherent TH-PPM systems, the proposed method provides a more accurate and comprehensive evaluation of the multiple access performance than the existing code correlation function based analysis.

In [12], we have investigated various MA codes to be applied for a non-coherent TH-PPM system. It can be concluded that for a moderate number of users, optical orthogonal codes (truncated Costas codes, prime codes) with low code weights ensure a good multiple access performance while adding only a very small additional non-coherent combining loss.

4. Digital receiver implementations

Digital receiver implementations according to Fig. 9 have several advantages. First of all, they offer a superior interference rejection capability [11, 16], since user specific filtering can take place prior to the non-coherent signal processing. This restricts the non-coherent combining loss to the multipath arrivals (which exhibit stochastic path weights), whereas that part of the signal energy, which is already spread by a user-specific code at the transmitter, is coherently summed up. Furthermore, digital receiver implementations enable advanced modulations such as Walsh-modulation [15, 17] or advanced NBI-suppression strategies based on soft-limiting [13].

The block diagram of a receiver with a "digital code matched filter" (DCMF) is shown in Fig. 9. The ADC operates with a sampling rate, which is not smaller than the UWB signal bandwidth, where the ADC resolution has been chosen between 1 and 4 bits. Regarding the following results, we have always assumed TH impulse radio transmission.

Fig. 10 shows the E_b / N_0 improvement of a DCMF-based receiver as a function of N_s, where N_s depicts the number of non-zero elements of the user-specific code. As the DCMF combines the corresponding pulses coherently, the benefit compared to an analog receiver increases with N_s.

4.1. Applicability of low-resolution ADCs (single user case)

In [14] we have shown that in the 2-PPM case and under certain conditions, low-resolution ADCs can almost achieve the full resolution E_b / N_0-performance. One important condition is the number of pulse repetitions N_s within one modulated symbol, which should not be too small. For the one bit ADC case, $N_s = 8$ and $N_s = 20$ just correspond to quantization penalties of 2 dB and 1.5 dB, respectively, cf. Fig. 11(a). If the resolution is increased from 1 bit to 2 or 4 bits, the penalty may decrease, but only if the input level of the ADC is well controlled by an additional gain-control circuit. In [14] we have also proven that a 1 bit ADC with its inherent clipping characteristic offers a superior interference rejection capability.

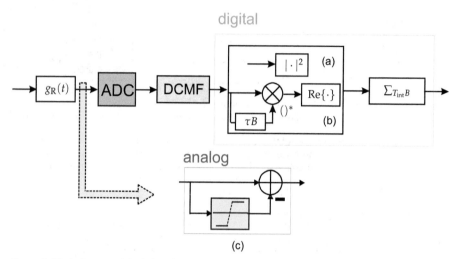

Figure 9. Block diagram of the DCMF-based non-coherent receiver shown in the complex baseband for (a) energy detection and (b) differential detection. (c) A NBI suppression scheme using a soft-limiter.

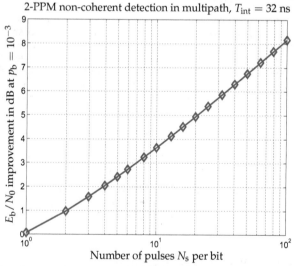

Figure 10. E_b/N_0 performance improvement of a fully digital 2-PPM non-coherent receiver compared to its analog counterpart for the multipath single-user case. The integration window extents over $T_{int} = 32$ ns. A full-resolution ADC is assumed.

We have also investigated the applicability of Sigma-Delta ADCs, especially if M-ary Walsh modulation is used, cf. Section 4.4. The results show that the full resolution performance can be obtained for an oversampling rate of 4. Since the power consumption of an ADC depends linearly on the sampling rate, but exponentially on the resolution [8], Sigma-Delta ADCs can thus be considered as attractive candidates.

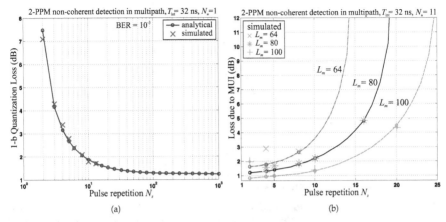

Figure 11. Power penalty due to 1-bit quantization (a) and multi-user interference (b), where random-codes are assumed. BER is 10^{-3}.

4.2. Multiple access codes for time hopping PPM

Fig. 11(b) shows the power penalty due to MUI for a network with 11 users, where perfect transmit power control and a full resolution ADC was assumed. Random codes were applied. For a given processing gain $N_s \cdot N_h$, the penalty depends strongly on the ratio of the parameters N_s (number of non-zero pulses) and N_h (number of hopping positions). Since N_s determines the ADC quantization induced penalty, too, we conclude that N_s on the order of 8 leads to a good trade-off between the quantization loss and the MUI penalty. In [14] we have shown that this rule does not only apply to random codes but also to optical orthogonal codes as suggested for analog receiver implementations.

4.3. Performance of simultaneously operating piconets

A test geometry of simultaneously operating piconets (SOP) is shown in Fig. 12, where a single co-channel interference is considered. The reference distance d_{ref} (desired piconet 1) is chosen such that the power at the receiver is 6 dB above the receiver sensitivity threshold. The interfering transmitter (uncoordinated piconet 2) operates at the same power as the transmitter of piconet 1, but at a distance d_{int} to the reference receiver. We have considered the IEEE 802.15.4a channel model 3 (indoor LOS) and the channel model 4 (indoor non-LOS) [9], where random TH codes with $N_s = 10$ and $N_h = 8$ are applied altogether with forward error correction (Reed Solomon code with a rate of 0.87). The results shown in Table 1 prove clearly that digital receiver implementations outperform analog ones. Furthermore, 1-bit ADCs are desirable.

Channels	Analog	DCMF (full)	DCMF (1 bit)
CM3	1.53	0.64	0.32
CM4	1.05	0.71	0.45

Table 1. d_{int}/d_{ref} at 1% packet error ratio for SOP tests

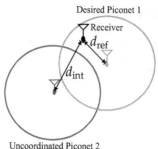

Figure 12. The SOP test geometry with a single co-channel interference

4.4. Power efficient Walsh-modulation

In [17], we have proposed two advanced (low data rate) transmission schemes based on M-ary Walsh-modulation, namely repeated Walsh (R-Walsh) and spread Walsh (S-Walsh). For both schemes, the fast Walsh Hadamard transformation can be used to efficiently implement the demodulator. Whereas the more implementation friendly R-Walsh transmission is favorable for data rates of up to 180 kbps ($M = 8$ or $M = 16$), S-Walsh transmission with $M \geq 32$ is an option for higher data rates.

In [15], we have compared R-Walsh transmission with M-PPM. It has been shown that R-Walsh works well with a 1-bit quantization. Fig. 13 shows that in the case of Walsh modulation, the quantization loss (compared to the full resolution case) is only about 1.5 dB — independently of M and N_s. However, if a strong near-far effect is present, M-PPM outperforms R-Walsh with respect to the MA-performance.

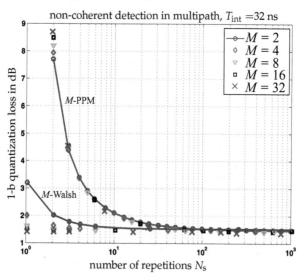

Figure 13. Quantization loss due to a one-bit ADC versus the number of repetitions N_s at $p_b = 10^{-3}$ for M-Walsh and M-PPM in multipath channels (analytical estimation).

4.5. Advanced narrowband interference suppression schemes

In [13] we have presented a new NBI-mitigation technique, which is shown in Fig. 9c). It is based on a soft limiter, where the soft limiter itself was originally proposed to suppress impulse interference [2]. The thresholds of the soft limiters are adjusted according to NBI power.

We have shown that the proposed receiver can effectively mitigate the NBI, if the threshold factor and the input level of the subsequent ADC are chosen appropriately. Furthermore, the performance improves if the ADC resolution is increased. In the presence of the OFDM interference, the proposed scheme could also be used, but it is required to adjust the threshold dynamically. It should be noted that the performance also depends on the frequency of the interference, since the DCMF has a frequency dependent transfer function.

5. Conclusions

We have derived concepts for energy efficient impulse radio UWB systems with a low transceiver complexity. These concepts are especially suitable for wireless sensor networks operating at low data rates. The E_b/N_0-performance of non-coherently detected 2-PPM and DPSK is very similar. It differs by 3 dB in favour of DPSK. However, if the multipath combining should take place in the analog domain, i.e., by means a simple integrate and dump filter, the difficulty to realize analog broadband delays makes it almost impossible to use differential detection and thus DPSK. On the contrary, digital receiver implementations enable advanced modulation schemes and offer superior interference rejection capabilities. With low-resolution ADCs, only a small quantization loss is observed. Compared to the full-resolution case, a one-bit receiver shows a higher MUI suppressing capability. Sigma-Delta ADCs can be considered as attractive candidates for the analog to digital conversion. Our results show that the full resolution performance can be obtained for an oversampling rate of 4.

Author details

Nuan Song, Mike Wolf and Martin Haardt
Ilmenau University of Technology, Germany

6. References

[1] 05-58, F. [2005]. Petition for Waiver of the Part 15 UWB Regulations Filed by the Multi-band OFDM Alliance Special Interest Group, ET Docket 04-352.

[2] Beaulieu, N. C. & Hu, B. [2008]. Soft-Limiting Receiver Structures for Time-Hopping UWB in Multiple Access Interference, *IEEE Transactions on Vehicular Technology* 57.

[3] Commission, F. C. [2002]. First Report and Order: Revision of Part 15 of the Commission's Rules Regarding Ultra-Wideband Transmission Systems, ET Docker 98-153.

[4] ECC [2007]. ECC Decision of 24 March 2006 amended 6 July 2007 at Constanta on the harmonised conditions for devices using Ultra-Wideband (UWB) technology in bands below 10.6 GHz, amended ECC/DEC/(06)04.

[5] Hao, M. & Wicker, S. [1995].   Performance evaluation of FSK and CPFSK optical communication systems: a stable and accurate method, *Journal of lightwave technology* (13): 1613–1623.

[6] Kunisch, J. & Pamp, J. [2002]. Measurement Results and Modeling Aspects for the UWB Radio Channel, *IEEE Conference on Ultra Wide-Band Systems and Technologies* .

[7] Kunisch, J. & Pamp, J. [n.d.].   *IMST-UWBW: 1-11 GHz UWB Indoor Radio Channel Measurement Data*, IMST GmbH.   available   at http://www.imst.com/imst/en/research/radio-networks/funk_wel_dow.php.

[8] Le, B., Rondeau, T. W., Reed, J. H. & Bostian, C. W. [2005]. Analog-to-digital converters, *IEEE Signal Processing Magazine* 22(6): 69–77.

[9] Molisch, A. F. et al. [2005].   IEEE 802.15.4a channel model - final report, *Tech. Rep. Document IEEE 802.15-04-0662- 02-004a*.

[10] Song, N., Wolf, M. & Haardt, M. [2007].   Low-complexity and Energy Efficient Non-coherent Receivers for UWB Communications, *Proc. 18-th Annual IEEE International Symposium on Personal Indoor and Mobile Radio Communications (PIMRC07)*, Greece.

[11] Song, N., Wolf, M. & Haardt, M. [2009a].   A Digital Code Matched Filter-based Non-Coherent Receiver for Low Data Rate TH-PPM-UWB Systems in the Presence of MUI, *IEEE International Conference on Ultra Wideband (ICUWB 2009)*, Vancouver, Canada.

[12] Song, N., Wolf, M. & Haardt, M. [2009c].   Performance of PPM-Based Non-Coherent Impulse Radio UWB Systems using Sparse Codes in the Presence of Multi-User Interference, *Proc. of IEEE Wireless Communications and Networking Conference (WCNC 2009)*, Budapest, Hungary.

[13] Song, N., Wolf, M. & Haardt, M. [2010a].   A Digital Non-Coherent Ultra-Wideband Receiver using a Soft-Limiter for Narrowband Interference Suppression, *Proc. 7-th International Symposium on Wireless Communications Systems (ISWCS 2010)*, York, United Kingdom.

[14] Song, N., Wolf, M. & Haardt, M. [2010b].   A *b*-bit Non-Coherent Receiver based on a Digital Code Matched Filter for Low Data Rate TH-PPM-UWB Systems in the Presence of MUI, *Proc. of IEEE International Conference on Communications (ICC 2010)*, Cape Town, South Africa.

[15] Song, N., Wolf, M. & Haardt, M. [2010d].   Time-Hopping M-Walsh UWB Transmission Scheme for a One-Bit Non-Coherent Receiver, *Proc. 7-th International Symposium on Wireless Communications Systems (ISWCS 2010)*, York, United Kingdom.

[16] Song, N., Wolf, M. & Haardt, M. [Feb., 2009b].   Non-coherent Receivers for Energy Efficient UWB Transmission, *UKoLoS Annual Colloquium*, Erlangen, Germany.

[17] Song, N., Wolf, M. & Haardt, M. [Mar., 2010c]. Digital non-coherent UWB receiver based on TH-Walsh transmission schemes, *UKoLoS Annual Colloquium*, Günzburg , Germany.

[18] TG4a, I. . [2007]. Part 15.4: Wireless Medium Access Control (MAC) and Physical Layer (PHY) Specifications for Low-Rate Wireless Personal Area Networks (WPANs), *IEEE Standard for Information Technology* .

[19] Wolf, M., Song, N. & Haardt, M. [2009].   Non-Coherent UWB Communications, *FREQUENZ Journal of RF-Engineering and Telecommunications* .   special issue on Ultra-Wideband Radio Technologies for Communications, Localisation and Sensor applications.

Chapter 4

Pulse Rate Control for Low Power and Low Data Rate Ultra Wideband Networks

María Dolores Pérez Guirao

Additional information is available at the end of the chapter

1. Introduction

The growing request for low-to-medium data rate low cost networks is raising the interest in wireless sensor networking. For instance in the field of industrial and logistic applications, in order to improve the processes' efficiency the tight monitoring of goods, tools, and machinery -as to their state and position- is required. In response to this interest, IEEE approved in 2003 the IEEE 802.15.4 standard, this being the first one for low data rate, low complexity, and low power wireless networks.

The market success of wireless sensor networks (WSN) requires inexpensive devices with low power consumption. In order to satisfy this requirement, transmission technology, protocol as well as hardware design must give a common answer. UWB radio, particularly with impulse radio transmission (IR), is especially suitable for the development of WSN. IR-UWB is expected to allow low power, low complexity and low cost implementation as well as centimeter accuracy in ranging. The low complexity and low cost characteristics arise from the essentially baseband nature of the signal transmission. The high ranging accuracy results from the large absolute bandwidth, which must be at least 500 MHz. Indeed, the introduction of ranging functionality in low data rate networks was one of the main reasons for the IEEE 802.15.4a (2007) amendment, which added an IR-UWB physical layer to the original standard.

IEEE 802.15.4a allows for the use of non-coherent receivers, and defines ALOHA[1] as the mandatory medium access control (MAC) protocol. The use of a non-coherent receiver, such as an energy detector, helps to minimise power consumption and reduces implementation complexity. The choice of ALOHA is justified by the potential robustness of IR-UWB to multi-user interference (MUI) and by the low data rate nature of the applications envisioned.

In fact, the design of the MAC layer plays a very important role in order to materialize the benefits of IR-UWB in sensor networks. From a networking perspective, one potential

---

[1] Random medium access scheme that does not check whether the shared medium is already busy before transmission.

benefit of IR-UWB over narrowband radio technologies is the possibility to allow concurrent transmissions by using different pseudo-random, time hopping codes (THCs) as a multiple access (MA) method. However, TH codes are not perfectly orthogonal, and even if, Multi User Interference (MUI) is still a challenge due to the presence of multipath fading and the asynchronicity between sources. Beyond it, non-coherent receivers are less robust to MUI than coherent receivers; particularly, interference coming from close-by interferers can be very harmful. Thus, and specially if non-coherent receivers are used, additional interference mitigation features at the MAC layer are required.

This work is motivated by the fact that interference management at the MAC layer has not been extensively explored in the context of IR-UWB autonomous networks yet. The chapter is organized as follows. A general introduction into the field of IR-UWB radio technology and its relevant technical fundamentals is given in section 2. Additionally, a short overview into current research activities and basic principles of MAC protocol design for low to medium data rate IR-UWB networks is given. It follows a discussion about the use of game theory as a tool to model and analyse distributed MAC algorithms in wireless networks. The section ends with the description of the investigated scenario and the simulation model.

Section 3 introduces distributed Pulse Rate Control (PRC) as a novel approach for interference mitigation in autonomous IR-UWB networks. PRC enables concurrent transmissions at full power, allowing each source to independently adapt its pulse rate - measured in pulses per second (Pps)- to control the impact of pulse collisions at nearby receivers. This section shows that it is possible to incite autonomous users to decrease their impulsive emissions and thus, prevent network resource break-down. Finally, section 4 summarizes the achievements of the work presented and gives directions for future research.

2. Theoretical background

For the understanding of this chapter it is essential to have a good foundation in IR-UWB technology, as well as a general background knowledge of wireless autonomous networks[2] (AN) and game theory. The purpose of this section is to provide a short overview on these three topics. Furthermore, this section describes the scenario and the simulation model used in the investigations.

2.1. Impulse Radio Ultra Wideband (IR-UWB)

IR-UWB is a form of UWB transmission in which data is transmitted using sequences of extremely short pulses with a duration of less than 1ns and a large pulse repetition period (PRP). Due to the extreme short duration of the pulses, IR-UWB is capable of delivering high data rates in the order of several hundred Mbps, but at the expense of reduced transmission range due to the power restrictions.

Inherent to IR-UWB signaling is a high temporal resolution that enables accurate multipath resolution and ranging capabilities. Other interesting features related to the pulsed nature of IR-UWB are its robustness against fading, as well as its low power, low complexity and low cost implementation possibilities. In this respect, IR-UWB is a key technology for providing wireless networks with joint communication and ranging capabilities.

---

[2] Wireless autonomous networks can be considered as a special subclass of wireless ad hoc and sensor networks with reinforced self-organizing character

This chapter assumes a generic Time-Hopping IR-UWB (TH-UWB) physical layer as described in [21]. Time is divided into frames of length T_f and each user transmits one pulse of length T_p per frame. Furthermore, by dividing the frames into non-overlapping chips of length T_c, multi-access capability is provided. Each user transmits its pulse in a randomly chosen chip, according to a pseudo-random TH sequence (THS). Data modulation follows a pulse position modulation (PPM) scheme. Thus, the signal emitted by the k-th TH-PPM transmitter consists of a sequence of pulse waveforms shifted to different times.

$$s^{(k)}(t) = \sum_{j=-\infty}^{\infty} w_{tr}(t - jT_f^{(k)} - c^{(k)}[j]T_c - \eta b_{\lfloor j|N_s \rfloor}^{(k)} - \tau^{(k)}), \tag{1}$$

A typical expression is given in 1, where $w_{tr}(t)$ represents the transmitted pulse waveform and T_f is the average frame time, which is also denoted as the mean pulse repetition period. The inverse of the mean pulse repetition period is referred to as the mean pulse repetition frequency, $T_f = \frac{1}{prf}$. The expression $b_{\lfloor j|N_s \rfloor}^{(k)}$ represents that each data symbol $b^{(k)}$ can be transmitted by N_s identical pulses to enhance the quality of reception. The symbol duration equals then $T_s = N_s T_f$. The TH code value for pulse j is given by $c^{(k)}[j]$. The constant term η represents the time shift step introduced by the PPM modulator. Usually, this shift is much smaller than the one due to the TH code (T_c). The time shift $\tau^{(k)}$ represents the relative delay time between the instants at which user k and a reference user i start their transmission; it can be considered as a realization of a random process determined by the actions of the users.

Figure 1 illustrates some of the mentioned parameters; in the example: $N_s = 1$, $c^{(k)} = (2,3,4)$, $b^{(k)} = (1,1,0)$.

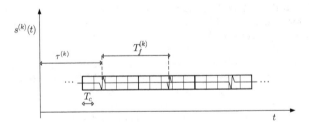

Figure 1. TH-PPM signal structure.

For more detailed information about UWB technology, the author recommends the book [3], which offers an easy-to-read, but complete introduction to the field. Concerning IR-UWB, see [21].

2.2. MAC layer design for low power low data rate IR-UWB networks

The design space for the MAC layer is large; it embraces several dimensions such as multiple access (MA), interference management, resource allocation and power saving. This section summarises the most relevant research findings concerning the design of the MAC layer for low power, low data rate (LDR) IR-UWB networks.

The unique characteristics of the IR-UWB physical layer provide both challenges and opportunities for the MAC layer design. A concrete challenge is the impossibility of carrier sensing, as practised in narrowband systems, since an IR-UWB signal has no carrier. Great opportunities such as robustness against MUI and multipath fading derive from IR-UWB's high temporal resolution. This makes uncoordinated access to the spectrum possible, provided that the local offered load is low compared with the available system bandwidth. For instance, at moderate pulse rates, the "dead time" between pulses allows several uncoordinated, concurrent transmissions to be time interleaved. As a result, ALOHA emerges as the most straightforward MA approach for low data rate IR-UWB networks [3, 9]. The inherent resilience of IR-UWB to MUI can be further increased if different links employ different pseudo-random THC in order to determine the temporal position of the transmitted pulses. The combination of ALOHA with TH coding leads to Time-Hopping Multiple-Access (THMA) [21], which is the most representative MA scheme for low data rate IR-UWB networks.

However TH codes are not perfectly orthogonal, and even if, due to asynchronicity between sources and multipath fading, THMA is still sensitive to impulsive interference. As for CDMA systems, interference coming from a near-by interferer, the so called *near-far* effect, can be very harmful and must be managed in order to avoid performance degradation. For instance, at the physical (PHY) layer multi-user receivers[3] can efficiently address the *near-far* effect at the cost of moderate to high additional hardware [19]. At the link layer, based on the estimation of the wireless link quality, it is possible to adapt the transmission rate to the level of interference experienced at the receiver; the goal can be the improvement of the data rate while satisfying a minimum BER requirement [12].

Within the MAC layer the adaptation of transmission parameters corresponds to a resource allocation task. In the technical literature some approaches on THMA resource allocation can be found, for instance in [18]. A broadly accepted result concerns the optimal power allocation in terms of *proportional fairness*[4]. Note that in autonomous networks with random topology, *proportional fairness* satisfyingly combines fairness and efficiency and outperforms other known performance metrics such as *max-min fairness* or *max total capacity*. This result indicates that the only necessary power control in a IR-UWB network, whose physical layer is in the linear regime[5], is the scheduling function $0 - P_{max}$. In other words, each node should either transmit with full power or not transmit at all [18]. Another common finding is the existence of an *exclusion region* around every destination such that the reference source and nodes within this area cannot transmit simultaneously. This is similar to the IEEE 802.11 CSMA/CA strategy. Certainly, the *exclusion region* vanishes granted that the system has infinite bandwidth ($W \rightarrow \infty$). However, in a practical IR-UWB network the bandwidth is always finite and the size of the *exclusion region* may become a critical issue forcing the MAC designer to concentrate on the scheduling task.

It can be concluded that for IR-UWB networks, and when maximizing rates is the design objective, near-far effects should be tackled by combining scheduling and rate adaptation. Particularly, for low power networks the optimal MAC layer design follows an "all at once" scheduling while it adapts the transmission rates to interference [18].

---

[3] Those which can receive on M channels concurrently.

[4] The power allocation that maximises the sum of the log of source rates, which is a concave objective function.

[5] When the rate of a link can be approximated by a linear function of the signal-to-interference-and-noise ratio.

2.3. Basics of game theory

Game theory has been applied in the recent past to model complex interactions among radio devices that have possibly conflicting interests. For the designer of wireless communication systems game theory is a powerful tool to analyse and predict the behaviour of distributed algorithms and protocols. Respected reference books are [7]. A short overview focusing on the application of game theory in the field of wireless communications can be found in [5].

A resource allocation problem can be naturally modeled as a game, in which the players are the radio devices willing to transmit or receive data. In general, there is an interest conflict since the players have to cope with a limited transmission resource such as power, bandwitdh or pulse load. In order to resolve this conflict they can make certain decisions (or take certain actions) such as changing their transmission parameters. The most familiar game form is the *strategic* form, which models a single-shot, simultaneous interaction among players. It is worth to mention that "simultaneous" is not used here in its strict temporal meaning. It does not imply that players have to choose their actions at the same point of time, but much more that no player is aware of the choice of any other player prior to making his own decision.

A *strategic* form game $\Gamma \langle \mathbf{N}, \mathbf{A}, \mathbf{u} \rangle$ is defined by the following elements[6]:

- A set of players, $\mathbf{N} = \{1, 2, ..., |N|\}$ with cardinality $|N|$.
- The possible actions that the players can choose. Assume that player i can choose among $|A_i|$ possible actions (strategies). Then, the set of possible actions of player i is $\mathbf{A_i} = \{a_{i,1}, a_{i,2}, ..., a_{i,|A_i|}\}$, and the game's global action space $\mathbf{A}$ is given by the cartesian product of the action set of each player $\mathbf{A} = A_1 \times A_2 \times ... \times A_{|N|}$. If player i chooses strategy $a_i \in \mathbf{A_i}$ in a game move, the action profile chosen by all players is denoted as the vector $\mathbf{a} = (a_1, a_2, ..., a_{|N|}) = (a_i, a_{-i})$, with $a_{-i} = (a_1, a_2, ..., a_{i-1}, a_{i+1}, a_{|N|})$ representing the strategies of all players except of player i.
- The utility function, which describes the satisfaction level or payoff of player i given a certain action profile $\mathbf{a}$. The vector of utilities is denoted as $\mathbf{u}(\mathbf{a}) = \{u_1(\mathbf{a}), u_2(\mathbf{a}), ..., u_{|N|}(\mathbf{a})\}$.

To model asynchronous, continuous interactions as those which characterise resource allocation problems in AN the *strategic* form game model is however insufficient. The author agrees with the opinion advanced in [14] and considers *asynchronous, myopic, repeated* games as the most appropriate game model to analyse those problems. At this point it has to be stressed that the work in [14] provides a key result for the application of game theory in the analysis and modeling of AN, since it formalises the relation between the AN's *steady states* and the game behaviour characterised by the Nash equilibria (NE). Theorem 4.1 in [14] proofs that the *steady states* of an AN modeled by an *asynchronous, myopic, repeated* game with stage game $\Gamma \langle \mathbf{N}, \mathbf{A}, \mathbf{u} \rangle$, where all players are rational and act autonomously, coincide with the Nash equilibria of the stage game $\Gamma \langle \mathbf{N}, \mathbf{A}, \mathbf{u} \rangle$.

Notice that an NE is a strategy combination $\mathbf{a}^*$, where no player can improve his utility by individually deviating from its strategy. The existence of an NE for a game significantly

---

[6] We use bold capital letters to denote sets and bold small letters for vectors; for the sake of simplicity we have omitted the superscript T for transposed vectors

depends on the choice of the utility function; specifically on its mathematical properties. The most common existence result for a game's NE is given by the Glicksberg-Fan-Debreu fixed point theorem [7].

2.4. Investigated scenario

This chapter assumes an autonomous sensor, positioning, and identification network (SPIN) as the example system in all investigations. A SPIN is a system characterized by a medium to high node density (up to 2-3 nodes per m^2) in industrial factories or warehouses. Nodes transmit low to medium rate data (up to 1Mbps) combined with position information (position accuracy under 1m) over medium to long distances (typically less than 30 m) to a common receiver.

Concretely, we consider a cluster of up to a hundred IR-UWB sensor nodes that transmit data packets to one common receiver, called the cluster head (CH). The network operation model is based on the beacon-enabled mode of the IEEE 802.15.4a standard [8]. Each sensor node (SN) is considered a source; a link is formed by a transmitting node (source) and the cluster head (CH). Users are asynchronous among themselves. We investigate two scenarios:

- **Scenario 1- Continuous transmission**: Up to 10 UWB sensor nodes are equidistantly situated to the CH, but not necessarily to themselves, along a circle of radius 10 m. They continuously send packets to the CH.

- **Scenario 2- Factory hall**: This scenario accounts for a square simulation field with dimensions 30m×30m. In this field, 100 IR-UWB sensor nodes are considered: 5 are collocated at fixed positions, 75 move along a production line at 5m/s while the rest randomly moves at the same speed within the simulation field. It is assumed that each sensor node generates packets following an exponential distributed packet inter-arrival process and that the exponential processes of the individual sensor nodes are statistically independent. The maximum information data rate per sensor node is 50 Kbps.

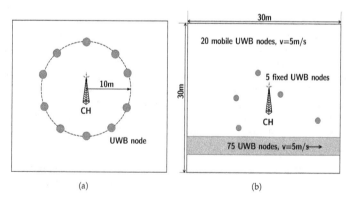

(a)   (b)

Figure 2. Investigated scenarios: (a) Continuous transmission vs. (b) Factory hall.

2.5. Simulation model

The dynamic simulation model has been developed with the discrete event simulation system OMNeT++ [15]. The CH and each SN comprise a PHY layer, a DLC layer and an application layer instance; network and transport layer operation is transparent. For the air interface the superframe structure described in Section 5.4.1 of the IEEE 802.15.4a standard [8] has been selected.

At the application layer each SN generates data packets according to an exponential distributed packet inter-arrival process. Packets are addressed to the CH; its size has been chosen to be $L_p = 400$ bits. The exponential processes are statistically independent from each other, and a maximum information data rate of 1 Mbps is considered. The DLC layer implementation corresponds with the basic "data transfer model to a coordinator" in the standard IEEE 802.15.4a [see 8, Section 5.5.2.1]. For each DLC packet a packet error rate (PER) is calculated as a function of the received power, interference from concurrent transmissions and thermal noise. At the MAC layer a link adaptation function has been implemented which aims at optimising link/system capacity under several channel and interference conditions. The development and analysis of this function is the main achievement of the work presented in this chapter and is covered in section 3.

It is assumed that the network has a fixed chip duration, T_c, so that all changes at the PHY layer transmission parameters are induced by instructions coming from the MAC/DLC layer. The selected set of PHY layer parameters remains constant for one MAC packet transmission, but can be changed from packet to packet according to the time variant channel and interference conditions.

3. MUI Mitigation by distributed pulse rate control

Interference mitigation is a fundamental problem in wireless networks. In CDMA systems, a well-known technique for this is to control the nodes' transmit powers [22]. The work in [17] has shown that for wireless networks in the *linear regime*, and that allow fine-grained rate adaptation, the optimal power allocation is to let nodes either transmit at full power or do not transmit at all. IR-UWB conforms to both attributes, and thus, according to [17], the MAC layer should concentrate on alternative interference mitigation techniques such as scheduling and rate adaptation.

The term rate adaptation embraces all technical means in a system to adapt the transmission speed (rate) to the current quality of the radio link. In IR-UWB networks, rate control can be achieved by adapting the channel coding rate, the modulation order or the processing gain. In order to adapt these parameters, the link's transmitter must have an estimate of the level of interference at its intended receiver. In autonomous networks, most approaches make use of feedback information from the receiver to the transmitter, for example within ACK packets. This information can take various forms; conventionally, it is a function of the signal-to-noise-plus-interference-ratio (SNIR). However, measuring the SNIR is difficult in practice due to the very low transmit power of UWB signals. Therefore, recent approaches [9] rely on information provided by the channel decoder, namely on BER estimations. This work follows theses approaches and also considers BER instead of SNIR feedbacks.

The processing gain of IR-UWB is twofold: the number of pulses per symbol (N_s) and the average frame duration (T_f). The adaptation of IR-UWB processing gain impacts both rate and average emitted power. Adapting the processing gain in IR-UWB systems was first suggested by Lovelace et. al. in [10]. Lovelace proposed a technique for adjusting the number of pulses per symbol in a single-hop link affected by uncoordinated near pulse interferers. This technique requires a particular type of receiver, capable of selectively and passively blank large interfering pulses from symbol decisions. Also, the approach assumes a system with a large number of pulses per symbols, so that blanking a few of them has only a minimal impact on the resulting BER. The joint adaptation of both types of processing gain was first studied in the context of IEEE 802.15.3a WPANs to reduce the mutual interference among uncoordinated, collocated WPANs [23], and was lately extended to cluster based wireless sensor networks in [20]. The basic observation in [23] was that the larger the number of collocated WPANs the larger the average frame time, T_f, has to be in order to reduce the amount of impulsive interference.

The first, general, theoretical considerations about the performance tradeoff between the two types of IR-UWB processing gain were published by Fishler and Poor in [6]. Fishler and Poor examined this tradeoff as a function of the BER in a system with fixed processing gain over flat-fading and frequency-selective channels. The study concludes that in a coded system transmitting over a flat-fading channel, the BER is independent of the ratio between the two types of processing gain. In contrast, in an uncoded system over a flat-fading channel and in frequency-selective channels there is a trade-off. This trade-off favours systems with a low number of pulses per symbol, as the system BER considerably degrades as the number of pulses per symbol increases. Thus, regarding processing gain adaptation, and assuming that the energy per transmitted pulse is the same for all users, it is preferable to extend the signal's average frame time (reduce the signal's duty cycle) than to increase the number of pulses per symbol. Moreover, using large frame times help to reduce the system complexity since a lower sampling rate can be used.

With the exception of the work in [10], which requires a particular receiver technique, the distributed adaptation of IR-UWB processing gain in autonomous networks has not been addressed in the literature before. The remaining approaches referenced in this section rely on the presence of a coordinator node that implements the adaptation algorithm and instructs other nodes on how to scale their parameters. This work focuses on autonomous networks; although hierarchical structures are not ruled out here, they are not required and therefore adaptation schemes cannot rely on the presence of coordination entities, but must be distributed.

The author claims that in autonomous IR-UWB networks, due to its self-organising and asynchronous character, and due to the monotonically increasing throughput for increasing pulse rates, the system's local pulse load can become so high that bit errors may not be longer tractable with error coding schemes. Based on this assumption, and following previous theoretical considerations on processing gain adaptation, this work develops a novel mechanism -distributed Pulse Rate Control (PRC)- to coordinate the links' pulse rate levels to optimise the overall network performance, measured in terms of total network logarithmic utility (*proportional fairness*), while satisfying a minimum per link BER requirement.

3.1. Distributed pulse rate control

Pulse rate control (PRC) can be realised in form of a link adaptation function whose goal is the improvement of the network throughput while satisfying a minimum per link BER requirement.

While adaptive modulation and channel coding are local decisions to a sender-receiver pair, PRC involves a cooperation among different links since the average[7] probability of pulse collision at the i-th receiver ($P_{coll,i}$) does not depend on its own link's pulse rate, but on the pulse rates of transmitters in its vicinity [20]. Since the collision probability is an indirect measurement of the BER, it can be further assumed that the BER at the i-th receiver, $P_{e,i}$, does not directly depend on its link's pulse rate, but on the pulse rate of the neighbouring links.

In the following we model, analyse and evaluate the distributed PRC approach. We first formulate PRC as a network logarithmic utility maximisation problem with quality of service (QoS) constraints. In order to solve the problem in a distributed manner, PRC is reformulated to a non-cooperative game with pricing. A distributed asynchronous algorithm is proposed which converges to the globally optimal solution of the original problem.

3.1.1. Problem formulation

The objective of the PRC approach is to determine the maximum pulse rate allocation, such that the QoS demands - in terms of BER- of all network links are fulfilled. This goal can be expressed by equation 2.

$$\max_{\mathbf{prf}} \sum_{i=1}^{|\mathbf{N}|} log\,(r_i(\mathbf{prf}))$$

subject to: $\hspace{9cm}$ (2)

$$P_{e,i}(\mathbf{prf}_{-i}) \leq \beta_i, \forall\, i \in \mathbf{N},$$

$$prf_i \in \mathbf{PRF}_i = [prf_i^{min}, prf_i^{max}], \forall\, i \in \mathbf{N}$$

PRC assumes that each IR-UWB node can autonomously adapt its average pulse repetition frequency, prf - that is the inverse of the average frame duration ($prf = \frac{1}{T_f}$). Additionally, it assumes that the energy per transmitted pulse, E_p^{tx}, is fixed and equal for all users, independent of the modulation scheme, and is chosen so that the IR-UWB node with the highest data rate completely exploits the FCC requirements in terms of EIRP and peak power. Controlling the source's prf is equivalent to controlling it's data rate (r) in terms of pulses per second; with fixed E_p^{tx}, it is also equivalent to controlling the source's average transmitted power.

QoS constraints in equation 2 limits the set of feasible pulse rate allocations. Since a higher pulse rate level from one transmitter increases the collision probability -and in turn the BER

---

[7] The average is taken over the links' asynchronism. In fact, the collision probability between two transmitters, for instance transmitter j and the reference transmitter i, depends on their pulse rates and on the relative delay time between the instants at which both transmitters start their transmissions. This relative delay is a random variable determined by the action of the users; it is represented by the time shift τ_j.

(P_e)- at neighbouring receivers, there may not be any feasible pulse rate allocation to satisfy the requirements of all users.

The logarithmic utility function in equation 2 captures the link's desire for higher data transmission rate. In an IR-UWB network, the raw data rate per link can be controlled by adapting the channel coding rate (R_i), the modulation order (m_i) or the processing gain, that is the number of pulses per symbol (N_s^i) and/or the average frame duration (T_f^i). Equation 3 depicts the dependency of the raw data rate on these parameters.

$$r_i^{raw} = \frac{1}{N_s^i \cdot T_f^i} \cdot R_i \cdot log_2(m_i) \text{ [bit/s]} \tag{3}$$

Following [6], this work focuses on systems with low number of pulses per symbol. For the sake of simplicity and without loss of generality we consider hereafter $N_s^i = 1, \forall i \in \mathbf{N}$. The useful (net) data rate per link is given in equation 4.

$$r_i(\mathbf{prf}) = r_i^{raw} \cdot \left(1 - P_{e,i}(\mathbf{prf}_{-i})\right) \text{ [bit/s]} \tag{4}$$

Since the bit error rate, P_e, is a nonlinear, and neither convex nor concave function of the links' pulse rates, the pulse rate optimisation problem is in general a nonlinear optimisation problem. The classical optimisation theory has no effective method for solving the general nonlinear optimisation problem, but several different approaches such as geometric programming (GP). Each of these approaches involves some compromise [1]; for instance, GP is limited to algorithms with a central single point of computation [2]. Game theory represents an alternative to GP and it is used in the next section to model the PRC problem (in equation 2) in a distributed manner.

3.1.2. Pulse rate control game

From the author's point of view, distributed PRC can be interpreted as a resource allocation mechanism that regulates the link's number of transmitted pulses per second. Hence, the framework of non-cooperative game theory can be applied to model and analyse the problem that searches for the network's maximum pulse rate allocation that satisfies a certain set of per-link BER constraints. Next we show how a game theoretical formulation helps to provide the UWB devices with incentives to minimise impulsive emissions when the cumulative system pulse load excesses certain limits, which are determined by some QoS constraints.

In PRC an increase in a link's average *prf* directly and negatively affects the probability of pulse collision, and consequently the BER, of neighbouring links [20]. A game theorist would refer to this fact by saying that there are *negative externalities* in the system. In order to deal with QoS constraints in the presence of negative externalities cooperation among the autonomous users must be enforced. Pricing is one of the most commonly used incentives to regulate selfish user behaviour and establish cooperation. Keeping this in mind, the original logarithmic utility function $u_i(\mathbf{prf}) = log(r_i(\mathbf{prf}))$ in equation 2 is modified by adding a linear pricing function of the link's *prf*. The new utility function is given in equation 5.

$$v_i(\mathbf{prf}) = log(r_i(\mathbf{prf})) - \pi_i(\mathbf{prf}) \cdot prf_i \tag{5}$$

The original logarithmic utility function reflects the level of a user's satisfaction from consuming the resource *prf* (directly related to the transmission data rate). The pricing factor, $\pi_i(\mathbf{prf})$, reflects the cost per unit of resource charged to user i. Hence, the new utility function can be interpreted as if each user i maximises the difference between its old net utility and a payment to other users in the network due to the interference (pulse collisions) it generates.

With the new utility function $v_i(\mathbf{prf})$ a non-cooperative PRC game with pricing, denoted by $\Gamma_{PRC} = \langle \mathbf{N}, \mathbf{PRF}, \mathbf{v} \rangle$, is developed. For Γ_{PRC} the set of players $\mathbf{N} = \{1, 2, ..., |N|\}$ corresponds with the set of active links (users) in the network, so that the terms "player" and "user" are used as synonym. The vector of utilities corresponds to $\mathbf{v}(\mathbf{prf}) = \{v_1(\mathbf{prf}), v_2(\mathbf{prf}), ..., v_{|N|}(\mathbf{prf})\}$, and the set of actions that players can choose, $\mathbf{PRF} = PRF_1 \times PRF_2 \times ... \times PRF_{|N|}$, is compact and convex.

As described in 2.4 this work considers a network model with a single centralised receiver (CH) and several uncoordinated sources. With a common pricing factor provided by the CH, $\pi_i(\mathbf{prf}) = \pi_j(\mathbf{prf}) = \pi_{CH}(\mathbf{prf})$, $\forall i, j \in \mathbf{N}$, each user in the network can be guided by the altruistic goal of maximising the cumulative network throughput at the CH, while keeping their average bit error rate, $P_{e,i}$, as close as possible to a target $\beta_{CH} = \beta_i, \forall i \in \mathbf{N}$. In this setting, the pricing term acts as a control parameter employed by the CH to discourage the overuse of the wireless resource *prf* and to keep the interference sustainable. It is expected that the choice of a common pricing factor for all links degenerates the *proportional-fairness* character of the original problem formulation into a *max-min fairness* solution.

Generally, an average cumulative $P_{e,CH} >> \beta_{CH}$ suggests a congestion situation caused by an overload in the local pulse density. In contrast, $P_{e,CH} << \beta_{CH}$ suggests an underload situation in which the local pulse density is below the sustainable load for the given QoS criteria. Accordingly, the CH measures the cumulative bit error rate at each superframe and continuously tracks its deviation from the target value in the variable $\triangle P_{e,CH}^s$

$$\triangle P_{e,CH}^s = \sum_{k \in \mathbf{AL}^s} P_{e,k} - \beta_{CH}, \tag{6}$$

where s represents the superframe index and $\mathbf{AL}^s$ is the set of active links during superframe s. Based on $\triangle P_{e,CH}^s$, the CH computes a congestion cost for superframe $s + 1$ and feedbacks it to the UWB nodes in the next beacon frame. The computation rule for the congestion cost is very simple. If there is congestion in the current superframe s, the congestion cost for the next superframe $s + 1$ must be increased, in contrast, if there is underload in the current superframe s, the congestion cost for the next superframe $s + 1$ can be decreased. The congestion cost represents a common price factor for all players, $\pi_{CH} = \pi_i, \forall i \in \mathbf{N}$. The UWB nodes regulate their *prf*, and therewith their data rate, in response to the congestion cost feedback from the CH.

Specifically, the following computation rule is proposed

$$\pi_{CH}^{s+1} = \begin{cases} (1.0 - \delta)\pi_{CH}^s + \mu\pi_{CH}^s\delta, & \triangle P_{e,CH}^s > 0 \\ (1.0 - \delta)\pi_{CH}^s - \frac{\pi_{CH}^s\delta}{\mu}, & \triangle P_{e,CH}^s < (\beta_{CH}\omega|AL|^s) \end{cases}, \tag{7}$$

where μ is a weight factor and δ is the smoothing factor of a weighted exponential-moving-average (EMA) algorithm [4]. In order to improve game convergence a tolerance region for the cluster congestion level has been defined in which no adaptation is done. The tolerance range is defined by $|AL|^s$, the number of active links in superframe s, and a constant $\omega \leq 0$.

3.1.3. Pulse rate control algorithm

In a realistic distributed environment, at the start of a game it is not possible that a player $i \in N$ has the complete price information (adjacent channel gains, link qualities) that is necessary to discover an NE inmediately. However, the player can make a guess, denoted by its selected action prf_i, regarding its equilibrium average prf denoted by prf_i^*. Then, assuming that the actions of the other players, $\mathbf{prf}_{-i}$, and its own price factor, π_i, remain constant while it makes a decision, player i improves its guess by selecting a new action which maximises its utility function. This new guess results in a new approximation to prf_i^*. When the deviations in all players' actions become negligibly small, the game can be assumed to have converged to an NE.

The PRC algorithm implements the adaptive behaviour described above and distributively solves the pulse rate allocation maximisation problem in equation 2.

PRC Algorithm

- **Step 1: Initialisation**

  For each user $i \in N$ choose some $prf_i(0) \in PRF_i$ and a price factor $\pi_i(0) \geq 0$.
- **Step 2: Price update**

  At each iteration $t \in T_{i,\pi}$, user i updates its price according to equation 7.
- **Step 3: Pulse rate update**

  At each iteration $t \in T_{i,prf}$, user i updates prf_i according to a best response decision rule:

$$prf_i^*(t+1) = \mathrm{BR}_i\left(\mathbf{prf}_{-i}(t)\right) = \left[\frac{1}{\pi_i(t)}\right]_{prf^{min}}^{prf^{max}} \tag{8}$$

where the notation $[x]_a^b$ means $\max\{a, \min\{b, x\}\}$ and BR_i is the set of best responses of player i to the strategy profile $\mathbf{prf}_{-i}(t)$.

Notice that the price and prf adaptations do not need to happen at the same time. The price update instants are determined by the superframe beacon raster, while the prf update instants depend on the beacon raster as well as on the source traffic model and can therefore be asynchronous across users.

In the practical implementation of the algorithm, a discretisation of the action space ($\mathbf{PRF}_i$) is unavoidable. The granularity of this discretisation process represents a trade-off for the convergence properties of the algorithm. An infinitely small granularity equals an infinite action space and guarantees that the NE action profile (if there is one) is considered in the search process; however, it increases the computation cost of the algorithm. With an increasing granularity the action space becomes finite, so that the search space for the algorithm shrinks and the computational cost is reduced. A brief discussion about the existence of Nash equilibria in discrete games can be found in [7]. In general it holds that, if the game with continuous action space has a stable equilibrium, the discrete pendant also has an equilibrium.

3.2. Game analysis

In this section, the utility function of Γ_{PRC} is analysed in terms of existence and uniqueness of Nash equilibria. The fact that Γ_{PRC} fits the framework of potential games ([13, 14]) significantly facilitates the analysis.

A potential game is characterised by the existence of a function, denoted as the potential function, $\Phi : A \to \Re$, such that the change in the utility function of a player when it unilateraly deviates in its strategy ($\triangle u_i$) is reflected in a change in value of the potential function ($\triangle \Phi$). If for all unilateral deviations, $\triangle \Phi = \triangle u_i$ the game is referred to as an *exact* potential game (EPG). If the relationship between the potential function and the utility functions is relaxed so that only sign changes are preserved, $sgn(\triangle \Phi) = sgn(\triangle u_i)$, the game is called an *ordinal* potential game (OPG).

3.2.1. Existence of an equilibrium

In order to prove the existence of an NE for game Γ_{PRC} a definition and a powerful result concerning the identification of ordinal potential games are leveraged. Definition 1, theorem 1 and 2 and their respective proofs can be found in [14].

Definition 1. *Better-Response Equivalence*

A game $\Gamma = \langle \mathbf{N}, \mathbf{A}, \mathbf{v} \rangle$ is said to be better response equivalent to game $\tilde{\Gamma} = \langle \mathbf{N}, \mathbf{A}, \tilde{\mathbf{v}} \rangle$, if $\forall i \in \mathbf{N}, a \in \mathbf{A}, v_i(a_i, \mathbf{a}_{-i}) > v_i(b_i, \mathbf{a}_{-i}) \Leftrightarrow \tilde{v}_i(a_i, \mathbf{a}_{-i}) > \tilde{v}_i(b_i, \mathbf{a}_{-i}).$

Theorem 1. *Identification of Ordinal Potential Games*

A game $\Gamma = \langle \mathbf{N}, \mathbf{A}, \mathbf{v} \rangle$ is an ordinal potential game if and only if it is better response equivalent to an exact potential game.

Theorem 2. *NE of Better-Response Equivalent Games*

If a game $\Gamma = \langle \mathbf{N}, \mathbf{A}, \mathbf{v} \rangle$ is better response equivalent to game $\tilde{\Gamma} = \langle \mathbf{N}, \mathbf{A}, \tilde{\mathbf{v}} \rangle$, then the Nash Equilibria of Γ, if any exist, coincide with the Nash equilibria of $\tilde{\Gamma}$.

With these results in mind, game $\tilde{\Gamma}_{PRC} = \langle \mathbf{N}, \mathbf{A}, \tilde{\mathbf{v}} \rangle$ with utility function

$$\tilde{v}_i(\mathbf{prf}) = log(prf_i) - \pi_{CH}prf_i, \tag{9}$$

is introduced. Since $log(xy) = log(x) + log(y)$ and through definition 1, there is a trivial better response equivalence relationship between Γ_{PRC} and $\tilde{\Gamma}_{PRC}$. Hence, from theorem 2 we know that by analysing the set of Nash equilibria of game $\tilde{\Gamma}_{PRC}$ we are at the same time solving for the Nash equilibria of Γ_{PRC}. Next, we prove that $\tilde{\Gamma}_{PRC}$ is an EPG and, consequently thanks to theorem 1, Γ_{PRC} is an OPG. Further, in [14] we find a powerful result concerning all potential games with a compact action space and a continuous potential function: The existence of at least one NE.

From [14], a sufficient condition for the existence of a potential function in game $\Gamma = \langle \mathbf{N}, \mathbf{A}, \mathbf{u} \rangle$ is

$$\frac{\partial^2 u_i(\mathbf{a})}{\partial a_i \partial a_k} = \frac{\partial^2 u_k(\mathbf{a})}{\partial a_k \partial a_i}, \forall i, k \in \mathbf{N}, \forall \mathbf{a} \in \mathbf{A} \tag{10}$$

In $\tilde{\Gamma}_{\text{PRC}}$ the utility functions are given by:

$$\tilde{v}_i(prf_i, \mathbf{prf}_{-i}) = log(prf_i) - \pi_{CH}prf_i \tag{11}$$

and

$$\tilde{v}_k(prf_k, \mathbf{prf}_{-k}) = log(prf_k) - \pi_{CH}prf_k. \tag{12}$$

In this network settings, users ignore any influence they may have on the price calculated at the CH. Hence, $\frac{\partial \pi_{CH}}{\partial prf_i} = 0 \,\forall\, i \in \mathbf{N}$, and it is easy to prove that

$$\frac{\partial^2 \tilde{v}_i(prf_i, \mathbf{prf}_{-i})}{\partial prf_i \partial prf_k} = \frac{\partial^2 \tilde{v}_k(prf_k, \mathbf{prf}_{-k})}{\partial prf_k \partial prf_i} = 0 \tag{13}$$

Characteristic for an EPG is the existence of a potential function Φ that exactly reflects any unilateral change in the utility of any player, that is $\triangle\Phi(a) = \triangle u_i(a)$. Hence, starting from any arbitrary strategy vector $\mathbf{a}$ any unilaterally player's adaptation that increases its utility $u_i(a)$ identically translates in an increase of the potential function $\Phi(a)$.

The potential function of game $\tilde{\Gamma}_{\text{PRC}}$ is given in equation 14.

$$\tilde{\Phi}_{\text{PRC}}(\mathbf{prf}) = \sum_{i \in N} log(prf_i) - \pi_{CH} \sum_{i \in N} prf_i \tag{14}$$

Based on the potential game definition given in [13], the proof that $\Gamma = \langle \mathbf{N}, \mathbf{A}, \mathbf{u} \rangle$ is an EPG requires that:

$$u_i(a_i, a_{-i}) - v_i(b_i, \mathbf{prf}_{-i}) = \Phi(a_i, a_{-i}) - \Phi(b_i, a_{-i}), \forall\, i \in \mathbf{N}, \forall\, \mathbf{a} \in \mathbf{A} \tag{15}$$

Equation 16 proves condition 15 for $\tilde{\Gamma}_{\text{PRC}}$. Note that all sum terms in the potential function that are independent of user i's strategy are constant and can be grouped in an extra term denoted as c. By the substraction of the potential functions, the term c dissapears leaving the difference of the potential functions identical to the difference of the utility functions.

$$\tilde{v}_i(prf_i, \mathbf{prf}_{-i}) - \tilde{v}_i(prf_i', \mathbf{prf}_{-i}) = \tilde{\Phi}_{\text{PRC}}(prf_i, \mathbf{prf}_{-i}) - \tilde{\Phi}_{\text{PRC}}(prf_i', \mathbf{prf}_{-i})$$

$$log(prf_i) - \pi_{CH}prf_i - log(prf_i') + \pi_{CH}prf_i' =$$

$$log(prf_i) - \pi_{CH}prf_i$$

$$+ \underbrace{\sum_{j \in N \,\wedge\, j \neq i} log(prf_j) - \pi_{CH} \sum_{j \in N \,\wedge\, j \neq i} prf_j}_{=c}$$

$$-log(prf_i') + \pi_{CH}prf_i' \tag{16}$$

$$- \underbrace{\sum_{j \in N \,\wedge\, j \neq i} log(prf_j) + \pi_{CH} \sum_{j \in N \,\wedge\, j \neq i} prf_j}_{=-c}$$

$$log(prf_i) - \pi_{CH}prf_i - log(prf_i') + \pi_{CH}prf_i' =$$

$$log(prf_i) - \pi_{CH}prf_i - log(prf_i') + \pi_{CH}prf_i'$$

$$\forall\, i \in \mathbf{N}, \forall\, prf \in \mathbf{PRF}$$

Equation 14 is continuous since it is the sum of continuous functions, furthermore the action space of $\tilde{\Gamma}_{PRC}$ is compact per definition. Hence, equation 16 verifies that (as any other EPG) $\tilde{\Gamma}_{PRC}$ has at least one NE. Finally, applying theorem 2 it is proven that Γ_{PRC} has at least one NE - in fact the same as $\tilde{\Gamma}_{PRC}$).

3.2.2. Uniqueness of the equilibrium

From [14], it is known that an EPG following an asynchronous, myopic, best response decision rule converges to a pure strategy NE that maximises the potential function. Furthermore, if the potential function is strictly concave, it has a unique global maximum which is then the unique NE of the EPG. Based on these results the following proposition can be stated.

Proposition 1. *If $\tilde{\Phi}_{PRC}$ in equation 14 is strictly concave, the proposed PRC algorithm will always converge to the unique NE of game $\tilde{\Gamma}_{PRC}$, which in turn, is the unique global maximum of $\tilde{\Phi}_{PRC}$.*

Proof. The PRC algorithm can be interpreted as the players employing asynchronous myopic best response (MBR) updates. Thus, to demonstrate proposition 1 it suffices to prove the strict concavity of $\tilde{\Phi}_{PRC}$ in equation 14. As explained in [1, Section 3.1.4], this can be verified with the Hessian matrix and the second-order conditions.

For the Hessian matrix in equation 18 the second derivatives of equation 14 are required.

$$H(\tilde{\Phi}) = \begin{pmatrix} \frac{\partial^2 \tilde{\Phi}_{PRC}}{\partial^2 prf_1} & \cdots & \frac{\partial^2 \tilde{\Phi}_{PRC}}{\partial prf_1 \partial prf_j} & \cdots & \frac{\partial^2 \tilde{\Phi}_{PRC}}{\partial prf_1 \partial prf_N} \\ \vdots & \ddots & \vdots & \ddots & \vdots \\ \frac{\partial^2 \tilde{\Phi}_{PRC}}{\partial prf_N \partial prf_1} & \cdots & \frac{\partial^2 \tilde{\Phi}_{PRC}}{\partial prf_N \partial prf_j} & \cdots & \frac{\partial^2 \tilde{\Phi}_{PRC}}{\partial^2 prf_N} \end{pmatrix} \tag{17}$$

$$\frac{\partial^2 \tilde{\Phi}_{PRC}}{\partial prf_i \partial prf_j} = \begin{cases} -\frac{1}{prf_i^2} & \text{if } i = j \\ 0 & \text{if } i \neq j \end{cases} \tag{18}$$

The matrix is negative definite, $x^T H(\tilde{\Phi}) x < 0$, since all diagonal elements are negative. Hence, the proof is complete. □

Finally, proposition 1 can be reformulated to proposition 2.

Proposition 2. *If $\tilde{\Phi}_{PRC}$ in equation 14 is strictly concave, the proposed PRC algorithm will always converge to the unique NE of game $\tilde{\Gamma}_{PRC}$, which, in turn, is the unique NE of game Γ_{PRC}.*

Proof. The proof of Proposition 2 results from combining the proof of Proposition 1 with Theorem 2. □

3.2.3. Optimality of the equilibrium

So far, there is no general result about the optimality of Nash equilibria in potential games or in any other more general class of games. However by designing the potential game in a way that its potential function complies with the network design function, a quite elegant way to demonstrate NE optimality is possible. In that case, any strategy that maximises the potential function (any NE) maximises as well the network design function.

In this sense, note that the utility function of game Γ_{PRC} in equation 5, combines the network utility function of the original resource allocation problem (cf. to equation 2) with a linear price function. By exploiting the linear space properties of EPGs, the potential function in equation 14 preserves the properties (such as concavity and uniqueness of the global maximisers) of the original network objective function $\sum_1^{|N|} log(r_i(\mathbf{prf}))$. The addition of the linear price term aims at adjusting the unique NE of $\tilde{\Gamma}_{PRC}$ so that the QoS constraint in the original problem is respected. Hence the NE is optimal from a network design perspective.

3.3. Performance evaluation

Results highlighted in this section have been computed with the simulation model presented in Section 2.5 in the scenarios depicted in Fig. 2. The aim of this section is to demonstrate the regulative character and the interference compensation functionality of the PRC approach. Therefore, its performance under increasing offered system load is compared to the one obtained with the ALOHA MAC protocol without feedback. Note that the IEEE 802.15.4a proposes ALOHA with optional feedback as the standard MAC protocol for IR-UWB physical layers.

Parameter	Value	Parameter	Value
E_p^{tx}	$2 \cdot 10^{-11}$ [Ws]	BW	1.5 [GHz]
$T_{f\,rcx}$	$5 \cdot 10^{-9}$ [s]	prf granularity	1 [kHz]
prf_i^{max}	1 [MHz]	prf_i^{min}	10 [kHz]
m-PPM	2	β_{CH}	$5 \cdot 10^{-4}$
L_p	400 bit	μ	2
δ	$1 \cdot 10^{-2}$	ω	$2.5 \cdot 10^{-3}$

Table 1. List of main simulation parameters.

3.3.1. Existence of an equilibrium

First a simplified setting with a constant number of players, as usually assumed in game theory, is considered. Then, a more realistic setting where the number of players is a random variable controlled by the traffic distribution is explored. For both settings simulation results[8] confirm the existence of an NE regardless of the algorithm's initialisation parameters.

Constant number of players This setting considers the continuous transmission scenario described in Section 2.4. Recall that in this scenario up to 10 nodes are collocated along a circumference of 10m radius with the CH in the centre. It is assumed that all nodes have always packets to send; this is assured by configuring the source traffic generator to an information data rate of 1Mbps.

Figure 3 confirms the existence of a stable equilibrium in game Γ_{PRC}, and that the PRC algorithm converges to it. Figure 3(a) depicts a symmetric prf allocation, which agrees with the analytically predicted NE; the equilibrium prf per link is approximately 220 kHz. In fact, a symmetric equilibrium was expected since the congestion cost is the same for each link and

---
[8] These results were obtained in [11].

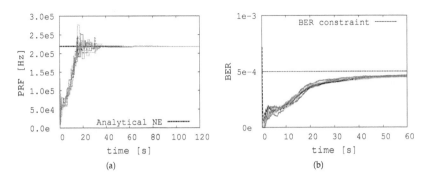

Figure 3. Existence of an equilibrium in Scenario 1 - Continuous transmission. Each link is represented with a different color: (a) Convergence to the unique equilibrium *prf* allocation; (b) Upper-bounded character of the link's BER.

the node topology is symmetric. Figure 3(b) confirms that the QoS constraint on the BER, $P_{e,i} \leq \beta_{CH}$, is fulfilled for all links.

Time-varying number of players This setting considers the factory hall scenario described in Section 2.4. There are a total of 100 heterogeneous nodes, some of them are moving along a production line, others are fixed in known positions and the rest is moving around the CH following a random waypoint movement model. The information data rate has been chosen low enough to guarantee that nodes do not always have packets to send; this results in a time-varying number of players[9].

Figure 4 presents results for two values of the information data rate which are 10 kbit/s and 15 kbit/s, respectively. With 10 kbit/s information data rate, the CH is exposed to a low system pulse load measured in pulses per second (Pps). Figure 4(a) suggests the existence of a stable equilibrium in game Γ_{PRC}, and that the PRC algorithm converges to it. The *prf* equilibrium allocation is again symmetric, with all links converging to the maximum possible *prf* of 1 MHz. Figure 4(b) confirms that the QoS constraint on the BER is satisfied for all links. Note that the BER curves of all links except one follow almost an identical course. The outlier curve corresponds to the sensor node which is located closest to the CH (at 3m) and therefore exhibits the best channel gain and the lowest BER.

By increasing the user data rate to 15 kbit/s, the system pulse load achieves a level which is not compliant with the problem's QoS constraint. The PRC algorithm should then reduce the links' pulse rates to a sustainable level, which is identified by the congestion cost factor π_{CH}. Figure 4(c) illustrates the regulative effect of the PRC algorithm. In spite of the heterogeneous network character, the *prf* equilibrium corresponds to a *max-min* socially fair allocation induced by the choice of a common pricing factor for all links. Besides, it can be observed that in Figure 4(d) some of the links violate the QoS constraint. Still, if we consider the average BER over all links (see Figure 8(b)) it is below the upper bound β_{CH}. This is

---

[9] Note that the number of players in the game coincide with the number of active links in each superframe, and this is a random variable controlled by the exponential traffic distribution.

due to the global altruistic behaviour of the congestion cost factor, π_{CH}, which works with an estimation of the average[10] BER at each stage of the game.

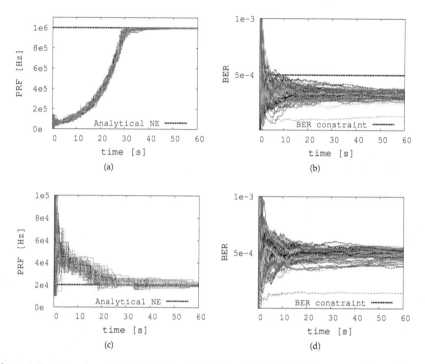

Figure 4. Existence of an equilibrium in Scenario 2 - Factory hall. Each link is represented by a different color: Graphics (a) and (c) show the convergence to the unique equilibrium *prf* allocation when the information data rate per user is 10 kbit/s and 15 kbit/s, respectively; Graphics (b) and (d) show the temporal evolution of the BER per link when the information data rate per user is 10 kbit/s and 15 kbit/s, respectively.

3.3.2. Variation of the offered load

The results in this clause show the effect of increasing offered system load on the aggregate network throughput and the average BER, both measured at the CH. Like in the previous clause, the game performance with a constant number of players is differentiated from the that with a time-varying number of players.

Constant number of players The information data rate is fixed and equal to 1 Mbps to ensure continuous packet transmission. In order to progressively increase the offered system pulse load, the number of sensor nodes collocated along the circumference of 10 m radius has been stepped up from 2 to 10.

---

[10] Over all links.

In general, and due to the increasing MUI, the aggregated network throughput is expected to drop as the number of nodes in the network grows. Figure 5(a) illustrates the aggregated network throughput. It can be observed that with four to ten source nodes the aggregated network throughput remains almost constant. These results suggest an interference compensation effect of the PRC algorithm; Figure 6 confirms this effect. Additionally, Figure 5(b) shows that the cumulative BER scratches the QoS upper bound in all cases except in the case of having only two nodes. In this special case, even with nodes sending with the maximum *prf* the offered system load is low enough to guarantee an average BER far below the QoS upper bound.

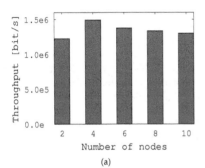
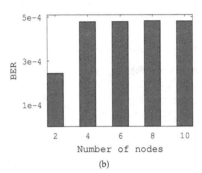

(a)   (b)

Figure 5. System performance in Scenario 1 - Continuous transmission: (a) Aggregated network throughput measured at the CH's application layer; (b) Average BER per link measured at the CH.

Figure 6 shows the mean[11] *prf* per link; where each link has been represented with a different color. It can be observed that the PRC algorithm relaxes the mean *prf* per link as the number of nodes in the network grows, since consequently the pulse density in the system increases. Recall that a higher pulse density raises the probability of pulse collisions and this, in turn, the probability of bit errors. The larger the average BER at the CH, the larger the congestion factor, π_{CH}, is. Finally, larger congestion factors lead to lower *prf* levels.

Time-varying number of players Since in Scenario 2 - Factory hall the number of sensor nodes is constant, in order to increase the offered system pulse load we progressively raise the information data rate per sensor node.

In Figure 7 two different phases can be identified. As long as the information data rate is kept below 10 kbit/s, the mean *prf* per link does not drop, but remains close to the maximum allowed *prf* value (1 MHz). When the information data rate reaches 15 kbit/s, the mean *prf* per link drops down to approximately 20 kHz and remains there despite growing information data rate. These results are consistent with the PRC algorithm's behaviour illustrated in Figure 4(a) and Figure 4(c).

In a similar way Figure 8 illustrates the regulative behaviour of the PRC algorithm. At information data rates up to 10 kbit/s the PRC algorithm converges to the maximum possible

---

[11] Over the whole simulation time.

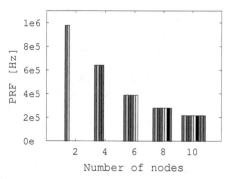

Figure 6. Average *prf* per link in Scenario 1 - Continuous transmission.

prf per link, since the offered system pulse load is low enough to guarantee the QoS constraint on the BER (see Figure 8(b)). With higher information data rates, the PRC algorithm has to relax the effective system pulse load per user to avoid network congestion, and to be able to satisfy the QoS constraint.

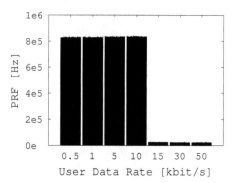

Figure 7. Average *prf* per link in Scenario 2 - Factory hall.

3.3.3. Performance comparison with IEEE 802.14.5a MAC

Finally, this clause is dedicated to compare the performance of the ALOHA MAC protocol, as described in IEEE 802.15.4a, with and without distributed PRC. For the simulations without PRC we have set the fixed *prf* per link to 1 MHz. IEEE 802.15.4a recommends ALOHA with optional feedback channel as the MAC protocol for low data rate sensor networks with IR-UWB physical layers. This work focuses on random access without feedback channel to keep the power consumption and the receiver complexity at the sensor nodes low. In exchange, the drawback has to be accepted that successful reception of data packets cannot be guaranteed since retransmission requests are not possible. Note that in the application field considered in this chapter, the relatively small throughput offered by the random access method without feedback channel is still satisfying.

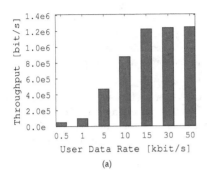
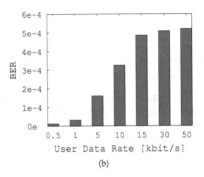

Figure 8. System performance in Scenario 2 - Factory hall: (a) Aggregated network throughput measured at the CH's application layer; (b) Average BER per link measured at the CH.

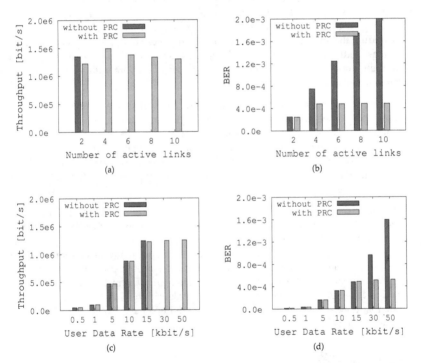

Figure 9. Performance comparison of ALOHA with and without PRC: (a) Aggregated network throughput in Scenario 1; (b) Average BER per link in Scenario 1; (c) Aggregated network throughput in Scenario 2; (d) Average BER per link in Scenario 2.

Figures 9(a) and 9(b) depict a performance comparison between ALOHA with and without PRC in the continuous transmission scenario with high offered pulse load and constant number of players. The network throughput obtained without PRC rapidly collapses[12] as the number of source nodes grows. This is comprehensible, since pulse collisions (and therewith bit errors, see Figure 9(b)) augment as the offered pulse load increases. In contrast, the interference compensation effect of the PRC approach under increasing offered load can be well observed. Notice how the PRC algorithm is able to limit the system pulse load to a level that ensures the QoS constraint.

Figures 9(c) and 9(d) depict results in the factory hall scenario with low to moderate pulse load and time-varying number of players. Notice that the regulative effect of the PRC approach (see Figure 7) limits the maximum possible cumulative network throughput, while ALOHA with fixed *prf* cannot guarantee the design QoS constraint and breaks down as the information data rate per link increases to 30 kbit/s.

4. Summary and conclusions

This chapter introduces a novel concept for impulsive interference management in low power, autonomous, IR-UWB networks. The concept enables concurrent transmissions at full power, while allowing each source to independently adapt its pulse rate (measured in pulses per second) to reduce the impact of pulse collisions at nearby receivers. The design is independent of a particular modulation scheme and can be applied to any IR-UWB physical layer. Beyond, it does not rely on any particular receiver technique and can work with a simple, low cost, non-coherent receiver.

The chapter formulates and evaluates the pulse rate allocation problem as a social rate optimisation problem with QoS constraints. It introduces a distributed algorithm implementation and analyses its performance via simulation. It has been analytically proven that the game Γ_{PRC} fits the framework of ordinal potential games, and that the NE is unique. In all considered scenarios, simulation results have confirmed the existence of an equilibrium for the game and that the distributed PRC algorithm converges to it, provided that the pricing parameters have been appropriately chosen.

We can conclude that distributed Pulse Rate Control is an appropriate means for impulsive interference management and network throughput optimisation with QoS constraints in highly loaded IR-UWB networks with a common central receiver. In [16] the author extends the work presented here and investigates the applicability of distributed PRC as well as its combination with adaptive channel coding in peer-to-peer networks, i.e. with multiple uncoordinated receivers and without any hierarchical infrastructure. Moreover she investigates a low-complexity, heuristic, alternative algorithm to the one proposed in this chapter that is more suitable for embedded hardware implementations, as preferred in the design of sensor networks.

Acknowledgements

The author thanks Prof. K. Jobmann, Prof. M. G. di Benedetto and Ralf Luebben for their valuable support to complete the work presented in this chapter.

---

[12] The design QoS constraint can only be guaranteed when the number of active links is below 4.

Author details

Pérez Guirao María Dolores
Institute of Communications Technology (IKT), Leibniz Universitaet Hannover (LUH), Hanover, Germany

5. References

[1] Boyd, S. & Vandenberghe, L. [2006]. *Convex Optimization*, Cambrigde University Press.

[2] Chiang, M., Tan, C. W., Palomar, D., O'Neill, D. & Julian, D. [July 2007]. Power Control by Geometric Programming, *Wireless Communications, IEEE Transactions on* 6(7): 2640–2651.

[3] di Benedetto, M. G. & Giancola, G. [2004]. *Understanding Ultra Wide Band Radio Fundamentals*, Prentice Hall PTR.

[4] Everette, S. & Gardner, J. [October-December 2006]. Exponential Smoothing: The State of the Art–Part II, *International Journal of Forecasting* 22, Issue 4: Pages 637–666.

[5] Felegyhazi, M. & Hubaux, J.-P. [2006]. Game Theory in Wireless Networks: A Tutorial, *Technical report*.

[6] Fishler, E. & Poor, V. [2005]. On the Tradeoff between two Types of Processing Gains, *IEEE Transactions on Communications* 53(9): 1744–1753.

[7] Fudenberg, D. & Tirole, J. [2001]. *Game Theory*, MIT Press.

[8] IEEE *Standard for Information Technology - Telecommunications and information exchange between systems - Local and metropolitan area networks - specific requirement Part 15.4: Wireless Medium Access Control (MAC) and Physical Layer (PHY) Specifications for Low-Rate Wireless Personal Area Networks* (WPANs) [2007]. *IEEE Std 802.15.4a-2007 (Amendment to IEEE Std 802.15.4-2006)* pp. 1–203.

[9] Le Boudec, J.-Y., Merz, R., Radunovic, B. & Widmer, J. [2004]. DCC-MAC: a Decentralized MAC Protocol for 802.15.4a-like UWB Mobile Ad-Hoc Networks based on Dynamic Channel Coding, *Broadband Networks, 2004. BroadNets 2004. Proceedings. First International Conference on*, pp. 396–405. TY - CONF.

[10] Lovelace, W. & Townsend, J. [16-19 Nov. 2003]. Adaptive Rate Control with Chip Discrimination in UWB Networks, *Ultra Wideband Systems and Technologies, 2003 IEEE Conference on* pp. 195–199.

[11] Luebben, R. [October 2007]. *Spieltheoretische Optimierung des Durchsatzes eines IR-UWB Sensornetzes*, Master's thesis, Institut für Kommunikationstechnik, Leibniz Universität Hannover, Germany.

[12] Merz, R., Widmer, J., Le Boudec, J.-Y. & Radunovic, B. [2005]. A joint PHY-MAC Architecture for Low-Radiated Power TH-UWB Wireless Ad Hoc Networks, *Wireless Communications and Mobile Computing Journal, Special Issue on UWB Communications, John Wiley & Sons Ltd* vol. 5: pp. 567–580.

[13] Monderer, D. & Shapley, S. [1996]. Potential Games, *Games and Economics Behaviour* 14: 124–143.

[14] Neel, J. [2006]. *Analysis and Design of Cognitive Radio Networks and Distributed Radio Resource Management Algorithms*, PhD thesis, Faculty of the Virginia Polytechnic Institute and State University.

[15] OMNeT++, *Discrete Event Simulation System.* [n.d.].
URL: *www.omnetpp.org*

[16] Perez-Guirao, M. [2009].   *Cross-Layer, Cognitive, Cooperative Pulse Rate Control for Autonomous, Low Power, IR-UWB Networks,* Shaker Verlag.

[17] Radunovic, B. & Le Boudec, J.-Y. [13-16 June 2005].   Power Control is Not Required for Wireless Networks in the Linear Regime, *World of Wireless Mobile and Multimedia Networks, 2005. WoWMoM 2005. Sixth IEEE International Symposium on a* pp. 417–427.

[18] Radunovic, B. & Le Boudec, J.-Y. [2004].   Optimal Power Control, Scheduling, and Routing in UWB Networks, *Selected Areas in Communications, IEEE Journal on* 22(7): 1252–1270.

[19] Tong, L., Naware, V. & Venkitasubramaniam, P. [Sept. 2004].   Signal Processing in Random Access, *Signal Processing Magazine, IEEE* 21(5): 29–39.

[20] Weisenhorn, M. & Hirt, W. [Sept. 2005].   Uncoordinated Rate-Division Multiple-Access Scheme for Pulsed UWB Signals, *Vehicular Technology, IEEE Transactions on* 54(5): 1646–1662.

[21] Win, M. & Scholtz, R. [2000].   Ultra-Wide Bandwidth Time-Hopping Spread-Spectrum Impulse Radio for Wireless Multiple-Access Communications, *Communications, IEEE Transactions on* 48(4): 679–689.

[22] Yates, R. [Sep 1995]. A Framework for Uplink Power Control in Cellular Radio Systems, *Selected Areas in Communications, IEEE Journal on* 13(7): 1341–1347.

[23] Yomo, H., Popovski, P., Wijting, C., Kovács, I. Z., Deblauwe, N., Baena, A. F. & Prasad, R. [Sep. 2003].   Medium Access Techniques in Ultra-Wideband Ad Hoc Networks, *in Proc. the 6th Nat. Conf. Society for Electronic, Telecommunication, Automatics, and Informatics (ETAI) of the Republic of Macedonia* .

Chapter 5

Coding, Modulation, and Detection for Power-Efficient Low-Complexity Receivers in Impulse-Radio Ultra-Wideband Transmission Systems

Andreas Schenk and Robert F.H. Fischer

Additional information is available at the end of the chapter

1. Introduction

Impulse-radio ultra-wideband (IR-UWB) is a promising transmission scheme, especially for short-range low-data-rate communications, as, e.g., in wireless-sensor networks. One of the main reasons for this is its potential to employ noncoherent, hence low-complexity, receivers even in dense multipath propagation scenarios, where channel estimation required for coherent detection would be overly complex due to the large signal bandwidth and hence rich multipath propagation [23].

Differential pulse-amplitude-modulated IR-UWB in combination with autocorrelation-based detection constitutes an attractive variant of noncoherent detection schemes [11]. The inherent loss in performance of traditional noncoherent autocorrelation-based differential detection (DD), as compared to coherent detection based on explicit channel estimation, can be alleviated by advanced autocorrelation-based detection schemes operating on the output of an extended autocorrelation receiver (ACR). This ACR delivers correlation coefficients of symbols separated by several symbol durations. In this context, block-based detection schemes, which partition the receive symbol stream into (possibly overlapping) blocks and thus process multiple symbols jointly, have proven to enable power-efficient, yet low-complexity detection in both uncoded and coded IR-UWB transmission systems [3, 6, 11, 12, 15–18].

In this chapter, a comprehensive review of block-based detection schemes is presented. Starting with an exposition of the operation in uncoded schemes, we discuss the generation of soft output, required in coded IR-UWB systems employing autocorrelation-based detection. For the design of such systems, an information theoretic performance analysis of IR-UWB transmission with autocorrelation-based detection delivers design rules for coded IR-UWB systems. In particular, optimum rates for the applied channel code are derived, which improve the overall power efficiency (i.e., required signal-to-noise ratio to guarantee a desired error rate) of the system. The chapter concludes with a brief summary.

2. IR-UWB system model with autocorrelation-based detection

2.1. IR-UWB system model

Throughout this chapter binary pulse-amplitude-modulated IR-UWB transmission in combination with bit-interleaved coded modulation (BICM), as shown in Fig. 1, is considered. Avoiding up-/downconversion due to operation at a carrier frequency, transmission takes place in the baseband; hence, all signals are real valued. The sequence of information bits (assumed to be equiprobable and independent, i.e., with maximum entropy) is encoded with a channel code of rate R_c. After symbolwise mapping from (interleaved) codeword bits c_k to binary information symbols $a_k \in \{\pm 1\}$, differential encoding is performed, yielding the transmit symbols $b_k \in \{\pm 1\}$, where $b_k = b_{k-1} a_k$ and $b_0 = 1$. The IR-UWB receive signal, after propagation through an UWB multipath channel, is given by [23]

$$r(t) = \sum_{k=0}^{+\infty} b_k p(t - kT) + n(t) \tag{1}$$

where T is the symbol duration and $p(t)$ denotes the overall receive pulse shape, resulting from the convolution of transmit (TX) pulse, receive (RX) filter, and channel (CH) impulse response; its energy is normalized to one, thus, the energy per information symbol[1] is given as $E_s = 1$. We assume the channel to remain constant within one codeword. $n(t)$ results from white Gaussian noise of two-sided power-spectral density $N_0/2$ passed through the RX filter. To preclude inter-symbol interference, the symbol duration T is chosen sufficiently large, such that each pulse has decayed before the next pulse is received. For clarity, we do not explicitly consider the typically applied frame structure used for time-hopping and code-division multiple access, as it can be averaged out prior to further receive signal processing [6].

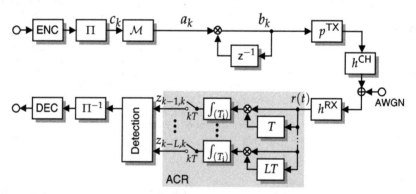

Figure 1. System model of coded IR-UWB transmission with autocorrelation-based detection.

For convenient representation and in view of an all-digital implementation of the receiver, we define the sampled receive signal of the kth symbol interval as $\tilde{r}_k = [r(kT), r(kT + T_s), \ldots, r(kT + (N_s - 1)T_s)]^T$, where $N_s = f_s T$ is the number of samples per symbol interval, and $f_s = 1/T_s$ is the sampling rate (greater than or equal to the Nyquist rate). With respective

---

[1] For long bursts, the energy for the reference symbol may be neglected.

definitions for the end-to-end impulse and the noise, assuming no inter-symbol interference, we compactly write

$$\bar{r}_k = b_k \bar{p} + \bar{n}_k .$$ (2)

The noise components in $\bar{n}_k$ are modeled as uncorrelated Gaussian random variables with variance $\sigma_n^2 = f_s N_0 / 2$, which is the case for a square-root Nyquist low-pass receiver front-end filter with two-sided bandwidth f_s.

2.2. Autocorrelation-based detection

Autocorrelation-based noncoherent detection of IR-UWB, cf., e.g., [2, 3, 6, 9, 15], requires to compute the correlation of the current symbol with up to L preceding symbols, as shown in Fig. 1. Significant gains are achieved by adopting the integration interval to the channel characteristics at hand [23], i.e., choosing $T_i < T$ in the order of the expected channel delay spread. Simplified, larger T_i lead to decreased performance, but become inevitable in case of only coarse synchronization or insufficient knowledge of the channel characteristics[2]. Defining the time-bandwidth product $N = f_s T_i$ and r_k as the part of $\bar{r}_k$ relevant for the ACR integration, i.e., (typically the first) N successive components out of N_s, in an digital implementation we have, for $l = 1, \dots, L$,

$$z_{k-l,k} = r_{k-l}^T r_k .$$ (3)

The correlation coefficients serve as input for various detection schemes, cf., Sec. 3 and [3, 6, 9, 10, 12, 15, 23]. E.g., symbolwise differential detection (DD) utilizes only the correlation coefficient of the current symbol and its predecessor, i.e., $L = 1$, and, since $b_{k-1} b_k = a_k$, the decision rule for the information symbols reads $a_k^{DD} = \text{sign}(z_{k-1,k})$.

We explicitly point out the major drawback of an autocorrelation-based receiver, namely the required accurate analog delay lines in an analog implementation, or the large sampling rate[3] in an all-digital implementation. Especially approaches based on the principle of compressed sensing (CS) promise to circumvent these problems [14, 24]. These approaches avoid sampling the receive signal at the (possibly prohibitively) large Nyquist rate by taking fewer measurements in a different domain (e.g., frequency or some transform domain). In [14] it has been shown, that a CS-front-end can readily be applied prior to an ACR, i.e., via direct correlation of the measurements, thus also avoiding the need for computationally complex CS-reconstruction algorithms. In combination with autocorrelation-based DD the inherent loss in performance of CS/ACR-based detection is proportional only to the square root of the compression ratio (number of measurements over N_s) [14].

2.3. Equivalent discrete-time system model

Based on the all-digital implementation, we introduce an equivalent discrete-time system model of ACR-based detection. The ACR-output can be written as

$$z_{k-l,k} = E_i x_{k-l,k} + \eta_{k-l,l}$$ (4)

---

[2] A typical setting for realistic IR-UWB scenarios, e.g., modelled by the IEEE channel models [7, 8] is $T_i = 33$ ns, whereas $T = 75$ ns to avoid inter-symbol interference. With $f_s = 12$ GHz, we have $N_s = 900$ and $N \approx 400$ [23].

[3] With the advance in micro electronics, one can expect that an all-digital implementation becomes realistic within no later than the next decade.

where $E_i = \boldsymbol{p}^\mathsf{T} \boldsymbol{p}$ denotes the captured pulse energy. It is composed of the phase transition from b_{k-1} to b_k, i.e., $x_{k-1,k} = b_{k-1} b_k$, and "information $\times$ noise" and "noise $\times$ noise" terms, summarized in the equivalent noise term

$$\eta_{k-1,l} = b_{k-1} \boldsymbol{p}^\mathsf{T} \boldsymbol{n}_k + b_k \boldsymbol{n}_{k-1}^\mathsf{T} \boldsymbol{p} + \boldsymbol{n}_{k-1}^\mathsf{T} \boldsymbol{n}_k \; . \tag{5}$$

A detailed analysis of the components of the equivalent noise term in (5) shows that already for moderate time-bandwidth products N it is reasonable to approximate the respective terms as uncorrelated Gaussian random variables [9, 10, 14]. In particular, the "information $\times$ noise" terms are zero-mean with variance σ_n^2, and the "noise $\times$ noise" term, as the sum of N products of independent Gaussian random variables, is zero-mean with variance $N(\sigma_n^2)^2$. Consequently, $\eta_{k-1,k}$ may be modeled as a zero-mean Gaussian random variable with variance $\sigma_\eta^2 = 2\sigma_n^2 + N(\sigma_n^2)^2$. Since each $\eta_{k-1,k}$ results from the multipication of different parts of noise and symbols, the equivalent noise samples at different time instances and ACR branches are uncorrelated.

This approximation is only valid under the following prerequisites, which typically are fulfilled in common IR-UWB systems: i) the symbol duration is chosen sufficiently large, such that no inter-symbol interference is present, ii) the integration interval of the ACR and the time-bandwidth product N are chosen sufficiently large, such that the Gaussian approximation holds, iii) the receiver front-end filter is a square-root Nyquist low-pass with two-sided bandwidth f_s to avoid correlations of the noise samples, and iv) the channel remains constant over the block of at least $L + 1$ symbols. We emphasize that this model not only enables the subsequent information theoretic analysis of ACR-based detection of IR-UWB, but also serves as a tool for efficient numerical simulations of the IR-UWB transmission chain.

3. Advanced detection schemes for IR-UWB

3.1. Multiple-symbol differential detection

One of the most powerful detection schemes is based on the principle of multiple-symbol differential detection (MSDD), cf., [1] and its modifications for IR-UWB detection [3, 6, 15]. In MSDD the stream of receive symbols is decomposed into blocks of $L + 1$ symbols (note that the blocks have to overlap by at least one symbol), and for each block the blockwise-optimal sequence of L information symbols is decided jointly based on the correlation coefficients corresponding to this block. The decision metric given a hypothesis of information symbols grouped into a vector $\tilde{\boldsymbol{a}}$ and the corresponding hypothesis of the ACR output $\tilde{\boldsymbol{x}}$, reads

$$\Lambda(\tilde{\boldsymbol{a}}) = \sum_{k=1}^{L} \left(\sum_{l=0}^{k-1} \left(|z_{l,k}| - \tilde{x}_{l,k} z_{l,k} \right) \right) \; . \tag{6}$$

The blockwise-optimal sequence $\boldsymbol{a}^{\mathrm{MSDD}}$ of hard-output MSDD is given as the sequence with minimum decision metric.

To fully exploit the benefits of channel coding, reliability information on the estimated codeword bits should be delivered to the subsequent channel decoder, i.e., so-called soft-output MSDD (SO-MSDD) should be performed. Sticking to the so-called max-log approximation, in terms of log-likelihood ratios (LLRs) reliability information corresponds to the (scaled) difference of the decision metric of the optimum sequence [12], i.e.,

$$\Lambda^{\text{MSDD}} = \Lambda(a^{\text{MSDD}}) = \min_{\tilde{a} \in \{\pm 1\}^L} \Lambda(\tilde{a}) \tag{7}$$

and the decision metric of the corresponding counter hypothesis, i.e., the minimum metric with the restriction $\tilde{a}_k = -a_k^{\text{MSDD}}$, i.e., for $k = 1, \ldots, L$,

$$\Lambda_k^{\overline{\text{MSDD}}} = \min_{\tilde{a} \in \{\pm 1\}^L, \tilde{a}_k = -a_k^{\text{MSDD}}} \Lambda(\tilde{a}) . \tag{8}$$

Finally, the reliablitiy of the kth symbol/codeword bit is proportional to

$$\text{LLR}_k \sim a_k^{\text{MSDD}} \left(\Lambda_k^{\overline{\text{MSDD}}} - \Lambda^{\text{MSDD}} \right) . \tag{9}$$

In the case of SO-DD ($L = 1$), the LLRs are directly given as the (scaled) ACR output, i.e., $\text{LLR}_k^{\text{DD}} \sim z_{k-1,k}$ [12].

Utilizing the triangular structure of the decision metric, an efficient solution to the MSDD search problem (7) is obtained by employing the sphere decoder algorithm [6, 18, 19]. In the case of SO-MSDD, incorporating modifications in the sphere decoder algorithm proposed for efficient soft-output detection in multi-antenna systems [21], the $L + 1$ search problems per block, (7) and (8), can be solved in a single sphere decoder run per block using the single-tree-search soft-output sphere decoder [12, 21]. Thus, SO-MSDD can be realized at only moderate complexity increase compared to hard-output MSDD.

3.2. Decision-feedback differential detection

A closely-related detection scheme is blockwise decision-feedback differential detection (DF-DD), cf., [5] and its modifications for IR-UWB detection [15], which decides the symbols within each block in a successive manner taking into account the feedback from already decided symbols within the block. The blockwise processing of the receive signal enables to optimize the decision order, such that in each step the most reliable symbol is decided next, resulting in almost the performance of MSDD at lower and in particular constant complexity.

Briefly sketched, following [15] and focusing on the first block, with $\hat{k}_0 = 0$, $b_0^{\text{DF-DD}} = 1$, the optimized decision order and the estimates are given by

$$\hat{k}_i = \operatorname*{argmax}_{k \in \{1,\ldots,L\}/\{\hat{k}_1,\ldots,\hat{k}_{i-1}\}} \left| \sum_{l=0}^{i-1} z_{\hat{k}_l,k} \, b_{\hat{k}_l}^{\text{DF-DD}} \right| \tag{10}$$

$$b_{\hat{k}_i}^{\text{DF-DD}} = \operatorname{sign} \sum_{l=0}^{i-1} z_{\hat{k}_l,\hat{k}_i} \, b_{\hat{k}_l}^{\text{DF-DD}} . \tag{11}$$

Basically, the optimized decision order forces reliable decisions for the first decided symbols, which then strongly influence the upcoming decisions. In contrast to the related detection scheme BLAST in multiple-antenna systems, sorting is done per block based on the actual receive symbols and previous decisions, rather than on the channel realization.

3.3. Low-complexity soft-output detection via combining multiple observations

The blockwise processing of the receive symbol stream enables a further possibility to improve the performance without increase of the maximum delay of the ACR [17]. This method utilizes an overlapping block-structure. Since multiple blocks thus contain the same symbol, processing of each block delivers (possibly different) beliefs on the same symbol, i.e., multiple observations are available. Suitably combining the observations obtained from processing of each block, results in a (possibly more reliable) final decision. Depending on the applied blockwise decision scheme (here SO-MSDD and DF-DD are considered), there are different options how to combine multiple soft/hard observations to deliver a final hard and/or soft decision for the respective symbol [17]. The most interesting option is to combine multiple hard decisions, e.g., obtained from DF-DD, of the same symbol to form a single soft decision. This method can be implemented by using the sum of the individual hard-decisions as (quantized and scaled) "soft-output"; it preserves the low complexity of blockwise DF-DD, yet enables to exploit the additional gain of soft- vs. hard-decision channel decoding.

3.4. Performance of advanced detection schemes for uncoded IR-UWB transmission

In Fig. 2 the presented ACR-based detection schemes are compared with respect to bit error rate of uncoded IR-UWB transmission and a time-bandwidth product of $N = 400$. This parameter setting is based on the reasoning in Footnote 2; the Gaussian approximation as described in Sec. 2.3 is employed assuming that the integration interval captures the entire pulse energy, i.e., $E_i = E_s$). It can be observed, that i) with increasing blocksize performance improves over traditional DD (the significant loss compared to coherent detection is mainly caused by the squared original noise variance σ_n^2 in the equivalent noise variance σ_η^2) and approaches coherent detection with perfect channel estimation, ii) DF-DD with optimized decision order (dashed lines) achieves almost the performance of MSDD (solid lines, exactly the same performance for $L = 2$ with minimum overlap, and iii) combining multiple observations obtained by introducing a maximum block-overlap, but using the same ACR front-end (right hand side of Fig. 2) leads to significant gains over traditional blockwise processing without overlapping blocks (left hand side of Fig. 2), for both soft-output MSDD and hard-output DF-DD (except for $L = 2$) as blockwise detection scheme. This gain comes at the cost of an increased computational complexity (roughly proportional to L).

4. Design rules for coded IR-UWB systems

Based on an information theoretic performance analysis of IR-UWB in combination with ACR-based detection [16], in this section design rules for coded IR-UWB transmission systems are derived and verified by means of numerical results employing convolutional codes.

4.1. Capacity of IR-UWB with MSDD

In contrast to coded modulation using multi-level codes [22], common IR-UWB systems adopt the conventional serial concatenation of coding and modulation at transmitter, and detection and decoding at receiver side, as shown in Fig. 1, i.e., restrain to the BICM philosophy. This approach offers increased flexibility and robustness in fading scenarios. The BICM capacity of the overall transmission chain composed of mapping, differential encoding, and ACR-based

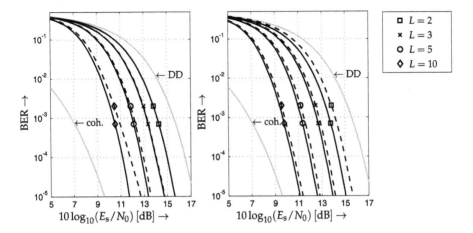

Figure 2. BER of uncoded BPSK IR-UWB transmission with autocorrelation-based detection with $L = 2$, 3, 5, and 10 (right to left). Solid: MSDD, dashed: DF-DD, left: processing of non-overlapping blocks, right: combining of multiple observations obtained by processing of maximum-overlapping blocks. Gaussian approximation with time-bandwidth product $N = 400$.

detection, is depicted in Fig. 3 (using the same parameter setup as in Sec. 3.4). Since an exact evaluation of the BICM capacity of the IR-UWB system at hand is overly complex, the equivalent discrete-time channel model and the Gaussian approximation, as derived in Sec. 2.3, have been applied [16]. Soft-output MSDD with $L = 2, 5$, and 10 (solid black), DF-DD with $L = 2$ and 5 (dashed black), and soft- and hard-output DD (solid gray/dashed gray) are shown; for comparison the capacity of BPSK with coherent detection is included.

In line with the BER results, the ACR operation causes a significant gap compared to coherent detection; the capacity improves with increasing blocksize. As expected for noncoherent detection schemes, cf., e.g., [20], the capacity curves of IR-UWB with ACR-based detection plotted vs. E_b/N_0, with E_b denoting the energy per information bit, have a C-like shape. Thus, as opposed to coherent detection, the minimum ratio E_b/N_0, which still guarantees reliable transmission, is obtained at non-zero rates (indicated with markers). At the operating point of minimum E_b/N_0 and optimum rate, both options, decreasing and increasing the code rate, lead to operating points which do not allow reliable transmission. Consequently, as known from other noncoherent detection schemes [20], also in realistic BICM IR-UWB systems the code rate should be carefully selected. Especially for increasing L this minimum gets more and more pronounced, and higher code rates should be favored compared to the probably more common choice of $R_c = 0.5$ [4]. In all cases, the optimum rate for the hard-output schemes DD and DF-DD is larger than the respective optimum rate of soft-output MSDD.

These effects are also observed for noncoherent detection (energy detection) of pulse-position modulation [20]. However, in this case already the application of BICM in combination with coherent detection leads to optimum operating points at non-zero code rates [13].

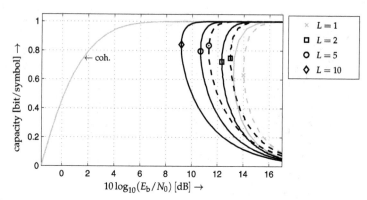

Figure 3. Capacity of BICM BPSK IR-UWB for soft/hard-output DD (solid gray/dashed gray), soft-output MSDD with $L = 2, 5,$ and 10 (solid black), and DF-DD with $L = 2$ and 5 (dashed black). Gaussian approximation with time-bandwidth product $N = 400$.

In addition, a more detailed analysis shows that in non-fading scenarios an interleaver is not required for BICM IR-UWB [16].

4.2. Performance of advanced detection schemes for coded IR-UWB transmission

Finally, the design rules derived above are verified by means of numerical simulations. Fig. 4 depicts the BER of coded IR-UWB transmission using convolutional codes with optimized code rate compared to the default rate choice of $R_c = 0.5$. We apply the same channel model as in Sec. 3.4), and nonrecursively nonsystematically encoded maximum-free-distance convolutional codes with constraint length $\nu = 4$. For soft-output DD, the optimum rate

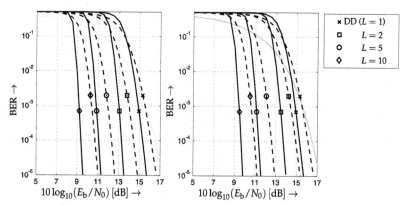

Figure 4. BER of convolutionally-coded BICM BPSK IR-UWB with autocorrelation-based detection with $L = 1, 2, 5,$ and 10 (right to left). Solid: optimum code rate ($R_c = 2/3$ for DD with $L = 1$ and $R_c = 3/4$ for $L = 2, 5$ and 10), dashed: $R_c = 1/2$, gray: DD uncoded, left: soft-output MSDD, right: DF-DD, both using multiple-observations combining with maximum overlap. Gaussian approximation with time-bandwidth product $N = 400$.

is quantized to $R_c = 2/3$. Note that due to the increased decoder complexity of high-rate convolutional codes, for MSDD/DF-DD $R_c = 3/4$ is selected for all L, although higher rates are suggested by Fig. 3. ACR-based detection using soft-output MSDD (left) and DF-DD (right) with multiple observations combining is applied. It can clearly be observed that the performance is significantly improved with an optimized choice of the code rate, although the optimum code rates are larger than the default setting of $R_c = 0.5$ for all L—of course the relations are exactly opposite for coherent detection. As expected from the shape of the curves in Fig. 3, this effect is emphasized for larger blocksizes, yielding gains of almost 1 dB for $L = 10$ compared to $R_c = 0.5$.

Similar results are obtained for different coding schemes, such as LDPC codes with belief-propagation decoding [16], and also for coded IR-UWB pulse-position modulation in combination with energy detection.

5. Summary and conclusions

In this chapter we have presented a comprehensive review of coding, modulation, and detection for IR-UWB binary phase-shift keying. We conclude that noncoherent autocorrelation-based receivers in combination with blockwise detection constitute a power-efficient low-complexity reference for uncoded, as well as coded transmission. We derived and verified design rules for coded IR-UWB systems, in particular optimum code rates, which take into account the noncoherent detection at the receiver side.

Author details

Andreas Schenk
Lehrstuhl für Informationsübertragung, Friedrich-Alexander-Universität Erlangen-Nürnberg, Germany

Robert F.H. Fischer
Institut für Nachrichtentechnik, Universität Ulm, Germany

6. References

[1] Divsalar, D. & Simon, M. K. [1990]. Multiple-Symbol Differential Detection of MPSK, *IEEE Trans. Commun.* 38(3): 300–308.

[2] Farhang, M. & Salehi, J. A. [2010]. Optimum Receiver Design for Transmitted-Reference Signaling, *IEEE Trans. Commun.* 58(5): 1589–1598.

[3] Guo, N. & Qiu, R. C. [2006]. Improved Autocorrelation Demodulation Receivers Based on Multiple-Symbol Detection for UWB Communications, *IEEE Trans. Wireless Commun.* 5(8): 2026–2031.

[4] *IEEE Std 802.15.4a-2007, IEEE Standard for PART 15.4: Wireless MAC and PHY Specifications for Low-Rate Wireless Personal Area Networks* [2007].

[5] Leib, H. & Pasupathy, S. [1988]. The Phase of a Vector Perturbed by Gaussian Noise and Differentially Coherent Receivers, *IEEE Trans. Inf. Theory* 34(6): 1491–1501.

[6] Lottici, V. & Tian, Z. [2008]. Multiple Symbol Differential Detection for UWB Communications, *IEEE Trans. Wireless Commun.* 7(5): 1656–1666.

[7] Molisch, A. F., Cassioli, D., Chong, C.-C., Emami, S., Fort, A., Kannan, B., Karedal, J., Kunisch, J., Schantz, H. G., Siwiak, K. & Win, M. Z. [2006]. A Comprehensive Standardized Model for Ultrawideband Propagation Channels, *IEEE Trans. Antennas Propag.* 54(11): 3151–3166.

[8] Molisch, A. F., Foerster, J. R. & Pendergrass, M. [2003]. Channel Models for Ultrawideband Personal Area Networks, *IEEE Wireless Commun. Mag.* 10(6): 14–21.

[9] Pausini, M. & Janssen, G. J. M. [2007]. Performance Analysis of UWB Autocorrelation Receivers Over Nakagami-Fading Channels, *IEEE J. Sel. Topics Signal Process.* 1(3): 443–455.

[10] Quek, T., Win, M. Z. & Dardari, D. [2007]. Unified Analysis of UWB Transmitted-Reference Schemes in the Presence of Narrowband Interference, *IEEE Trans. Wireless Commun.* 6(6): 2126–2139.

[11] Schenk, A. & Fischer, R. F. H. [2010a]. Multiple-Symbol-Detection-Based Noncoherent Receivers for Impulse-Radio Ultra-Wideband, *2010 International Zürich Seminar on Communications (IZS)*, Zürich, Switzerland, pp. 70–73.

[12] Schenk, A. & Fischer, R. F. H. [2010b]. Soft-Output Sphere Decoder for Multiple-Symbol Differential Detection of Impulse-Radio Ultra-Wideband, *2010 IEEE International Symposium on Information Theory (ISIT)*, Austin (TX), U.S.A., pp. 2258–2262.

[13] Schenk, A. & Fischer, R. F. H. [2011a]. Capacity of BICM Using (Bi-)Orthogonal Signal Constellations in the Wideband Regime, *2011 IEEE International Conference on Ultra-Wideband (ICUWB)*, Bologna, Italy.

[14] Schenk, A. & Fischer, R. F. H. [2011b]. Compressed-Sensing (Decision-Feedback) Differential Detection in Impulse-Radio Ultra-Wideband Systems, *2011 IEEE International Conference on Ultra-Wideband (ICUWB)*, Bologna, Italy.

[15] Schenk, A. & Fischer, R. F. H. [2011c]. Decision-Feedback Differential Detection in Impulse-Radio Ultra-Wideband Systems, *IEEE Trans. Commun.* 59(6): 1604–1611.

[16] Schenk, A. & Fischer, R. F. H. [2011d]. Design Rules for Bit-Interleaved Coded Impulse-Radio Ultra-Wideband Modulation with Autocorrelation-Based Detection, *2011 International Symposium on Wireless Communication Systems (ISWCS)*, Aachen, Germany, pp. 146–150.

[17] Schenk, A. & Fischer, R. F. H. [2012]. Low-Complexity Soft-Output Detection for Impulse-Radio Ultra-Wideband Systems Via Combining Multiple Observations, *2012 International Zürich Seminar on Communications (IZS)*, Zurich, Switzerland, pp. 119–122.

[18] Schenk, A., Fischer, R. F. H. & Lampe, L. [2009a]. A New Stopping Criterion for the Sphere Decoder in UWB Impulse-Radio Multiple-Symbol Differential Detection, *2009 IEEE International Conference on Ultra-Wideband (ICUWB)*, Vancouver, Canada, pp. 606–611.

[19] Schenk, A., Fischer, R. F. H. & Lampe, L. [2009b]. A Stopping Radius for the Sphere Decoder and its Application to MSDD of DPSK, *IEEE Commun. Lett.* 13(7): 465–467.

[20] Stark, W. [1985]. Capacity and Cutoff Rate of Noncoherent FSK with Nonselective Rician Fading, *IEEE Trans. Commun.* 33(11): 1153 – 1159.

[21] Studer, C., Burg, A. & Boelcskei, H. [2008]. Soft-Output Sphere Decoding: Algorithms and VLSI Implementation, *IEEE J. Sel. Areas Commun.* 26(2): 290–300.

[22] Wachsmann, U., Fischer, R. F. H. & Huber, J. B. [1999]. Multilevel Codes: Theoretical Concepts and Practical Design Rules, *IEEE Trans. Inf. Theory* 45(5): 1361–1391.

[23] Witrisal, K., Leus, G., Janssen, G., Pausini, M., Troesch, F., Zasowski, T. & Romme, J. [2009]. Noncoherent Ultra-Wideband Systems, *IEEE Signal Process. Mag.* 26(4): 48–66.

[24] Zhongmin W., Arce, G. R., Paredes, J. L. & Sadler, B. M. [2007]. Compressed Detection for Ultra-Wideband Impulse Radio, *2007 IEEE 8th Workshop Signal Processing Advances in Wireless Communications (SPAWC)* , pp. 1–5.

Chapter 6

Interference Alignment for UWB-MIMO Communication Systems

Mohamed El-Hadidy, Mohammed El-Absi, Yoke Leen Sit, Markus Kock, Thomas Zwick, Holger Blume and Thomas Kaiser

Additional information is available at the end of the chapter

1. Introduction

Due to the enormous occupied bandwidth even by a single pair of UWB users, one major scientific challenge in UWB communications is interference management. Recently, Interference Alignment (IA) has become popular not only to well manage the interference, but also to optimally exploit the possible capacity gain caused by multiple pairs of transmitters and receivers. Theoretically, IA scales the channel capacity by $K/2$, where K is the number of user pairs. This fact makes IA highly attractive for future communication systems with numerous pairs of users. However, in the literature it has been reported that IA is not robust [7] against imperfections such as channel estimation errors. Thanks to the interdisciplinary research of antenna, communication and hardware engineers within the UKoLoS project *Decimus*, we were able to jointly search for solutions to improve IA robustness while not suffering from too perfect simulation idealities or too unrealistic hardware requirements. We find that *antenna* or *pattern selection* is a promising approach to improving the robustness of IA while keeping the underlying algorithms at a reasonable complexity and feasibility for implementation. Our contribution is structured as follows: first, a brief introduction of IA tailored to UWB is given. Then, we propose an antenna selection algorithm with low complexity and demonstrate its performance. In the third chapter a general methodology of MIMO UWB antenna design for orthogonal channels maximizing the channel capacity is presented. A first outcome of this methodology is a multi-mode orthogonal antenna which has been used for the investigated antenna selection approach. At last, the hardware requirements of IA systems with the proposed antenna selection method are studied. A conclusion summarizes the outcomes of our contribution.

2. MIMO-UWB interference mitigation by interference alignment

IA is a promising technique which achieves the maximum degrees of freedom (DoF) for K users in interference channels [4]. This can be achieved by a combination of linear precoding at the transmitters and interference suppression at the receivers. IA permits to force interfering signals at each receiver in one subspace and the desired signal in another orthogonal subspace [5].

Consider a K-user UWB Multi Band Orthogonal Frequency Division Multiplexing (MB-OFDM) interference channel with M_j transmit antennas at transmitter j and N_i receive antennas at receiver i. All users transmit d_s streams using N sub-carriers. Every transmitter communicates with his desired receiver and causes interference to other pairs of transmitter and receiver. The discrete-time complex received signal over the nth subcarrier at the ith receiver over a flat channel is represented as[21],[28]:

$$\mathbf{y}_i^n = \sum_{j=1}^{K} \mathbf{H}_{ij}^n \mathbf{V}_j^n \mathbf{x}_j^n + \mathbf{z}_i^n = \mathbf{H}_{ii}^n \mathbf{V}_i^n \mathbf{x}_i^n + \sum_{j=1,j\neq i}^{K} \mathbf{H}_{ij}^n \mathbf{V}_j^n \mathbf{x}_j^n + \mathbf{z}_i^n \tag{1}$$

where $\mathbf{y}_i^n$ is the $N_i \times 1$ received vector at receiver i, $\mathbf{H}_{ij}^n$ is the $N_i \times M_j$ flat frequency domain channel matrix over nth subcarrier between jth transmitter and ith receiver, $\mathbf{V}_j^n$ is the $M_j \times d_s$ unitary precoding matrix which is applied for the transmitted $M_j \times 1$ vector $\mathbf{x}_j^n$ from the jth transmitter, and $\mathbf{z}_i^n$ is the $N_i \times 1$ zero mean unit variance circularly symmetric additive white Gaussian noise vector at receiver i. The Channel State Information (CSI) is assumed to be perfectly known at each node. To reconstruct the transmitted d_s signal at the ith receiver, the received signal is decoded using a unitary linear suppression interference matrix $\mathbf{U}_i^n$. The reconstructed data $\hat{\mathbf{y}}$ at receiver i is defined as:

$$\hat{\mathbf{y}}_i^n = \mathbf{U}_i^{n\,H} \mathbf{H}_{ii}^n \mathbf{V}_i^n \mathbf{x}_i^n + \sum_{j=1,j\neq i}^{K} \mathbf{U}_i^{n\,H} \mathbf{H}_{ij}^n \mathbf{V}_j^n x_j^n + \mathbf{U}_i^{n\,H} \mathbf{z}_i^n \tag{2}$$

For perfect interference alignment, the following conditions need to be fulfilled [1]:

$$\text{rank}\,(\mathbf{U}_i^{n\,H} \mathbf{H}_{ii}^n \mathbf{V}_i^n) = d_s \qquad \forall i \tag{3}$$

and

$$\mathbf{U}_i^{n\,H} \mathbf{H}_{ij}^n \mathbf{V}_j^n = 0 \qquad \forall j \neq i \tag{4}$$

According to (3) and (4), the received signal after processed by the linear suppression interference matrix is:

$$\hat{\mathbf{y}}_i^n = \mathbf{U}_i^{n\,H} \mathbf{H}_{ii}^n \mathbf{V}_i^n \mathbf{x}_i^n + \mathbf{U}_i^{n\,H} \mathbf{z}_i^n \tag{5}$$

2.1. Closed-form interference alignment

In order to achieve a closed-form IA solution, 3 users interference channel (K=3) has been considered, where each node has $M = 2d$ antennas, and each user wishes to achieve d degrees of freedom by applying the IA principles. The conditions of IA given in (3) and (4) are obtained by setting the precoding matrices as [4]:

$$\mathbf{V}_1 = \text{eign}(\mathbf{H}_{31}^{-1}\mathbf{H}_{32}\mathbf{H}_{12}^{-1}\mathbf{H}_{13}\mathbf{H}_{23}^{-1}\mathbf{H}_{21}) \tag{6}$$

$$\mathbf{V}_2 = (\mathbf{H}_{32}^{-1}\mathbf{H}_{31}\mathbf{V}_1) \tag{7}$$

$$\mathbf{V}_3 = (\mathbf{H}_{23}^{-1}\mathbf{H}_{21}\mathbf{V}_1) \tag{8}$$

and the interference suppression matrix for receiver i is given by:

$$\mathbf{U}_i = \text{null}([\mathbf{H}_{ij}\mathbf{V}_j]) \qquad \forall j \neq i \tag{9}$$

Figure 1. Transmitter block diagram of a MIMO UWB MB-OFDM communication system.

2.2. Artificial channel diversity algorithm

For successful applying the IA principle a sufficient orthogonality between all channels is required. In real-world indoor environments and MIMO UWB systems such orthogonality is not guaranteed; instead, the small distances between neighboring antennas and a possible low scattering could lead to high correlation among the several channels of the communication system. Therefore, the orthogonal component of the desired signal to the plane of the aligned undesired signals would be less pronounced, leading to worse overall system performance and robustness.

To overcome this problem, an artificial channel diversity technique is applied utilizing an antenna selection algorithm. The goal of this algorithm is to maximize the *orthogonality* of the desired signal on the plane of the aligned undesired ones. Here each transmitting node has Q_i antennas and only the best M_i antennas will be selected for maximizing the orthogonality of the desired component on the undesired signals plane. A brut force iterative process is carried out for all available combinations to choose the best selection that realizes this maximum orthogonality [8]. In the next section, the antenna selection criteria are illustrated in more detail.

3. Low complexity signal processing antenna selection algorithm

As mentioned before we propose an antenna selection algorithm in order to increase the required orthogonality directly leading to a more robust communication system in terms of minimum Bit Error Rate (BER). Note that antenna selection is a widely known approach in order to capture diversity and to improve the SNR of the communication systems [19],[14].

3.1. Antenna selection criterion

In the following we consider a K-user MIMO system with perfect IA. The selection algorithm consists of choosing the best M out of the L available transmit antennas. Denote by S_k the selected subset indices of the transmit antennas of k users [6]. The goal of the selection is to find S_k for all users $\{k = 1 : K\}$ which maximizes the average SNR_S for the multi-user system by increasing the projected desired signal into the interference-free space. This can be achieved by minimizing the principal angles between the desired signal subspace and the interference-free subspace. Minimizing the principal angles is equivalent to maximizing so-called *canonical correlations*.

Let ζ_1 and ζ_2 be subspaces in the complex plane $\mathbb{C}$. Considering the dimension of ζ_1 is smaller than or equal to the dimension of ζ_2 $(\dim \zeta_1 = d_1 \leqslant \dim \zeta_2 = d_2)$. The canonical correlations are defined as the cosines of the principal angles between any two linear subspaces, which can uniquely defined as [3]

$$\cos \theta_i = \max_{\mathbf{a}_i \in \zeta_1} \max_{\mathbf{b}_i \in \zeta_2} \mathbf{a}_i^H \mathbf{b}_i, \qquad i = 1, ..., d_1, \tag{10}$$

where $\mathbf{a}_i$ and $\mathbf{b}_i$ are principal vectors of ζ_1 and ζ_2 respectively, subject to $\mathbf{a}_i^H \cdot \mathbf{a}_i = \mathbf{b}_i^H \cdot \mathbf{b}_i = 1$ and $\mathbf{a}_i^H \cdot \mathbf{a}_j = \mathbf{b}_i^H \cdot \mathbf{b}_j = 0, i \neq j$.

If $\mathbf{Q}_1$ and $\mathbf{Q}_2$ are orthonormal bases of the two subspaces ζ_1 and ζ_2 respectively, the canonical correlations are obtained as singular values of $\mathbf{Q}_1^H \mathbf{Q}_2 \in \mathbb{C}^{d_1 \times d_2}$ as follows [3]

$$\mathbf{Q}_1^H \mathbf{Q}_2 = \mathbf{P}_1 \Lambda \mathbf{P}_2^H \quad , \tag{11}$$

where $\mathbf{P}_1$ is a $d_1 \times d_1$ unitary matrix and $\mathbf{P}_2$ is a $d_2 \times d_2$ unitary matrix, Λ is is $d_1 \times d_2$ diagonal matrix with nonnegative real numbers on the diagonal. Therefore, $\Lambda = \mathrm{diag}(\alpha_1, .., \alpha_{d_1})$ and $\alpha_1, .., \alpha_{d_1}$ are the canonical correlations of the subspaces.

Observe that the principal angles are given by

$$\theta_i = \cos^{-1}(\alpha_i), \qquad i = 1, .., d_1. \tag{12}$$

In order to maximize the SNR at the receiver, the antenna selection criterion relies on maximizing the canonical correlations between $\mathbf{U}_k$ and $\mathbf{H}_{kk} \mathbf{V}_k$ as follows

$$S_k = \arg \min \angle (\mathbf{U}_k; \mathbf{H}_{kk} \mathbf{V}_k) \qquad ; k = 1, .., K \tag{13}$$

$$S_k = \arg \max \cos (\angle (\mathbf{U}_k; \mathbf{H}_{kk} \mathbf{V}_k)) \qquad ; k = 1, .., K \tag{14}$$

$$S_k = \arg \max (\alpha_1, ..., \alpha_{d_s}) \qquad ; k = 1, 2, .., K, \tag{15}$$

where $(\alpha_1, ..., \alpha_{d_s})$ are the canonical correlations between subspace $\mathbf{U}_k$ and subspace $\mathbf{H}_{kk} \mathbf{V}_k$.

3.2. Relation between sum-rate and canonical correlations

The impact of the canonical correlations on the sum rate of a K-user MIMO system is given by [6]:

$$C = \sum_{k=1}^{K} \log \left| \mathbf{I}_N + \left(\sigma^2 \mathbf{I}_N + \sum_{l \neq k} \mathbf{W}_{kl} \right)^{-1} \mathbf{W}_{kk} \right|, \tag{16}$$

where C is the sum-rate, $\mathbf{W}_{kl} = \mathbf{H}_{kl} \mathbf{V}_l \mathbf{V}_l^H \mathbf{H}_{kl}^H$ denotes the $N \times N$ interference covariance matrix of the signal from the l-th transmitter to the k-th receiver, σ^2 is the variance of the additive white Gaussian noise, and $\mathbf{W}_{kk} = \mathbf{H}_{kk} \mathbf{V}_k \mathbf{V}_k^H \mathbf{H}_{kk}^H$ denotes the $N \times N$ covariance matrix of the desired signal. While perfect IA is assumed according to (3) and (4), the interference channel is equivalent to a set of parallel Gaussian MIMO channels, where the MIMO channel transfer function is given by $\overline{\mathbf{H}}_k = \mathbf{U}_k^H \mathbf{H}_{kk} \mathbf{V}_k$, for $k = 1, .., K$. Then the sumrate equation in (16) reduces to

$$C = \sum_{k=1}^{K} \log \left| \mathbf{I}_N + \frac{1}{\sigma^2} \mathbf{U}_k^H \mathbf{H}_{kk} \mathbf{V}_k \mathbf{V}_k^H \mathbf{H}_{kk}^H \mathbf{U}_k \right|. \tag{17}$$

Note that at high SNR, (17) can be approximated as

$$C \simeq \sum_{k=1}^{K} \log \left| \frac{1}{\sigma^2} \mathbf{U}_k^H \mathbf{H}_{kk} \mathbf{V}_k \mathbf{V}_k^H \mathbf{H}_{kk}^H \mathbf{U}_k \right|, \tag{18}$$

and by applying a thin QR decomposition

$$\mathbf{U}_k = \mathbf{Q}_{U_k} \mathbf{R}_{U_k}, \tag{19}$$

where

$$\mathbf{H}_{kk} \mathbf{V}_k = \mathbf{Q}_{V_k} \mathbf{R}_{V_k} \tag{20}$$

and $\mathbf{Q}_{U_k}, \mathbf{Q}_{V_k}$ are orthonormal $N \times d_s$ matrix and $\mathbf{R}_{U_k}, \mathbf{R}_{V_k}$ are $d_s \times d_s$ upper triangle matrix it follows

$$C = \sum_{k=1}^{K} \log \left| \frac{1}{\sigma^2} \left(\mathbf{Q}_{U_k} \mathbf{R}_{U_k} \right)^H \left(\mathbf{Q}_{V_k} \mathbf{R}_{V_k} \right) \left(\mathbf{Q}_{V_k} \mathbf{R}_{V_k} \right)^H \left(\mathbf{Q}_{U_k} \mathbf{R}_{U_k} \right) \right|. \tag{21}$$

Since $\mathbf{U}_k$ is a unitary matrix, meaning $\left| \mathbf{R}_{U_k} \mathbf{R}_{U_k}^H \right| = 1$ it furthermore follows

$$C = \sum_{k=1}^{K} \log \left(\left(\frac{1}{\sigma^2} \right)^2 \left| \mathbf{Q}_{U_k}^H \mathbf{Q}_{V_k} \right| \left| \mathbf{Q}_{V_k}^H \mathbf{Q}_{U_k} \right| \left| \mathbf{R}_{V_k} \mathbf{R}_{V_k}^H \right| \right). \tag{22}$$

Since $\mathbf{Q}_{U_k}$ and $\mathbf{Q}_{V_k}$ are the orthonormal basis of the two subspaces $\mathbf{U}_k$ and $\mathbf{H}_{kk} \mathbf{V}_k$ respectively, (22) can be linked to the principal angles between the two subspace using (11). Therefore, (22) can be written as

$$C = \sum_{k=1}^{K} \log \left(\left(\frac{1}{\sigma^2} \right)^2 \left| \mathbf{P}_{k1} \Lambda \mathbf{P}_{k2}^H \right| \left| \mathbf{P}_{k2} \Lambda \mathbf{P}_{k1}^H \right| \left| \mathbf{R}_{V_k} \mathbf{R}_{V_k}^H \right| \right) \tag{23}$$

· such that

$$\mathbf{Q}_{U_k}^H \mathbf{Q}_{V_k} = \mathbf{P}_{k1} \Lambda \mathbf{P}_{k2}^H,$$

where $\mathbf{P}_{k1}$ and $\mathbf{P}_{k2}$ are $d_s \times d_s$ unitary matrices and Λ is $d_s \times d_s$ diagonal matrix equals $\text{diag}(\alpha_1, .., \alpha_{d_s})$.

Thereafter, (22) can be re-formulated as

$$C = \sum_{k=1}^{K} \log \left(\left(\frac{1}{\sigma^2} \right)^2 \left(\prod_{i=1}^{d_s} \alpha_i \right)^2 \left| \mathbf{R}_{V_k} \mathbf{R}_{V_k}^H \right| \right), \tag{24}$$

where $(\alpha_1, ..., \alpha_{d_s})$ are the canonical correlations between subspace $\mathbf{U}_k$ and subspace $\mathbf{H}_{kk} \mathbf{V}_k$. From (23) it is shown that maximizing the canonical correlations increases C, but still does not result in the maximum C because the term $\left| \mathbf{R}_{V_k} \mathbf{R}_{V_k}^H \right|$ is linked to the matrix of coefficients participating in the linear combinations yielding the columns of $\mathbf{H}_{kk}$.

3.3. Simulation results analysis

All the following simulation results have been obtained based on real-world deterministic scenarios. The deterministic hybrid EM ray-tracing channel model was considered for the MIMO UWB channel [9]. This model considers the spatial channel and the environmental effects such as path-loss, frequency dependence, reflections, transmissions, and also diffractions. It considers as well the characteristics of the antennas as part of the effective channel such as directional gain, matching and polarization. A fair comparison has been carried out among three systems: the first uses two omnidirectional Half-Wave Dipole (HWD) antennas at each node, the second uses three directional antennas (horn antenna) at each transmitter node and two directional antennas (horn antenna) at each receiver node, in this system antenna selection (AS) technique is applied to select two antennas from the three at each transmitter. The third system uses two directional antennas at each node without AS (we choose the worst case in this manner). Fig. 2 shows a comparison between the three systems using the *average* BER vs. E_b/N_o for the whole multiuser system. As shown in Fig. 2 the artificial diversity technique improves the performance of the system significantly compared to the system that uses a HWD omnidirectional antennas and the other which uses the directional antennas without using AS. This figure proves that AS is a powerful technique to improve the BER performance of the system.

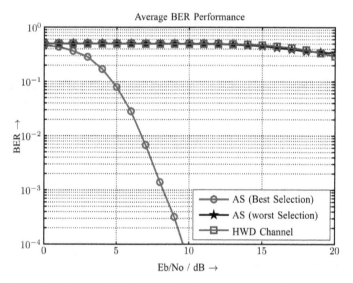

Figure 2. Comparison of average BER vs. Eb/No for a 3-user system by applying antenna selection (best and worst scenarios) and using omni-directional HWD antennas.

MIMO systems typically use antenna arrays and beamforming and spatial multiplexing. These beamforming methods do not result in (1) orthogonal channels nor (2) guarantee optimum data rates within the predefined environment e.g. indoors, outdoors, urban, suburban, etc. In designing optimal antenna systems, which fulfill the previously stated conditions, up to now, only heuristic methods have been employed. Presented here is a systematic framework to fulfill the above 2 conditions, that aids in reducing the number of real world antennas to be used. The goal is to synthesize an optimal UWB antenna system which can be used by both the transmitter and receiver at any location within the investigated scenario. This synthesis method is based on [24] and has been used to design antennas for narrowband systems in [25]. The method is now extended to the design of UWB antenna systems. In the following sections, the theory and concepts behind this systematic synthesis will be given, followed by the methodology of realization and the results.

4. Spatial sampling with sampling antennas

The concept of 'spatial sampling' is presented in greater detail in [24, 30], but can be simplified to this: 'Given a predefined overall antenna aperture confined to a limited volume V, there exists a maximum spatial capacity limit with transmission system parameters i.e. antenna aperture size and element spacing. This can be determined by sampling the transmit and receive volumes with a set of ideal sampling antennas'. In a realistic case sampling antennas possess an overall aperture size and occupy a certain spatial volume. Therefore three parameters have to be considered for their design. First, the antenna aperture size, which will approximate the size of the real world antennas. The larger the aperture size, the higher the capacity, but this size is limited by the physical size of the desired real world antennas. Second, the minimum distance between the antenna elements in order to decrease their correlation [[18, 22]]. This parameter also influences the number of sampling antenna elements within the selected aperture size. It should be noted that the more elements the aperture contains, the more time is needed for the synthesis algorithm. Third, the frequency dependency of the previously mentioned parameters.

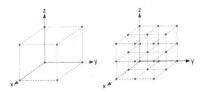

Figure 3. Sampling antenna configurations for spatial sampling, (left) minimum elements of $2\times2\times2$, (right) more elements added, $3\times3\times3$.

Here, a $5\times5\times5$ cm^3 aperture based on the configuration in Fig. 3 has been chosen to illustrate the sampling antenna design and antenna synthesis algorithm in section 4.1. This configuration is then used in the subsequent antenna synthesis steps and the changes to the resulting radiation pattern over frequency is noted. If the resulting radiation pattern varies too much over frequency, the number of sampling antenna elements is then increased and the synthesis algorithm is repeated. This procedure is iterated until the resulting radiation pattern over frequency appear similar.

4.1. Channel diagonalization

For a time-invariant system, the transfer function is a transfer coefficient for all transmit and receive antenna pairs and can be expressed in the form of an $N \times M$-dimentional matrix $\mathbf{H}$. Letting $\mathbf{y}_{out}$ be an $N \times 1$ matrix and $\mathbf{x}_{in}$ be an $M \times 1$ matrix with $\mathbf{n}$ as the noise vector, the received signal vector of a communication system can then be described in the frequency domain as $\mathbf{y}_{out} = \mathbf{H}\mathbf{x}_{in} + \mathbf{n}$. Using singular value decomposition (SVD) $\mathbf{H}$ can be decomposed into $\mathbf{H} = \mathbf{U}\mathbf{S}\mathbf{V}^\dagger$, where $\mathbf{S} = \mathbf{U}^\dagger\mathbf{H}\mathbf{V}$ is a diagonal matrix whose elements are non-negative square roots of the eigenvalues λ_i of the matrix $\mathbf{H}\mathbf{H}^\dagger$. $\mathbf{U}$ and $\mathbf{V}$ are unitary matrices, which fulfill the condition $(\mathbf{X}^{-1})^\dagger = \mathbf{X}$. Multiply the input vector $\mathbf{x}_{in}$ and the output vector $\mathbf{y}_{out}$ with the matrices $\mathbf{U}^\dagger$ and $\mathbf{V}$ respectively, and the original channel becomes and equivalent channel,

$$\hat{\mathbf{y}}_{\mathbf{out}} = \mathbf{U}^\dagger\mathbf{y}_{\mathbf{out}} = \mathbf{U}^\dagger(\mathbf{H}\mathbf{x}_{\mathbf{in}} + \mathbf{n}) = \mathbf{U}^\dagger(\mathbf{H}\mathbf{V}\hat{\mathbf{x}}_{\mathbf{in}} + \mathbf{n}) = \mathbf{S}\hat{\mathbf{x}}_{\mathbf{in}} + \hat{\mathbf{n}} \qquad (25)$$

where $\hat{\mathbf{x}}_{\mathbf{in}}$, $\hat{\mathbf{y}}_{\mathbf{out}}$ and $\hat{\mathbf{n}}$ are the equivalent input, output and noise vectors respectively. The diagonal matrix $\mathbf{S}$ now becomes the channel matrix of the equivalent channel where each Eigenmode is interpreted as an independent SISO (single-input-single-output) subchannel and the capacity of the system becomes a sum over these SISO capacities as expressed by [12]

$$C = \sum_{i=1}^{K} \log_2\left(1 + \frac{p_i\lambda_i}{\sigma_{\text{noise}}^2}\right) \qquad (26)$$

with $K = \min(M, N)$, which is the rank of the matrix $\mathbf{H}\mathbf{R}_{\mathbf{xx}}\mathbf{H}^\dagger$ with $\mathbf{R}_{\mathbf{xx}}$ being the covariance matrix of the transmit signal, its Eigenvalues $\lambda_i(i = 1, 2, ..., K)$ and power coefficients $p_i(i = 1, 2, ..., K)$. From (26) the capacity of a MIMO system can be seen as a sum of independent K SISO subchannels (Eigenmodes) represented by the Eigenvalues λ_i, where each Eigenmode corresponds to one orthogonal subchannel. More explanation can be found in [16, 24].

Employing the waterfilling algorithm in the case of channel state information (CSI) known to the transmitter will result in an optimum capacity solution for such a MIMO channel. The $\mathbf{U}$ and $\mathbf{V}$ matrices are called the 'beamforming matrices' as they determine the mapping and weighting of all the signals onto the antenna elements.

5. Scenario-based MIMO antenna synthesis

In order to apply the SVD technique to obtain parallel subchannels, the channel matrix $\mathbf{H}$ of the intended scenario must be provided. One of the most reliable and repeatable way of obtaining the SISO channel matrix $\mathbf{H}$ is through ray-tracing with the software developed by [10]. The ray-tracing simulations are done in parallel with the design of the sampling antenna configuration.

A typical indoor scenario (with glass windows, furniture, ceiling and floor) was built for simulation as shown in Fig. 4. The size of the room is $10 \times 10 \times 3$ m and simulations were done for transmitters and receivers at randomized positions in the room with an antenna height 1.5 m over the frequency band of 3.1 GHz to 10.6 GHz. Omni-directional antennas (dipoles) are used, along with the option of using both vertical and horizontal polarizations for an added degree of freedom for the design of the real world antenna. Around 45600 random transmitter-receiver points were simulated in order to acquire a synthesis result, that when the averaging strategies in section 6 have been applied, will be applicable from virtually any point in the intended scenario.

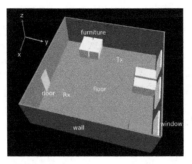

Figure 4. Indoor scenario for ray-tracing

5.1. SISO to MIMO extrapolation

A complete characterization of the MIMO channel matrix with ray-tracing requires $N \times M$ runs taking into account the sampling antenna configuration, which is computationally inefficient. [11] presents a method to reduce the calculation effort by assumming that the same plane wave impinges on all sampling antenna elements. Since the sampling antenna elements spacing is small and fixed, the difference of the incident wave at the origin of the sampling antenna configuration shown in Fig. 3 with the other antenna elements is only the phase difference expressed as

$$\Delta \varphi_i = -\beta \left(\Delta x_i \sin \vartheta \cos \psi + \Delta y_i \sin \vartheta \sin \psi + \Delta z_i \cos \vartheta \right) \tag{27}$$

where ϑ and ψ are the angles of arrival or departure of the incident wave in elevation and azimuth respectively. Hence the SISO to MIMO extrapolation reduces the computation of the MIMO **H** matrix to only one SISO run.

5.2. Antenna system simplification

In order to simplify the system, a plot of the eigenvalues (obtained after the SVD) versus the frequency is used to identify the channels with the strongest power. For instance, if only the first two subchannels were identified as having significant power as compared to the rest, the beamforming matrices **U** and **V** can be modified to contain only those two subchannels. With this, the system will now comprise only 2 inputs and 2 outputs.

6. Synthesis results

The resulting synthesized antenna radiation patterns for both the transmitter and receiver at one point for several different frequencies are as shown in Fig. 5. The figure shows the 3D plot of the radiation patterns for two subchannels and is computed using:

$$\vec{E}(d, \vartheta, \psi) = \vec{E}_{\text{single}}(d, \vartheta, \psi) \cdot \frac{e^{-j\beta d}}{d} \cdot \sum_{i=1}^{N_{\text{ant}}} a_i \, e^{-j(\beta(d_i - d) + \zeta_i)} \tag{28}$$

where $\vec{E}_{single}$ is the electric field of the sampling antenna used (a dipole in this case), N_{ant} is the total number of transmitter or receiver antennas (since they both use the same sampling antenna configuration), β is the wave number, d is the distance from the origin of the sampling antenna to a far-field observation point, $a_i \angle \zeta_i$ is the weighting from the **U** and **V** beamforming matrices, $d_i - d = \Delta\varphi$ in (27) and $\Delta x_i, \Delta y_i, \Delta z_i$ are the position of the individual elements in the array according to the Cartesian coordinate system.

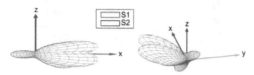

Figure 5. Resulting synthesized antenna radiation pattern for a transmit-receive pair (horizontal polarization) for subchannel 1 (S1), with line-of-sight propagation, and subchannel 2 (S2), with propagation paths reflected from the ceiling and floor, for the aperture size $5\times5\times5\,\mathrm{cm}^3$ with $5\times5\times5$ elements, (left) side view and (right) bottom view. (Image taken from [29])

Averaging strategy

Three averaging strategies were used, namely averaging over frequency, over location, and of transmitter and receiver radiation patterns. The first averages all radiation patterns obtained at frequency points between 3.1 to 10.6 GHz to obtain a pattern which is valid for the UWB. The second averages the radiation pattern obtained from random points around the scenario so that the resulting radiation pattern is valid for use in the whole scenario. The third averaging is done if the resulting transmitter and receiver radiation patterns look qualitatively similar, so that both can use the same antennas.

6.1. Capacity analysis

The capacity for the averaged synthesized patterns according to the number of sampling antennas across the ECC (Electronic Communications Committee) standard's UWB band is analyzed using (26). The term p_i is taken from the power spectral density levels of the ECC UWB spectral mask, λ_i is the Eigenvalue of the subchannel from the matrix **S** and $\sigma_{noise}^2 = kTB$, where k is the Boltzmann constant, $T = 297\,\mathrm{K}$ and $B = 100\,\mathrm{MHz}$. Fig. 6 shows the capacity of the synthesized radiation pattern using $2\times2\times2$ till $5\times5\times5$ sampling antenna elements within the defined $5\times5\times5\,\mathrm{cm}^3$ physical space. It can be seen that the higher the number of sampling antennas, the more the capacity increases, agreeing with the theory in [18]. Noting that the rise in the capacity is decreasing with the higher element configuration used, we conclude that the $5\times5\times5$ configuration is nearing the capacity saturation limit.

7. Real world antennas

The real world antennas which match the elevation characteristics over frequency of the optimized synthesized antennas can be found in [1]. Two dual orthogonal polarized antennas are used along with a 180° hybrid coupler to form the two subchannels. The comparison of the radiation pattern over frequency of the synthesized antennas and the real world antennas is shown in Fig. 7 .

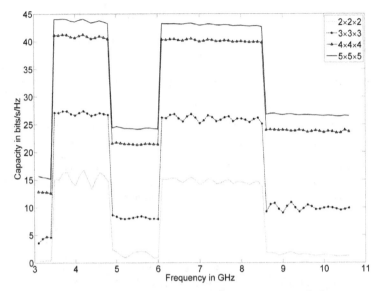

Figure 6. Capacity analysis for the aperture size $5\times5\times5\,\text{cm}^3$. (Image taken from [29])

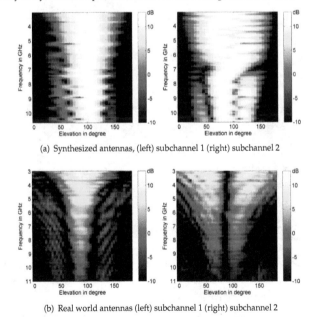

(a) Synthesized antennas, (left) subchannel 1 (right) subchannel 2

(b) Real world antennas (left) subchannel 1 (right) subchannel 2

Figure 7. Synthesized antennas vs. real world antennas: radiation pattern (elevation) over frequency

7.1. Orthogonal channels multimode antenna selection criteria

Before the above mentioned real world UWB antennas are used, the IA algorithm is tested with a narrowband multimode antenna with orthogonal channels. The narrowband antenna system is designed for use at 5.9 GHz to 6.15 GHz, consisting of four monopoles built on a finite ground plane as shown in Fig. 8(a). This antenna system is capable of radiating four different orthogonal modes based on the amplitude and phase of the excitation signals to the antenna ports. More details about this antenna can be found in [15].

Two modes as shown in Fig. 8(c) and (d) were chosen and the antenna system was simulated with a ray-tracing software within the scenario shown in Fig. 8(b). The simulation has been performed for 1000 different transmit and receive nodes locations. Simulation results shown in Fig. 9 illustrates that the overall BER system performance has been significantly improved with the multimode antenna system compared to the half wave dipoles (HWD). That is due to the additional path diversity to the communication system.

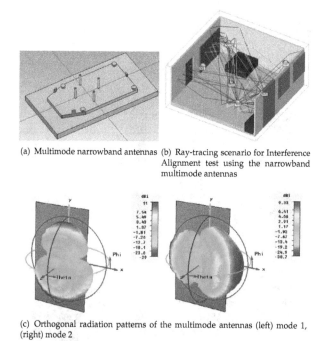

(a) Multimode narrowband antennas   (b) Ray-tracing scenario for Interference Alignment test using the narrowband multimode antennas

(c) Orthogonal radiation patterns of the multimode antennas (left) mode 1, (right) mode 2

Figure 8. Multimode narrowband antennas for preliminary Interference Alignment algorithm analysis

8. Hardware requirements of IA systems

High throughput wireless communication systems including LTE, ECMA-368 (WiMedia) and IEEE 802.11ac are built around sophisticated digital signal processing algorithms. Among the research goals for future communication standards are higher spectral efficiency, higher

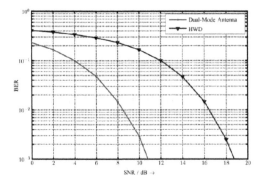

Figure 9. Comparison of the average BER vs. E_b / N_o of the IA system by applying multimode antenna systems and comparing to HWDs.

energy efficiencies towards the Shannon limit and increased data rates. Naturally, these benefits come at the price of higher computational complexity. The demand for flexible realtime hardware platforms capable of delivering the required huge number of operations per second at a severely limited power and silicon area budget has led to the development of specialized hardware platforms for software defined radio (SDR) applications. Techniques from FPGA-based ASIC verification and rapid prototyping are combined in this project for the design space exploration of highly optimized hardware architectures.

In a typical implementation scenario for complex designs, high level reference models are used. The choice of optimization-blocks is often based on profiling results, with those blocks contributing significantly to the overall resource requirements being chosen for optimization. This leads to a hybrid design consisting of a mixture of high level blocks and highly optimized blocks, running on hardware ranging from general purpose processors (GPP), application specific instruction set processors, FPGA-based rapid prototyping systems and dedicated hardware accelerators. The presented design space exploration framework reflects this structure and allows the designer to freely move processing blocks between the different layers of optimization.

The design space exploration framework created within this project is presented in Section 8.2. A hardware implementation case study of a closed-form 3-user IA algorithm has been selected for presentation in Section 8.3. Cost functions for an iterative IA algorithm are given in Section 8.4.

8.1. Wireless communication systems design space exploration

The process of designing complex digital electronic circuits offers a large variety of options to the designer. There are many valid possible implementations that fulfill the specification, but they differ in certain properties, e.g. silicon area, power efficiency, flexibility, testability and design effort. These properties span the so-called design space. A design space exploration establishes relations between possible points in the design space, ultimately leading to cost functions modeling the relation between the design properties and parameters. These models serve as a quantitative basis to make important design decisions in an early design phase.

Certain parameters are of special interest in the domain of wireless communication platforms. The limited power budget in mobile devices puts hard constraints on the power efficiency, requiring power optimization across all layers of algorithm development, design implementation and semiconductor technology.

Deriving comprehensive cost models using Monte-Carlo methods requires visiting a significantly larger number of points in the design space compared to existing heuristically driven parameter optimization approaches covered by existing FPGA-based simulation acceleration systems. The achievable simulation speedup is a key factor enabling the characterization and optimization of complex communication systems using Monte-Carlo approaches which are infeasible for pure software simulation due to the large required stimuli sets.

8.2. Development framework

The FPGA-based hybrid hardware-in-the-loop research and design space exploration (DSE) framework created in this work combines high-level tools (e.g. MATLAB/Simulink) and optimized hardware blocks [17]. Its application domain ranges from the design, optimization and verification of efficient and optimized signal processing blocks for computationally demanding next-generation wireless communication systems to system characterization and DSE.

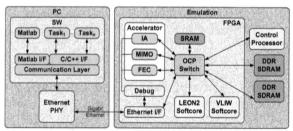

Figure 10. Emulation framework block diagram

The framework consists of a host PC, FPGA-based emulation systems, a generic fully synthesizeable VHDL SoC infrastructure, dedicated processors, processor softcores and a software library providing a transparent communication application programming interface (API). This allows signal processing blocks to be split and run distributed on a highly heterogeneous signal processing system. Software API libraries provide unified transparent communication between MATLAB, C/C+, embedded software and the hardware on-chip multilayer bus system. The same resources are accessible from all components, enabling a flexible partitioning and migration of processing task between high-level software, embedded software and dedicated hardware modules. The framework block diagram is shown in Figure 10. The properties of the optimized on-chip infrastructure template make it suitable for usage in final ASIC targets and thus enable the test, debugging and characterization of signal processing blocks in their target environment. Using standard FPGA design flows, new computationally intensive processing cores are directly implemented as the optimized hardware target modules. Instrumentation is used to enable dynamic, software controlled parameter adjustment. The remaining blocks may continue to run as high-level models, enabling a divide-and-conquer implementation and verification approach. The framework provides transparent data transport between the substituted MATLAB modules and multiple parallel instances of their FPGA hardware counterparts. The same interfaces are available for hardware simulation via the Modelsim foreign language interface (FLI), effectively also providing a verification and debugging environment at minimal extra effort.

The PC is connected to the emulation systems via Gigabit Ethernet. The generic FPGA infrastructure template comprises an OCP multilayer bus, the ethernet DMA interface, SDRAM controllers, on-chip memories and massively parallel parameterized softcore processors [20]. It has been adapted to and tested on a Xilinx Virtex-6 LX550T based BEE4 rapid prototyping system, the Xilinx Virtex-6 ML605 Evaluation Kit and the Virtex-5 LX220 based MCPA board [2] developed at IMS, see Fig. 11.

Figure 11. FPGA-based emulation system developed at IMS

8.3. Case study: 3-user antenna selection interference alignment

An implementation of the antenna selection interference alignment algorithm presented in Section 3 has been chosen as a case study using the development framework presented in Section 8.2. The proposed 3-user 2x2 MIMO zero-forcing IA antenna selection algorithm computes precoding matrices $\mathbf{V}$, decoding matrices $\mathbf{U}$ and a metric η based on [4]. Compared to the experimental testbed for fixed antenna patterns presented in [13], our implementation also chooses a subset of channels (i.e. antennas or radiation patterns) from the available channels. This leads to an increased channel orthogonality for the chosen channels at a reduced number of RF front ends.

The problem of finding the optimum antenna combination $\hat{i}$ from a set of I combinations can be formulated as

$$\hat{i} = \arg\max_{i=1...I} \sum_{k=1}^{K} \eta(\mathbf{V}_{k,i}, \mathbf{U}_{k,i}) \tag{29}$$

where $\eta(\mathbf{V}, \mathbf{U})$ is a function of the resulting SNR according to Section 3.1. $\mathbf{V}_{k,i}$ and $\mathbf{U}_{k,i}$ are the precoding and decoding matrices of user k for a given antenna combination i. Equation 29 is solved by visiting all I antenna combinations.

8.3.1. Computational complexity

The resource requirements of an optimized efficient integer implementation of the proposed novel antenna-selection IA algorithm is presented in this section, based on FPGA implementation results. Target systems include SDR platforms, FPGAs and ASICs.

This section focuses on the costs of the 3-user 2x2 MIMO processing consisting of matrix inversions, matrix multiplications, eigenvector computation and normalization, see Equations 6 to 9. The metric η is computed for both eigenvectors. All intermediate matrices can be independently scaled by arbitrary scalars without affecting the antenna decision or V and U. Exploiting this makes the cost of all involved 2x2 matrix inversions negligible and allows intermediate matrices to be block-normalized by shifting, i.e. extract a common power of 2 from all matrix elements. This results in reduced integer word lengths and thus reduced hardware costs. Table 1 summarizes the number of required real-valued mathematical base operations for antenna selection and the computation of V and U per antenna combination and subcarrier, without a final normalization step of V and U. Complex multiplications are composed of three real multiplications, three additions and two subtractions, INVSQRT denotes the reciprocal square root [26].

OP	ADD	MUL	SQRT	INVSQRT
Matrix mult.	696	348	0	0
Eigenvectors	15	8	3	0
Metric score	46	82	6	2
#OPC	757	438	9	2

Table 1. Operation counts for the computation of η per antenna combination i and subcarrier

To keep the total transmit power constant, the chosen antenna combination's precoding matrices V need to be normalized, resulting in 3 ADD, 8 MUL and 1 INVSQRT additional operations #OPN per transmitter and subcarrier. The above analysis implies that in general, the implementation cost is dominated by the multiplications in terms of silicon area and power consumption.

For the case of $K = 3$ users with $M = 2$ active transmit antennas used out of $L = 3$ physical antennas per transmitter and $N = 2$ antennas per receiver, there are 27 antenna combinations to be visited per subcarrier.

For realtime operation, the maximum allowable latency is defined to be T_0. Assigning relative operation costs α_i to each operation type OP_i, the total computational cost C for S subcarriers becomes

$$C = \frac{S}{T_0} \cdot \left(n \cdot \sum_{i \in OP} \alpha_i \cdot \#OPC_i + K \cdot \sum_{i \in OP} \alpha_i \cdot \#OPN_i \right) \quad (30)$$

8.3.2. Hardware cost estimation

Using α as relative silicon area costs, the total silicon area implementation cost of an architecture without resource sharing can be estimated from Eq. (30). The relative area α of 16-bit arithmetic operations for an ASIC implementation based on [23] results in the values given in Table 2. The relative costs α_{MUL} of a multiplier are defined to be 1.

OP	ADD	MUL	SQRT	INVSQRT
α	0.108	1	1.73	3

Table 2. Relative silicon area costs of 16-bit arithmetic operations

For a system using antenna selection at the transmitter only with $L = 3$ antennas, $S = 128$ subcarriers and $T_0 = 1$ ms, the total IA costs are estimated to be $C = 1.875$ GOPS.

For the configuration above, the original MATLAB algorithm takes 3.63 s on an Intel Xeon 2.4 GHz CPU running MATLAB R2012a for the computation of the optimal antenna combination $\hat{i}$ and its corresponding precoding and decoding matrices V_k and U_k from a set of channel information H. The FPGA implementation created in this case study achieves realtime operation, requiring 380 μs at 100 MHz clock frequency on a Xilinx Virtex-6 LX550T FPGA in a BEE4 emulation system. Thus, the achieved speedup is 9553.

8.4. Cost functions for K-user IA

The implementation presented in the previous section is a based on a closed-form $K = 3$ user IA algorithm. There is no known closed-form solution for $K > 3$ users, but iterative algorithms exist. In this section, we present implementation complexity estimates of the minimum mean square error (MMSE) IA algorithm presented in [27].

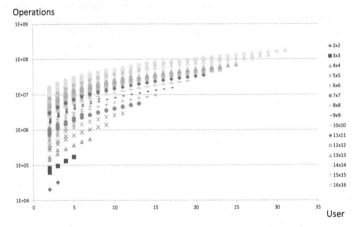

Figure 12. Number of operations for iterative MMSE interference alignment (4 iterations)

The MMSE-IA algorithm starts with arbitrary precoding matrices $\mathbf{V}_k$, then iteratively updates the decoding and precoding matrices $\mathbf{U}_k$ and $\mathbf{V}_k$ according to Eq. (31) and (32) until convergence. The Lagrange multiplier $\lambda_k \geq 0$ is computed to satisfy $\|\mathbf{V}_k\|_2^2 \leq 1$ by Newton iteration.

$$\mathbf{U}_k = \left(\sum_{j=1}^{K} \mathbf{H}_{kj} \mathbf{V}_j \mathbf{V}_j^H \mathbf{H}_{kj}^H + \sigma^2 \mathbf{I} \right)^{-1} \mathbf{H}_{kk} \mathbf{V}_k \qquad (31)$$

$$\mathbf{V}_k = \left(\sum_{j=1}^{K} \mathbf{H}_{jk}^H \mathbf{U}_j \mathbf{U}_j^H \mathbf{H}_{jk} + \lambda_k \mathbf{I} \right)^{-1} \mathbf{H}_{kk}^H \mathbf{U}_k \qquad (32)$$

The number of required iterations is data-dependent. Each iteration step requires the following operations to be executed: matrix multiplication, pseudo-inverse, Newton iterations. Figure 12 summarizes the estimated number of operations for the computation of a set of $\mathbf{V}$ and $\mathbf{U}$ matrices, based on well-known optimized hardware implementations. Comparing the iterative approach computational complexity to the the closed-form 2x2 IA implementation presented in Section 8.3, the number of operations is increased by a factor of approximately 60.8.

9. Conclusion

IA is a promising approach for communications with numerous pairs of users. In our contribution we have investigated the usefulness of IA for MIMO UWB communication systems. Beside apart of the significant power processing needed for the high data rate applications, the MIMO UWB antenna design remains a challenge. The antenna synthesis presented here can be viewed as synthesizing an antenna system with optimal radiation pattern catered towards an intended scenario. This antenna system radiates orthogonalized channels (after the averaging strategies) with sufficient power and has fixed beamforming (direction optimized according to the scenario and with averaging over various positions) at the transmitter and receiver antenna systems. Also, the resulting system has been simplified to 2 inputs and 2 outputs based on the subchannels with the strongest power. The whole system has been simulated by an indoor ray tracing tool and the corresponding MIMO UWB base band modulation schemes and detection techniques. Moreover, an antenna selection method is proposed in order to increase the robustness of IA in real environments. It is demonstrated that by using orthogonal multimode antennas a significant gain can be obtained. In the third and last part of our contribution the hardware efforts of IA algorithms are studied in more detail. It is worked out that highly challenging system blocks like IA can be elaborated today only by the help of suitable hardware emulation platforms which are typically FPGA-based. Therefore, a generic methodology has been elaborated and implemented which allows to explore the implementation design space. The hybrid hardware-in-the-loop research and design space exploration (DSE) framework created in this work combines high-level tools (e.g. Matlab/Simulink) and optimized hardware blocks. The properties of the elaborated optimized on-chip infrastructure template make it suitable for usage in final ASIC targets and thus enable the test, debugging and characterization of signal processing blocks in their target environment. This DSE framework has been used to derive cost models for K-user iterative IA algorithms. Estimates for the implementation effort (e.g. in terms of operation counts in dependency of the number of users) have been derived. Because of this project, a generic DSE framework is available and can be used to work out suitable architectures for further challenging building blocks.

Author details

Mohamed El-Hadidy, Mohammed El-Absi and Thomas Kaiser
Duisburg-Essen University, Institute of Digital Signal Processing (DSV), Germany

Yoke Leen Sit and Thomas Zwick
Karlsruhe Institute of Technology, Institut für Hochfrequenztechnik und Elektronik (IHE), Germany

Markus Kock and Holger Blume
Leibniz Universität Hannover, Institute of Microelectronic Systems (IMS), Germany

10. References

[1] Adamiuk, G. [2010]. *Methoden zur Realisierung von Dual-orthogonal, Linear Polarisierten Antennen für die UWB-Technik*, PhD thesis, Karlsruhe.
URL: *http://digbib.ubka.uni-karlsruhe.de/volltexte/1000019874*

[2] Banz, C., Hesselbarth, S., Flatt, H., Blume, H. & Pirsch, P. [2012]. Real-Time Stereo Vision System using Semi-Global Matching Disparity Estimation: Architecture and FPGA-Implementation, *Transactions on High-Performance Embedded Architectures and Compilers, Springer* .

[3] Björck, Å. & Golub, G. H. [1973]. Numerical methods for computing angles between linear subspaces, *Math. Comp.* 27: 579–594.

[4] Cadambe, V. & Jafar, S. [2008]. Interference alignment and degrees of freedom of the k -user interference channel, *Information Theory, IEEE Transactions on* 54(8): 3425 –3441.

[5] Cadambe, V. & Jafar, S. [2009]. Reflections on interference alignment and the degrees of freedom of the k-user mimo interference channel, *IEEE Information Theory Society Newsletter* 54(4): 5 –8.

[6] El-Absi, M., El-Hadidy, M. & Kaiser, T. [2012]. Antenna selection for interference alignment based on subspace canonical correlation, *2012 International Symposium on Communications and Information Technologies (ISCIT)* .

[7] El Ayach, O., Peters, S. & Heath, R. [2010]. The feasibility of interference alignment over measured mimo-ofdm channels, *Vehicular Technology, IEEE Transactions on* 59(9): 4309 –4321.

[8] El-Hadidy, M., El-Absi, M. & Kaiser, T. [2012]. Articial diversity for uwb mb-ofdm interference alignment based on real-world channel models and antenna selection techniques, *2012 IEEE International Conference on Ultra-Wideband (ICUWB)* .

[9] El-Hadidy, M., Mohamed, T., Zheng, F. & Kaiser, T. [2008]. 3d hybrid em ray-tracing deterministic uwb channel model, simulations and measurements, 2: 1 –4.

[10] Fügen, T., Maurer, J., Kayser, T. & Wiesbeck, W. [2006]. Capability of 3-D Ray Tracing for Defining Parameter Sets for the Specification of Future Mobile Communications Systems, *Antennas and Propagation, IEEE Transactions on* 54(11): 3125 –3137.

[11] Fügen, T., Waldschmidt, C., Maurer, J. & Wiesbeck, W. [2003]. MIMO capacity of bridge access points based on measurements and simulations for arbitrary arrays, *5th European Personal Mobile Communications Conference*, pp. 467–471.

[12] Gesbert, G., Shafi, M., Shiu, D., Smith, P. J. & Naguib, A. [2003]. From Theory to Practice: An Overview of MIMO Space-Time Coded Wireless Systems, *IEEE Journal on Selected Areas in Communications* 21: 281–302.

[13] González, O., Ramírez, D., Santamaría, I., García-Naya, J. & Castedo, L. [2011]. Experimental validation of interference alignment techniques using a multiuser MIMO testbed, *Smart Antennas (WSA), 2011 International ITG Workshop on*, pp. 1 –8.

[14] Heath, R., Sandhu, S. & Paulraj, A. [2001]. Antenna selection for spatial multiplexing systems with linear receivers, *IEEE Commun. Letters* 5(4): 142 –144.

[15] Jereczek, G. [2010]. *Design of Capacity Maximizing MIMO Antenna Systems for Car-2-Car Communication*, Master's thesis, Karlsruhe Institute of Technology (KIT), Karlsruhe, Germany.

[16] Khalighi, M. I., Brossier, J., Jurdain, G. & Raoof, K. [2001]. Water Filling Capacity of Rayleigh MIMO Channels, *IEEE Transactions on Antennas and Propagation* 1: A155–A158.

[17] Kock, M., Hesselbarth, S. & Blume, H. [2013]. Hardware-accelerated design space exploration framework for communication systems, *Wireless Innovation Forum Conference*

on Wireless Communications Technologies and Software Defined Radio (SDR-WInnComm 2012), Washington, DC, USA. Accepted for publication.

[18] Loyka, S. & Mosig, J. [2006]. *Information Theory and Electromagnetism: Are They Related?*, CRC Press.

[19] Molisch, A., Win, M. & Winters, J. [2001]. Capacity of mimo systems with antenna selection, *IEEE International Conference on Communications (ICC)* (4): 570 –574.

[20] Paya-Vaya, G. & Blume, H. [2012]. TUKUTURI: A dynamically reconfigurable multimedia soft-processor for video processing applications, *Poster presentation at the 7th International Conference on High-Performance and Embedded Architectures and Compilers, HiPEAC'12, Paris, France*, Vol. USB Proceedings.

[21] Peters, S. & Heath, J. [2009]. Interference alignment via alternating minimization, *Int. Conf. on Acoust. Speech and Signal Processing,(ICASSP)* .

[22] Petersen, D. P. & Middleton, D. [1962]. Sampling and Reconstruction of Wave-limited functions in N-dimensional Euclidean Spaces, *Information and Control* 5: 279 – 323.

[23] Pirsch, P. [1998]. *Architectures for Digital Signal Processing*, John Wiley & Sons, Inc.

[24] Pontes, J. [2010]. *Optimized Analysis and Design of Multiple Element Antenna for Urban Communication*, Karlsruher Institut fuer Technologie (KIT).

[25] Reichardt, L., Pontes, J., Sit, Y. & Zwick, T. [2011]. Antenna optimization for time-variant mimo systems, *Antennas and Propagation (EUCAP), Proceedings of the 5th European Conference on*, pp. 2569 –2573.

[26] Salmela, P., Burian, A., Järvinen, T., Happonen, A. & Takala, J. H. [2011]. Low-complexity inverse square root approximation for baseband matrix operations, *ISRN Signal Processing* vol. 2011.

[27] Schmidt, D., Shi, C., Berry, R., Honig, M. & Utschick, W. [2009]. Minimum mean squared error interference alignment, *Signals, Systems and Computers, 2009 Conference Record of the Forty-Third Asilomar Conference on*, pp. 1106 –1110.

[28] Shen, M., Zhao, C., Liang, X. & Ding, Z. [2011]. Best-effort interference alignment in ofdm systems with finite snr, *IEEE International Conference on Communications (ICC)* 2(4): 5 –9.

[29] Sit, Y. L., Reichardt, L., Liu, R., Liu, H. & Zwick, T. [2012]. Maximum Capacity Antenna Design for an Indoor MIMO-UWB Communication System, *10th International Symposium on Antennas, Propagation and EM Theory, Xian, China*.

[30] Wallace, J. W. & Jensen, M. A. [2002]. Intrinsic Capacity of the MIMO Wireless Channel, *IEEE Proceedings of the 56th Vehicular Technology Conference VTC 2002-Fall*, pp. 701–705.

Chapter 7

Power Allocation Procedure for Wireless Sensor Networks with Integrated Ultra-Wide Bandwidth Communications and Radar Capabilities

Gholamreza Alirezaei, Rudolf Mathar and Daniel Bielefeld

Additional information is available at the end of the chapter

1. Introduction

In this chapter, we analyze the problem of power allocation for a distributed wireless sensor network with sensor nodes based entirely on ultra-wide bandwidth (UWB) technology. The network is used to perform object detection as well as object classification, where the absence, the presence, or the type of an object is observed by the sensors independently. UWB signals can be used for data communication between the sensor nodes as well as for radar applications. The approach of misemploying the communication sensors as radar sensors, such that the data transmission is misused as a radar beam in order to detect or to classify a target object, helps in realizing an energy-efficient radar system with compact and cheap sensor nodes. A further advantage of such radar systems is the fulfillment of major requirements of wireless sensor networks. This exploitation presupposes that the integration of sensing functionality into usual UWB sensors is implementable easily without the usage of any additional hardware units. Since the compact and low complexity UWB sensors are limited in power and communication capabilities, the detection and classification performance of a single sensor is restricted compared to that of a common complex radar system. To obtain an appropriate overall system performance, we consider the case of distributed detection and classification, where the local observations of the sensors are fused into a reliable global decision. Due to noisy communication channels and differences in distances between the object and the sensor nodes, both, the observations and their transmissions are unequally interfered. One simple way to suppress noise interference is to increase the power of each sensor node. But if the total power of the entire network is limited, then power allocation procedures are needed in order to increase the overall detection and classification probabilities. In general, it is challenging to evaluate the detection and the classification probabilities analytically, if possible at all. This particularly holds for the detection probability under a Neyman-Pearson-hypotheses-test criterion as well as for the classification probability under a Bayesian-hypotheses-test criterion [5]. This limits the usability of these criterions for analytical optimization of power allocation. Bounds, such as the Bhattacharyya bound [8], are also difficult to use for optimizing multidimensional problems. Therefore, we employ an information theoretic approaches [3], which help to solve the power allocation problem with a lower mathematical complexity. This approach yields a simple however suboptimal analytical solution for the power allocation problem. Furthermore, the proposed technique enables the consideration of object detection and classification at the same time. This is a further advantage of this method, which enables the usage of the same allocation algorithm in both cases. Hence a sensor network, which is used to classify target objects, can also be used to detect the absence or the presence of a target object with equal system settings. Therefore we only describe the case of object classification, which also includes the case of object detection, in the following sections.

The origin of research on distributed detection has been the attempt to fuse signals of different radar devices [10]. Currently, distributed detection is usually discussed in the context of wireless sensor networks, where the sensor unit of the nodes might be based on radar technology [7, 9, 14]. Other applications for UWB radar systems, which require or benefit from the detection and classification capabilities, are for example localization and tracking [6] or through-wall surveillance [4]. The physical layer design for an integrated UWB radar network that utilizes OFDM technology was analyzed in [11]. In [2] the case of object detection is considered, where for the problem of power allocation an approach based on the maximization of the Kullback-Leibler distance is used. In a recent publication [1] another approach is discussed, where the bit-error probability of data communication is used in order to allocate the transmission power and to increase the overall detection probability.

This chapter is divided into the following three sections except the introduction. First, the system model of the wireless network including sensor nodes and the fusion center is described. Here, all system parameters and assumptions with detailed mathematical formulations are introduced. Furthermore, the global classification rule in the fusion center as well as the local decision rules in the nodes are motivated. In the second section, we present a novel approach for power allocation in order to increase the overall classification probability, following which, the solution of this optimization approach is briefly discussed. The last section shows some results and demonstrates the feasibility of object classification by using the proposed power allocation method in UWB signaling systems. This chapter concludes with an interpretation of the achieved system performance.

2. System model

Throughout this chapter we denote the set of natural, real, and complex numbers by $\mathbb{N}$, $\mathbb{R}$, and $\mathbb{C}$, respectively. Note that the set of natural numbers does not include the element zero. Furthermore, we use the subset $\mathbb{F}_N \subseteq \mathbb{N}$ which is defined as $\mathbb{F}_N := \{1, \ldots, N\}$ for any given natural number N. The mathematical operations $|z|$ and $|\mathbf{z}|$ denote the absolute value of a real or complex-valued number z and the Euclidian length of a real or complex vector $\mathbf{z}$, respectively.

Distributed *target object* classification can be formally modeled by a multiple hypotheses testing problem with hypotheses $H_k \forall k \in \mathbb{F}_K$ for a specified number $K \in \mathbb{N}$ of different objects. We assume that all objects have the same size, shape, alignment, and position. They only differ in material and are classified by their complex-valued reflection coefficients $r_k \in \mathbb{C}$, which are ordered in a strictly increasing manner $0 \leq |r_1| < \cdots < |r_K| \leq 1$. Therefore the reflection coefficients are the only recognition features in this work. Generally, this assumption is not realistic, but, this case describes an ideal scenario for increasing the classification probability by performing a power allocation and is not really suitable for analyzing the problems of manifoldness.

Power Allocation Procedure for Wireless Sensor Networks with Integrated Ultra-Wide Bandwidth
Communications and Radar Capabilities

155

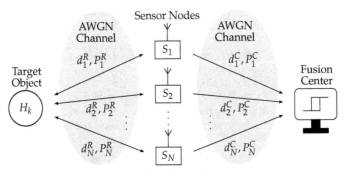

Figure 1. System model of the distributed wireless sensor network.

At any instance of time, a network of $N \in \mathbb{N}$ independent and spatially distributed sensors, as shown in Figure 1, obtains random observations $X_1, \ldots, X_N \in \mathbb{R}$. In the case of energy classification, X_n models the received signal at the receiver of the nth sensor. If a target object is present, then the received energy is a part of the radiated energy of the same sensor, which is reflected from the object's surface and is weighted by its reflection coefficient. We refer to this communication channel, between the sensors and the target object, as the *first* communication link and denote all dedicated parameters by the superscript R. The random observations $X_1, \ldots, X_N$ are assumed to be conditionally independent for each of the underlying hypotheses, i.e., the joint conditional probability density function of all the observations factorizes according to

$$f^R(X \mid H_k) := \prod_{n=1}^{N} f_n^R(X_n \mid H_k), \ \forall k \in \mathbb{F}_K, \tag{1}$$

where X denotes the sequence of random variables $X_1, \ldots, X_N$. In general, the observations are not identically distributed because the sensor nodes have different distances d_n^R from the target object and their radiated powers P_n^R are also different. Therefore, the signal-to-noise ratio (SNR) varies between the sensor nodes. Due to the distributed nature of the problem, the nth sensor S_n performs independent measurements and processes its respective observation X_n by generating a local decision $U_n := \theta_n(X_n) \forall n \in \mathbb{F}_N$, which depends only on its own observation and not on the observations of other sensor nodes. After deciding locally, each sensor transmits its decision to a fusion center located at a remote location. The communication between the sensor node and the fusion center is determined by the corresponding distance d_n^C as well as by the transmission power P_n^C of the same sensor node. We refer to this communication channel, between the sensor nodes and the fusion center, as the *second* communication link and denote all dedicated parameters by the superscript C. Furthermore, we assume that both communication channels are non-fading channels and that all data transmissions are affected only by additive white Gaussian noise (AWGN). We disregard time delays within all transmissions and assume synchronized data communication. We use two distinct pulse-shift patterns for each sensor node in order to distinguish its first and second communication link from the communication links of other sensor nodes as described in [13]. Each pattern has to be suitably chosen in order to suppress inter-user interference at each receiver. Hence the N received signals at the fusion center are uncorrelated and are assumed to be conditionally independent for each of the underlying hypotheses. These received random signals correspond to the local decisions $U_1, \ldots, U_N$ and are mapped to $\tilde{\mathbf{X}}_1, \ldots, \tilde{\mathbf{X}}_N \in \mathbb{R}^K$. Their joint conditional probability density function factorizes according to

$$f^C(\tilde{\mathbf{X}} \mid H_k) := \prod_{n=1}^{N} f_n^C(\tilde{\mathbf{X}}_n \mid H_k), \ \forall k \in \mathbb{F}_K, \tag{2}$$

where $\tilde{\mathbf{X}}$ denotes the sequence of random vectors $\tilde{\mathbf{X}}_1, \ldots, \tilde{\mathbf{X}}_N$. In general, these observations are – similar to the observations $X_1, \ldots, X_N$ – not identically distributed, because of variation in distances d_n^C as well as that of the radiated powers P_n^C. Unlike the local decision rules, the global decision rule $U_0 := \theta_0(\tilde{\mathbf{X}}_1, \ldots, \tilde{\mathbf{X}}_N)$ depends on all observations in order to increase the overall classification probability.

All described assumptions are necessary in order to obtain an ideal framework suited for analyzing the power allocation problem without studying problems of different classification methods in specific systems and their settings.

2.1. Local classification rules

The local decision and classification rules θ_n are mappings of the kind $\theta_n : \mathbb{R} \to \mathbb{F}_K, \ \forall n \in \mathbb{F}_N$. In this work, hard-decision rules are used for performing local classification given by

$$\theta_n(X_n = x_n) = k, \text{ if } \tau_{n,k} < x_n \leq \tau_{n,k+1}, \ k \in \mathbb{F}_K, \forall n \in \mathbb{F}_N, \tag{3}$$

where the thresholds $\tau_{n,k} \in \mathbb{R}$ are suitably chosen. The thresholds must be calculated separately for every sensor node in order to perform optimal classification. They depend on the prior probabilities of the hypotheses. Their values can be calculated by a suboptimal approach which is described in Section 3.1. In this way, every sensor node has a local probability of correct decision given by

$$\text{Prob}(U_n = k \mid H_k) = \text{Prob}(\tau_{n,k} < X_n \leq \tau_{n,k+1} \mid H_k), \ \forall k \in \mathbb{F}_K, \forall n \in \mathbb{F}_N \tag{4}$$

and a local probability of false decision given by

$$\text{Prob}(U_n \neq k \mid H_k) = 1 - \text{Prob}(U_n = k \mid H_k), \ \forall k \in \mathbb{F}_K, \forall n \in \mathbb{F}_N. \tag{5}$$

2.2. Fusion of local decisions and the global classification rule

The local decisions $U_1, \ldots, U_N$ at the sensor nodes are conditionally independent due to uncorrelated and independent noisy communication channels. By applying the Bayesian-hypotheses-test criterion the optimal fusion rule at the fusion center is given by

$$U_0 = \theta_0(\tilde{\mathbf{X}} = \tilde{\mathbf{x}}) = \underset{k \in \mathbb{F}_K}{\operatorname{argmax}} \left(\pi_k f^C(\tilde{\mathbf{x}} \mid H_k) \right), \tag{6}$$

where $\pi_k := \text{Prob}(H_k)$ with $\sum_{k=1}^{K} \pi_k = 1$ denotes the prior probability of hypothesis H_k. We use this formula to classify the target object. However, in order to optimize the allocation of the total power to the sensor nodes, we have to consider the overall classification probability. Therefore, we consider K pairwise disjoint regions $\mathcal{R}_1, \ldots, \mathcal{R}_K$ with

$$\mathcal{R}_k := \left\{ \tilde{\mathbf{x}} \in \mathbb{R}^{K \times N} \mid \pi_k f^C(\tilde{\mathbf{x}} \mid H_k) \geq \pi_l f^C(\tilde{\mathbf{x}} \mid H_l), \forall l \in \mathbb{F}_K, l \neq k \right\}, \ \forall k \in \mathbb{F}_K. \tag{7}$$

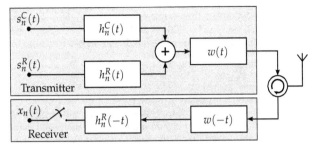

Figure 2. System model of the nth sensor node with circulator and antenna.

According to [5], the expected value of correct classification is given by

$$P_c := \sum_{k=1}^{K} \text{Prob}(\tilde{\mathbf{x}} \in \mathcal{R}_k, H_k), \tag{8}$$

which in general cannot be analytically evaluated. Therefore, the previous formula cannot be used to optimize the allocation of the total power analytically. Consequently, we choose a different approach for the optimization, which is described in Section 3.3.

2.3. Ultra-wide bandwidth sensor nodes

In Figure 2 the system model of the considered impulse-radio UWB (IR-UWB) sensor nodes with pulse position modulation (PPM) is shown. The transmitter generates two streams of data symbols $s_n^C(t)$ and $s_n^R(t)$.

The symbol stream s_n^C is used to transmit the local decisions $u_n(i) \in \mathbb{F}_K$ at the time index i to the fusion center, which are generated by the algorithm defined in (3). We describe the data symbols by Dirac delta functions $\delta(t - [u_n(i) - 1]\Delta)$, which are shifted pulses on the time axis. Their alignment is determined by the modulation index Δ. We assume that the product $K\Delta$ is much smaller than the symbol duration. Thus K different data symbols can be transmitted to the fusion center. The transmission power P_n^C of this stream is variable in order to adjust transmission power and to enable distributed power allocation.

The symbol stream s_n^R establishes the radiation to the target object and uses always the same data symbol. Its transmission power P_n^R is also variable. In order to increase the available power range at every sensor node, time-division multiple-access (TDMA) method is used to separate both streams into different time slots and to periodically share the same power amplifier.

In order to eliminate collisions due to multiple access, each user stream is assigned to a distinctive time-shift pattern after passing through the blocks $h_n^C(t)$ and $h_n^R(t)$. Their transfer functions are based on time-hopping sequences [13].

After superposition of both streams, a monocyclic pulse shape filter $w(t)$ limits the bandwidth of the signal. This filter has to fulfill the Nyquist intersymbol interference (ISI) criterion in order to avoid the intersymbol interferences.

When this superposition is transmitted, a part of the radiated signal s_n^R will be reflected from the target surface back to the antenna. The received signal will pass through the matched-filter

$w(-t)$ and will be decoded from its time-hopping sequence by $h_n^R(-t)$. The additive noise signal $b_n^R(t)$ will pass as well through both filters at the receiver. We denote the corresponding noise power by P_{noise}. If all receiver components are linear, then we can describe the received power by

$$\tilde{P}_{n|k}^R := P_n^R \frac{\alpha_n^R |r_k|^2}{g^2(2d_n^R)}, \quad \forall k \in \mathbb{F}_K, \forall n \in \mathbb{F}_N, \tag{9}$$

where the transmitted power is weighted by the product of the factors $\alpha_n^R > 0$, $|r_k|^2$, and $g^{-2}(2d_n^R)$. The factor α_n^R includes the radar cross section, the influence of the antenna, the impacts of the filters, and all additional attenuation of the transmitted power. Due to the reflection coefficient r_k of the target object the received power depends on the underlying hypothesis. The path loss function g depends on the assumed multipath propagation channel and is usually an increasing function of the distance between transmitter and receiver. Here, the factor of two in the distance results from that back and forth transmission between the transceiver and the object. The ratio of $\tilde{P}_{n|k}^R$ and P_{noise} is the observed conditional SNR at the receiver and is given by

$$\gamma_{n|k}^R := \frac{P_n^R}{P_{\text{noise}}} \cdot \frac{\alpha_n^R |r_k|^2}{g^2(2d_n^R)}, \quad \forall k \in \mathbb{F}_K, \forall n \in \mathbb{F}_N. \tag{10}$$

Due to the Gaussian distribution of the noise, each sample is also a Gaussian random variable, which is conditionally distributed according to

$$f_n^R(X_n = x_n \mid H_k) := \frac{1}{\sqrt{2\pi P_{\text{noise}}}} \exp\left(-\frac{\left(x_n - \sqrt{\tilde{P}_{n|k}^R}\right)^2}{2P_{\text{noise}}}\right), \quad \forall k \in \mathbb{F}_K, \forall n \in \mathbb{F}_N. \tag{11}$$

The local decision probabilities $\text{Prob}(U_n = l \mid H_k)$, see (4) and (5), can be computed by solving the integral

$$\tilde{\pi}_{n,l|k} := \text{Prob}(U_n = l \mid H_k) = \int_{\tau_{n,l}}^{\tau_{n,l+1}} f_n^R(x_n \mid H_k)\, dx_n$$

$$= \frac{1}{2}\left[\text{erf}\left(\frac{\sqrt{\tilde{P}_{n|k}^R} - \tau_{n,l}}{\sqrt{2P_{\text{noise}}}}\right) + \text{erf}\left(\frac{\tau_{n,l+1} - \sqrt{\tilde{P}_{n|k}^R}}{\sqrt{2P_{\text{noise}}}}\right)\right] \tag{12}$$

for all $k, l \in \mathbb{F}_K$ and for all $n \in \mathbb{F}_N$. Here, the mapping $\text{erf}(z)$ denotes the error function of z.

2.4. Fusion center

After radiation of the stream s_n^C by the sensor node S_n, the signal is attenuated depending on the distance and it reaches the antenna at the fusion center as depicted in Figure 3. The received signal is matched-filtered and decoded from its time-hopping sequence. Then a data splitter $v(t)$ is used to split the received signal into a K-dimensional vector space. This is necessary in order to retain the Euclidian distances between all transmitted symbols and achieve a higher classification probability. This filter is mathematically implemented as $\sum_{k=1}^{K} e_k \delta(t - (k-1)\Delta)$, where e_k is the standard basis vector of the K-dimensional space that points in the kth direction. Therefore the received signals $\tilde{X}_1, \dots, \tilde{X}_N \in \mathbb{R}^K$ are K-dimensional vectors. This new approach extends the method given by [13].

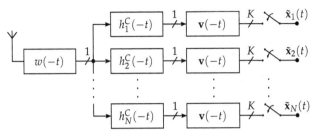

Figure 3. System model of the fusion center.

In case of additive zero-mean noise and due to the assumptions of $w(t)$, each vector sample of the received signal has the expected value of

$$\mathbf{m}_{n|l} := \mathrm{E}\big(\tilde{\mathbf{X}}_n \mid U_n = l\big) = \sqrt{P_n^C \frac{\alpha_n^C}{g^2(d_n^C)}} \cdot \mathbf{e}_l, \ \forall l \in \mathbb{F}_K, \forall n \in \mathbb{F}_N, \tag{13}$$

which depends on the transmitted symbol $U_n = l$. Thus the received power from the nth sensor node is given by

$$\tilde{P}_n^C := P_n^C \frac{\alpha_n^C}{g^2(d_n^C)}, \ \forall n \in \mathbb{F}_N, \tag{14}$$

where we assume that the path loss function is the same as for the first communication link. The power $\tilde{P}_n^C$ is independent of the underlying hypothesis because the data stream s_n^C has the same power for all kinds of transmitted data symbols.

The additive noise signal $b_n^C(t)$ will also pass through all the filters. We assume that the noise spectral density at the fusion center is the same as at the sensor nodes. Due to similarity in architecture of the fusion center and the sensor nodes, the noise power in each dimension of each stream is equal to P_{noise}. Because of the whiteness of noise, the interferences are uncorrelated in each dimension of each stream. Therefore the noise covariance matrix is determined by the product $P_{\text{noise}} \cdot \mathbf{I}_K$. Here $\mathbf{I}_K$ denotes the identity matrix of size K.

Similar to (10) we define an observed SNR for each data stream at the fusion center and denote it by

$$\gamma_n^C := \frac{P_n^C}{P_{\text{noise}}} \cdot \frac{\alpha_n^C}{g^2(d_n^C)}, \quad \forall n \in \mathbb{F}_N. \tag{15}$$

Due to the Gaussian distribution of noise, each vector sample is a Gaussian random vector, which is conditionally distributed according to

$$f_n^C(\tilde{\mathbf{X}}_n \mid H_k) := \sum_{l=1}^{K} \frac{\tilde{\pi}_{n,l|k}}{(2\pi P_{\text{noise}})^{K/2}} \exp\left(-\frac{(\tilde{\mathbf{x}}_n - \mathbf{m}_{n|l})^{\mathrm{T}}(\tilde{\mathbf{x}}_n - \mathbf{m}_{n|l})}{2P_{\text{noise}}}\right), \ \forall k \in \mathbb{F}_K, \forall n \in \mathbb{F}_N, \tag{16}$$

where the operator $\mathbf{z}^{\mathrm{T}}$ denotes the transpose of any vector $\mathbf{z}$.

Because of the convex superposition of multivariate Gaussian distributions, it is difficult to use (16) with the properties of (2) to optimize the distributed power allocation. Bounds, such as the Bhattacharyya bound [8], are also difficult to use due to multidimensional nature of (2) and (16). Therefore we propose an applicable technique which is motivated by concepts of information theory and is described in the next section.

3. Allocating power to the radar and to the communication task

In this section, we motivate and present an approach to suboptimally allocate transmission power to the radar and to the communication task. The objective is to maximize the overall classification probability, given a limited total transmission power P_{tot} that can be arbitrarily allocated to the radar task as well as to the communication task. A direct solution to this problem does not exist, since no analytical expression for the overall classification probability (8) is available. Instead, we independently maximize the mutual information of both communication channels to increase the information flow and in order to determine the power allocation. The motivation for this approach is the separation of the power allocation problem from the object classification procedure. Because in this case the data communication does not affect the classification of the target object.

Note that this theoretical concept is not realistic. However, we apply this concept as a heuristic method in this work.

3.1. Threshold calculation

For the optimization of the thresholds in Section 2.1, in order to increase the overall classification probability, the analytic evaluation of (8) is needed. Due to the fact that this explicit form for the overall classification probability is unknown and due to the separation of the data communication from the classification task we propose the following simple approach to calculate the thresholds.

We increase the probability of correct decision of each sensor node independently to achieve suboptimal values for the thresholds. Thus, the overall classification probability should be increased as well. According to equations (4) and (12) the local probability of correct decision, which has to be maximized, is given by

$$\sum_{k=1}^{K} \text{Prob}(H_k)\,\text{Prob}(U_n = k \mid H_k) = \sum_{k=1}^{K} \frac{\pi_k}{2}\left[\text{erf}\left(\frac{\sqrt{\tilde{P}_{n|k}^R} - \tau_{n,k}}{\sqrt{2P_{\text{noise}}}}\right) + \text{erf}\left(\frac{\tau_{n,k+1} - \sqrt{\tilde{P}_{n|k}^R}}{\sqrt{2P_{\text{noise}}}}\right)\right]. \quad (17)$$

Its solution can be found explicitly by using differential calculus. The corresponding result is identical to the one obtained by using the Bayesian-hypotheses-test criterion. It is given by

$$\tau_{n,k} = \begin{cases} \inf(\mathbb{I}_{n,k}) & \text{if} \quad \mathbb{I}_{n,k} \neq \{\}, \, k \in \mathbb{F}_K, \\ \tau_{n,k+1} & \text{if} \quad \mathbb{I}_{n,k} = \{\}, \, k \in \mathbb{F}_K, \\ \infty & \text{if} \quad k = K+1, \end{cases} \quad (18)$$

for all $n \in \mathbb{F}_N$, where the function $\inf(\mathbb{I}_{n,k})$ is the infimum of the interval $\mathbb{I}_{n,k}$ that is defined by

$$\mathbb{I}_{n,k} := \left\{x \in \mathbb{R} \mid \pi_k f_n^R(x \mid H_k) > \pi_l f_n^R(x \mid H_l), \forall l \in \mathbb{F}_K, l \neq k\right\}, \, \forall k \in \mathbb{F}_K, \forall n \in \mathbb{F}_N. \quad (19)$$

3.2. Limitation of transmission power

We assume that both the radar and the communication signal use the same bandwidth and are uncorrelated to each other, due to separation of the sensing task and the communication task into different time slots (see Section 2.3). In this case and for each new classification process, the limited total transmission power P_{tot} is an upper bound for the sum

$$\sum_{n=1}^{N} \underbrace{P_n^R}_{\text{Radar sensing}} + \underbrace{P_n^C}_{\text{Data communication}} \leq P_{\text{tot}}. \qquad (20)$$

$$\underbrace{\hspace{4cm}}_{\text{Total transmission power of one sensor for a single observation}}$$

By using this restriction, we present the power allocation procedure in the next section. But, we will have a look at some special cases previously.

In real applications the transmission power of each sensor node is also limited. Consider the case in which all sensor nodes have the same power limitation $P_{\max}$ with $\frac{P_{\text{tot}}}{N} \leq P_{\max} < P_{\text{tot}}$. If the power regulation, which is described in the next section, wants to allocate a higher power to $P_n^R > P_n^C$ of the nth sensor node than its limitation, then we set the transmission power P_n^R equal to its highest possible limitation given by $P_{\max}$, recalculate P_n^C which is given in terms of $P_n^R = P_{\max}$, discard this n^{th} sensor node from the list of unallocated sensor nodes, decrease the given total transmission power P_{tot} by $P_{\max} + P_n^C(P_{\max})$, and reallocate the remaining total power $P_{\text{tot}} - P_{\max} - P_n^C(P_{\max})$ recursively to the remaining sensor nodes by the same procedure described in the next section. In a case, where the power P_n^C instead of $P_n^R > P_n^C$ will be regulated higher than $P_{\max}$, we can reverse the roles of both transmission powers and repeat this reallocation method until no more sensor nodes are left which exceed their power limitation. Therefore, the described limitation of the total transmission power is the generalized case which includes the limitation of the transmission power of each sensor node.

Note that this procedure is applicable for individual power constraints per node as well. Furthermore, note that in each iteration more than one node can be discarded from the list of unallocated sensor nodes in order to decrease the computation complexity.

3.3. Mutual information-based power allocation

For the maximization of the information flow we set the mutual information of both communication channels equal. This leads to the same symbol error probabilities on both sides for low SNR values. For each sensor node an upper bound for the mutual information of its first and second link can simply be calculated. The identity of obtained bounds

$$\frac{1}{2}\log\left[1 + \frac{P_n^R \alpha_n^R (|r_K| - |r_1|)^2}{4P_{\text{noise}}g^2(2d_n^R)}\right] = \frac{K}{2}\log\left[1 + \frac{P_n^C \alpha_n^C (K-1)}{P_{\text{noise}}g^2(d_n^C)K^2}\right] \qquad (21)$$

has to be computed in order to find the relationship between the powers for all $n \in \mathbb{F}_N$. After calculation and usage of the simple approximation

$$\sqrt[K]{1+x} \approx 1 + \frac{x}{K} \qquad (22)$$

for any small values of x we obtain the analytical relationship

$$P_n^C = P_n^R \cdot \frac{\alpha_n^R}{\alpha_n^C} \frac{g^2(d_n^C)}{g^2(2d_n^R)} \frac{K}{K-1} \frac{(|r_K| - |r_1|)^2}{4}, \ \forall n \in \mathbb{F}_N. \qquad (23)$$

In the next step, we increase the overall information flow by maximization of the cumulative mutual information subject to the given total power of the sensor network. Then the optimization problem is given by

$$\underset{P_1^R,\dots,P_N^R}{\text{maximize}} \sum_{n=1}^{N} \tfrac{1}{2}\log\left[1 + \tfrac{P_n^R \alpha_n^R (|r_K|-|r_1|)^2}{4P_{\text{noise}}g^2(2d_n^R)}\right] \quad \text{subject to} \quad \sum_{n=1}^{N} P_n^C + P_n^R \le P_{\text{tot}}. \quad (24)$$

It has to be considered that the sum of concave functions is also concave and that the arguments of the logarithms are linear functions of the powers. Furthermore, the domain of the feasible set is a closed convex set and, therefore, only one global maximum of the problem exists. This maximum can be explicitly calculated by using the method of Lagrange multipliers which is equivalent to the water-filling power allocation result [3]. The result is given by

$$P_n^R = P_{\text{noise}} \tfrac{g^2(2d_n^R)}{\alpha_n^R} \tfrac{4}{(|r_K|-|r_1|)^2} \cdot \max\left(0, \tfrac{\lambda}{\beta_n} - 1\right), \forall n \in \mathbb{F}_N, \quad (25)$$

where the factor β_n is defined by

$$\beta_n := \tfrac{g^2(2d_n^R)}{\alpha_n^R} \tfrac{4}{(|r_K|-|r_1|)^2} + \tfrac{g^2(d_n^C)}{\alpha_n^C} \tfrac{K}{K-1}, \forall n \in \mathbb{F}_N. \quad (26)$$

For the following equations, we assume that the factors β_n are ordered in an increasing manner. Then the water-filling level λ is a value specified by the inequality

$$\beta_{\tilde{N}} < \lambda \le \tfrac{1}{\tilde{N}}\left[\tfrac{P_{\text{tot}}}{P_{\text{noise}}} + \sum_{n=1}^{\tilde{N}} \beta_n\right], \quad (27)$$

where the number $\tilde{N}$ with $1 \le \tilde{N} \le N$ is a suitably chosen integer value for which the inequality

$$\sum_{n=1}^{\tilde{N}} (\beta_{\tilde{N}} - \beta_n) < \tfrac{P_{\text{tot}}}{P_{\text{noise}}} \quad (28)$$

holds. From (23) and (25) the allocated power for the second channel is determined as

$$P_n^C = P_{\text{noise}} \tfrac{g^2(d_n^C)}{\alpha_n^C} \tfrac{K}{K-1} \cdot \max\left(0, \tfrac{\lambda}{\beta_n} - 1\right), \forall n \in \mathbb{F}_N. \quad (29)$$

This allocation has the following interpretation. The sensor node S_n with the lowest β_n gets the largest part of the total power because its communication channels are possibly the best due to the low distances. Therefore the observation of the target object is less interfered by noise and consequently results in better data communication. Sensor nodes with higher distances get smaller parts of the total power and some of them do not get any power at all. The last ones participate neither in the data communication nor in the classification of the target object. Their information reliability is too poor to be considered for data fusion. More and more sensor nodes will become active by increasing the total power. Then the overall classification probability increases because more correct information is provided by the observations.

Note that we have used the approximation (22) in order to simplify the maximization problem and to find analytical solutions for all equations. Without any approximation the maximization problem yields the *Lambert's trinomial equation*, which still does not have any analytical solutions. Although the above approximation is only valid for low transmission powers, we use the same solution for high transmission powers, too. If instead another approximation is used, the results are indeed different, but the behavior of solutions remains generally valid. However, this study is not the subject of this work.

3.4. Computational effort

In order to calculate the transmission powers (25) and (29) the computation of β_n, λ, and $\tilde{N}$ is necessary. The parameters K, N, P_{tot}, P_{noise}, r_k, α_n^R, and α_n^C are fixed system parameters which are known to the computation unit. The distances d_n^R and d_n^C depend on the position of the target object and are therefore unknown. They can be estimated for example by a tracking algorithm. If these values are also determined, then the equations (25) to (29) can be calculated with little effort, because of simple mathematical operations such as summation and multiplication. The only difficulty is the evaluation of the path loss function g, which can include complex mathematical operations. Its complexity depends on the given multipath channel.

However, the computation effort of the equations (25) to (29) is less complex than the evaluation of the classification algorithm such as (6). If one can find simpler algorithms than (6) (see, for example [12]), then the assessment of the calculation effort becomes important and should be considered in detail.

4. Numerical results and conclusions

In this section we present some numerical results obtained by applying the proposed optimization method from Section 3. We simulate target objects with equal prior probabilities $\pi_k = \frac{1}{K}$ $\forall k \in \mathbb{F}_K$ in sensor networks with different settings as described in Section 2. In all results, we consider three different kinds of target objects with reflection coefficients chosen as $|r_1| = 0$, $|r_2| = \frac{1}{2}$, and $|r_3| = 1$. Furthermore, the path loss function is modeled as line-of-sight propagation. The ratio SNR $= 10dB \log\left(\frac{P_{tot}}{P_{noise}}\right)$, instead of *received* SNRs, is depicted on the abscissa of all figures.

The verification of the proposed power allocation between both communication links of a single sensor node is shown in Figure 4. The overall error probability of the classification increases for higher SNR values for the case where the allocated power of one link is reduced by 10% and at the same time the power of the other link is stepped up by this 10%. When we reallocate a power amount of 10% − 30% to both links in an inverse manner, then the classification probability remains almost valid. This result shows that the proposed method allocates the given total power nearly optimal to both communication links, especially for higher SNR values.

In Figure 5 another verification of the proposed power allocation is shown, where a network of two sensor nodes is considered. The overall error probability of the classification decreases if we decrease the allocated power of the sensor node, which has the smallest part of the total power, by 10% and allocate this amount of power to the other sensor node. This result shows that the proposed method assigns the given total power suboptimal to the sensor nodes. The curves disperse, because of the approximation (22) which has been used for the equation (23).

As shown in Figure 6 the proposed method yields a better classification probability in comparison to a uniform power allocation where a network of ten sensor nodes is considered.

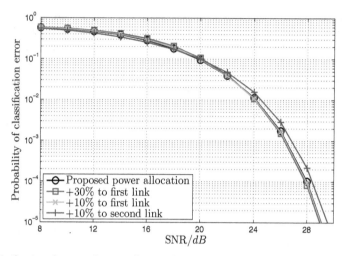

Figure 4. Verification of proposed power allocation between the two communication links of a single sensor node network.

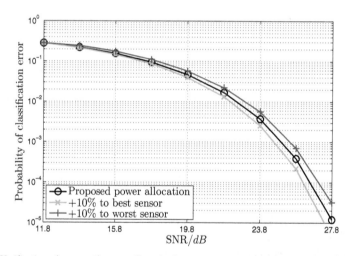

Figure 5. Verification of proposed power allocation between two sensor nodes.

In particular, it is shown that the same overall classification probability can be achieved with much lower transmission power, especially for low SNR values, by using an efficient power allocation method. Furthermore, the symbol-error probability of the sensor node with the highest part of the total power is also shown. The classification accuracy is better than the best symbol-error probability for higher SNR values, which affirms the gain of data fusion and illustrates the feasibility of object classification in this kind of distributed sensor networks.

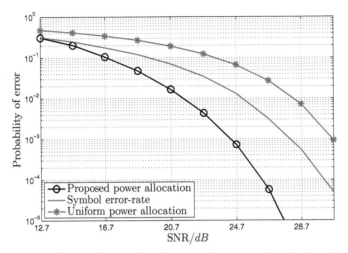

Figure 6. Comparison of proposed power allocation to a uniform power allocation in a network of ten sensor nodes.

Author details

Gholamreza Alirezaei, Rudolf Mathar and Daniel Bielefeld
Institute for Theoretical Information Technology, RWTH Aachen University, D-52056 Aachen, Germany

5. References

[1] Alirezaei, G. [2012]. Channel capacity related power allocation for ultra-wide bandwidth sensor networks with application in object detection, *IEEE ICUWB 2012 - International Conference on Ultra-Wideband*, Syracuse, NY, USA.

[2] Bielefeld, D., Fabeck, G., Zivkovic, M. & Mathar, R. [2010]. Optimization of cooperative spectrum sensing and implementation on software defined radios, *Proc. Int. Workshop Cogn. Radio Advanced Spectr. Management CogART*, pp. 1–5.

[3] Cover, T. M. & Thomas, J. A. [2006]. *Elements of Information Theory*, 2nd edn, John Wiley & Sons, Inc.

[4] Debes, C., Riedler, J., Zoubir, A. M. & Amin, M. G. [2010]. Adaptive target detection with application to through-the-wall radar imaging, *IEEE Trans. Signal Process.* Vol. 58(No. 11): 5572–5583.

[5] Duda, R. O., Hart, P. E. & Stork, D. G. [2000]. *Pattern Classification*, 2nd edn, John Wiley & Sons, Inc.

[6] Gezici, S., Tian, Z., Giannakis, G. B., Kobayashi, H., Molisch, A. F., Poor, H. V. & Sahinoglu, Z. [2005]. Localization via ultra-wideband radios: A look at positioning aspects for future sensor networks, *IEEE Signal Process. Mag.* Vol. 22: 70–84.

[7] Hume, A. L. & Baker, C. J. [2001]. Netted radar sensing, *Proc. IEEE Int. Radar Conf.*, pp. 23–26.

[8] Lapidoth, A. [2009]. *A Foundation in Digital Communication*, Cambridge University Press.

[9] Pescosolido, L., Barbarossa, S. & Scutari, G. [2008].   Radar sensor networks with distributed detection capabilities, *Proc. IEEE Int. Radar Conf.*, pp. 1–6.

[10] Srinivasan, R. [1986]. Distributed radar detection theory, *IEE Proceedings-F* Vol. 133(No. 1): 55–60.

[11] Surender, S. C. & Narayanan, R. M. [2011]. Uwb noise-ofdm netted radar: Physical layer design and analysis, *IEEE Trans. Aerosp. Electron. Syst.* Vol. 47(No. 2): 1380–1400.

[12] Varshney, P. K. [1997]. *Distributed Detection and Data Fusion*, 1st edn, Springer-Verlag

[13] Win, M. Z. & Scholtz, R. A. [2000].   Ultra-wide bandwidth time-hopping spread-spectrum impulse radio for wireless multiple-access communications, *IEEE Trans. on Communications* Vol. 48(No. 4): 679–691.

[14] Yang, Y., Blum, R. S. & Sadler, B. M. [2010]. Distributed energy-efficient scheduling for radar signal detection in sensor networks, *Proc. IEEE Int. Radar Conf.*, pp. 1094–1099.

Chapter 8

Antennas and Propagation for On-, Off- and In-Body Communications

Markus Grimm and Dirk Manteuffel

Additional information is available at the end of the chapter

1. Introduction

The ultra-wideband technology seems very attractive to be transferred to the challenging field of body centric communications. This technology involves the potential to establish robust communication links or high resolution localization systems. All these applications require a characterization of the propagation channel and the influence of the corresponding user to the system performance. Due to the inevitable interaction between the antenna and the related propagation channel a separation of both characteristics via traditional antenna theory methods is hardly applicable. The scope of this study is to establish a so called antenna de-embedding i.e. to separate the antenna form the underlying channel.

Traditional antenna parameters (e.g. directivity, gain, effective area) are based on free space propagation conditions. Underlying is the well known model of an isotropic radiator which enables the separation of channel, transmitter and receiver. It will be shown that this theory can be adapted to deduce approximations of equivalent antenna parameters for body centric communications. Key factor of this approach is the development of equivalent far field models of the corresponding in- and on-body scenarios. For off-body scenarios the propagation direction points away from the human body. The matching and the radiation pattern of the respective antennas may change due to the interaction with the human body but in general no modifications of the far field model are necessary. Therefore, the traditional theory is applicable with just minor restrictions and will not be discussed in further detail here.

The study is structured in two sections. The first part focuses on an in-body link i.e. the main propagation path of the electromagnetic wave leads through the tissue of the human body. Typical applications for this scenario are medical implants like wireless endoscopy or the RF breast cancer detection systems. The second part characterizes an on-body link. This means that the propagation path is defined along the body surface and the antenna is located in close proximity of the human body. The universality of this theory will be shown for the characterization of an UWB teardrop antenna.

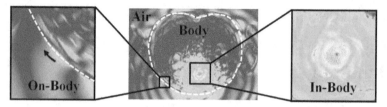

Figure 1. Distribution of the electric field of an implant located within the human abdomen; Left: On-body scenario showing surface waves guided by the body curvature; Right: In-body scenario characterized by circular shaped attenuation within the body.

2. Antenna de-embedding for in-body applications

From an electromagnetic point of view the human body consists of a large number of lossy dielectric materials of various combinations and spacial arrangements. In case of an in-body scenario the antenna is integrated into this complex dielectric structure and along an arbitrary propagation path various electromagnetic propagation effects occur. However, the most dominant effect is the attenuation of the propagating wave due to the lossy character of human tissues, see Figure 1. Furthermore, it has been shown that the average attenuation through the inhomogeneous tissue structure can be described for some scenarios by an equivalent homogeneous medium with appropriate properties [1]. In this case the analogy between the in-body and free space scenario enables an evident description of the related antenna far field. For this purpose the basic assumption of an isotropic point source is generalized for a lossy dielectric medium.

2.1. Model of an infinite homogeneous lossy dielectric media

The description of this model is based on the solution of the Helmholtz equation for a spherical wave which propagates in a homogeneous dissipative dielectric media. Due to this approach the absolute value of the electric field decreases proportionally to the reciprocal distance and shows an additional exponential attenuation with progressing distance compared to a lossless media. Therefore, the absolute value of the electric field E can be approximated by

$$E(d) \propto \frac{U_n}{d} e^{-\alpha d}, \tag{1}$$

where d, α and U_n denote the distance to the antenna, the attenuation constant of the media and a normalization factor related to the equivalent source of the spherical wave. For radiating elements, other than point sources, the accuracy of this model depends on the spacial current distribution of the source and is therefore a function of the distance. Due to the analogy of this dependency to the free space scenario the standard far field criteria seems applicable with the usual phase restrictions [2]. Related to the design of antennas which operates within the UWB frequency range [3] the radiating component is electrically large at the upper band edge frequency. In this case the appropriate far field criterium leads to following formulation:

$$d_{ff} = \frac{2D_{max}^2}{\lambda_i}, \tag{2}$$

where d_{ff} denotes the approximated far field distance, λ_i the wave length of a TEM wave in the dielectric media at the band edge frequency and D_{max} the largest diameter of the antenna. Note that D_{max} may be increased by passively excited components in the near field of the antenna which may contribute to the radiation, like the PCB of an implant or its encapsulation. Based on this far field model it is possible to deduce the equivalent gain of an in-body antenna. Doing this it seems logically consistent to refer the normalization constant U_n to an isotropic radiator. In this case the general definition of the gain G [2] is altered to

$$G = \frac{S\,(d \geq d_{ff})}{S_{\mathrm{iso,lossy}}\,(d \geq d_{ff})}, \tag{3}$$

where S denotes the power density of the antenna and $S_{\mathrm{iso,lossy}}$ denotes the power density of an isotropic source in a lossy medium at the same distance. Please note, due to the fact that the propagation medium itself contains losses and the directivity is by definition a lossless quantity, the definition of the directivity is not appropriate in this case. In order to achieve a constant normalization ratio versus the distance, the losses expressed by the exponential term of equation 1 have to be taken into account. Therefore, the absolute value of the lossy isotropic power density is given by

$$S_{\mathrm{iso,lossy}} = \frac{P_{\mathrm{rad}}}{4\pi d^2}e^{-2\alpha d}. \tag{4}$$

In equation 4, P_{rad} denotes the radiated power of the antenna. For an antenna whose losses are restricted to the surrounding medium P_{rad} is equal to the power on the antenna P_{ant}. Due to the dissipative nature of the tissue the antenna efficiency η decreases exponentially with progressing far field distance and can be calculated by

$$\eta\,(d_{ff}) = \frac{1}{P_{\mathrm{ant}}} \oiint_A \mathbf{S} \cdot d\mathbf{A}, \tag{5}$$

where A is the enclosure of the antenna at the distance d_{ff}. In order to calculate the path loss between two in-body antennas, the receive properties of such antennas have to be characterized as well. Due to the fact that the equivalent tissue medium is source free, linear and isotropic the definition of the effective antenna area A_{eff} [2] is also applicable for the in-body scenario. With respect to the definition of the gain in equation 3, the effective antenna area yields

$$A_{\mathrm{eff}} = G\frac{\lambda_i^2}{4\pi}. \tag{6}$$

Note, that the theory given above is based on an intrinsic far field model. Aim of this theory is an approach to give an intuitive formulation for the antenna design and handy path loss estimations. It raises no claim to give a closed analytical solution of the given problem. Despite this fact the model enables even estimations of theoretic problematic scenarios, such as a totally immersed antenna which is not insulated from the surrounding tissue media. As shown in [4], a theoretical formulation of this problem would lead to an inexpressible formulation. Nevertheless, to find a description between source and far field it is suggested to use an antenna which is electrically insulated from the surrounding media. Moreover the applicability of the model depends on the specific in-body scenario. The following chapter addresses the quality of the proposed model.

2.1.1. Validation of the in-body model on the example of an UWB teardrop antenna

Using the example of [1] the validity of the proposed in-body far field approach has been discussed by the evaluation of UWB localization of deep brain implants. As result it has been shown that the attenuation of an electromagnetic pulse within the inner human brain structure can be modeled by a homogeneous tissue with dielectric properties equivalent to grey matter. Despite the complex structure of the focused brain region time domain analyses indicate a radial wave conservation within the inner brain structure. These investigations indicate an average propagation velocity in arbitrary directions of the head which emphasize to the applicability of the suggested model.

The following analysis summarizes the results of an in-body scenario within the human trunk. In this case the representative media parameters are set to homogeneous muscle tissue, calculated by [5], and the validity of the approach is shown on the example of an UWB teardrop antenna. The antenna has been designed for the ultra wideband frequency range from 3.1 GHz to 10.6 GHz which is specified by [3]. Within this frequency range the return loss of the immersed teardrop antenna has been optimized to be lower than 10 dB. The capsulation of the antenna has been designed for the center frequency of 6.85 GHz by a lossless dielectric cylindrical insulation. The related permittivity of the capsulation has been set to $\varepsilon_r = 49.9$ to achieve an impedance matching between the insulation and the tissue. The validation of the model has been performed by numerical calculations with the FDTD simulation software EMPIRE XCcelTM [6].

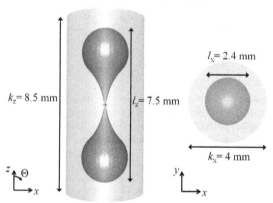

Figure 2. Insulated teardrop antenna designed for the focused UWB frequency band.

According to equation 2, the minimum far field distance for the upper band edge frequency ($f_u = 10.6$ GHz) is $d_{ff} = 3$ cm. Here, due to the insulation, the maximum antenna size D_{max} has been set to the maximal capsule dimension. The quality of the proposed far field model has been verified by the absolute difference ΔS between the calculated power density values S_{FDTD} and the far field model S_{ff}. For the angle $\Theta = 90°$ the difference ΔS has been calculated at the far field distance d_{ff} by the following equation:

$$\Delta S = \frac{|S_{FDTD} - S_{ff}|}{S_{FDTD}}. \tag{7}$$

Table 1 shows the calculated values for the lower, upper and center frequency of the UWB frequency range investigated. The error rises with increasing frequency and is still within the typical range of the equivalent free space far field considerations.

	$f = 3.1\,\text{GHz}$	$f = 6.85\,\text{GHz}$	$f = 10.6\,\text{GHz}$
$\Delta S[\%]$	1.1	3.7	6.9
$G(\Theta = 90°)$	1.9	2.3	3.3
$\eta_{\text{eff}}(d_{\text{ff}})$	$3.8 \cdot 10^{-2}$	$8.4 \cdot 10^{-5}$	$1.5 \cdot 10^{-8}$
$A_{\text{eff}}[m^2]$	$2.75 \cdot 10^{-5}$	$7.22 \cdot 10^{-6}$	$4.74 \cdot 10^{-6}$

Table 1. Calculated equivalent in-body antenna parameters of the UWB teardrop antenna at $d_{\text{ff}} = 3$ cm.

Table 1 also shows the gain calculated by the equations 3 and 4. As stated, additional derivations have shown that the value is nearly constant for distances greater than d_{ff}. Nevertheless, the value is not constant within the observed frequency range and rises with increasing frequency due to the variation of the electrical length of the antenna. To characterize the losses within the near field of the antenna the efficiency has been calculated by equation 5 at the minimum far field distance d_{ff}, see Table 1. As it might be expected, the efficiency decreases drastically with increased frequency due to the higher power consumption of the tissue medium. For greater distances the efficiency decreases exponentially with increasing distance. The effective antenna area, as shown in Table 1, is calculated using equation 6. The derived values of the gain and the effective antenna area enable the approximation of the path gain for arbitrary distances greater than the minimum far field distance. This ratio of transmitted to received power (path gain) is shown for the related frequencies in Figure 3.

As shown above the model of an homogeneous dissipative medium enables the definition of antenna parameters to establish an antenna de-embedding for in-body scenarios. Therefore, the consideration of the whole system is not necessary to achieve a path loss estimation. Moreover this assumption enables a basis for a purposeful antenna development.

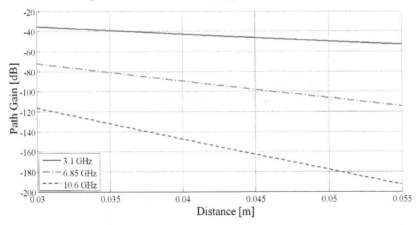

Figure 3. Calculated path gain of the UWB teardrop antenna.

2.1.2. Limitations of the in-body model

As shown in [1], boundaries between high water content tissues (muscle tissue) and air filled regions (paranasal sinus, frontal sinus) or low water content tissues (fat tissue, bone tissue) lead to various electromagnetic interactions which reduce the accuracy of the approach presented above. The average attenuation of the electromagnetic field may still be characterized by the proposed model but especially with respect to time domain analysis the multi path behavior of these structures leads to insufficient results. Moreover, the specific anatomical location of the RF application may have a strong influence on the corresponding radiation characteristic of the antenna which cannot be adequately described by a homogeneous tissue model. Despite this fact the generality of the approach enables an extension of the theory by including anatomic realistic human models in the antenna design process to derive the resulting radiation characteristic. In this case the corresponding far field distance has to be enlarged related to the anatomical structure but the average description of the far field may still be given by the proposed homogeneous model.

3. Antenna de-embedding for on-body applications

Encouraged by [7], the didactic first step to deduce a de-embedding for a wide class of on-body applications is the deduction of an adequate far field propagation model of a radiating source near a planar tissue like surface.

3.1. Model of an infinite homogeneous lossy plane surface

The propagation mechanisms of an electric doublet near a dissipative infinite homogenous plane has been investigated in [8]. These results provide a description of the electric field components as a function of the given geometry. Therefore the absolute value of the electric field E can be calculated depending on the dipole current distribution i, its effective height h, the frequency f, the complex dielectric parameters of the media ε and a tangential to the surface defined distance d, see Figure 4.

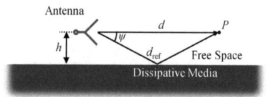

Figure 4. Principle geometry of the on-body scenario.

Included in this theory is the separation of the total electric field in its space and surface wave. As defined by Figure 4 the space wave consists of the wave components which propagates along the direct path d and the ground reflected path d_{ref}. A surface wave component which is excited by the dissipative nature of the tissue is guided by the air-media boundary. It only contributes to the related electromagnetic field if the antenna is located in close proximity to the surface. This means that the effective antenna high h is typically lower than a few wave lengths. Otherwise, this wave component is negligibly small compared to the space wave. The electric field of any observation point P along a parallel surface path at the effective antenna high h can be described by Norton's formulation N as follows:

$$E \propto \frac{U_n}{d} N \left(i_{v,h}, d, h, f, \varepsilon\right),$$ (8)

where the normalization value U_n depends on the antenna excitation. The term N is a function of the current distribution of the source $i_{v,h}$, the distance d, the frequency f and the complex tissue permittivity ε. The absolute value of the electric field is given by the superposition of the space wave E_{sp} and surface wave E_{sw} as follows:

$$E(d) = E_{sp} + E_{sw},$$ (9)

where E is given by

$$E(d) \propto \frac{U_n}{d} \left[\underbrace{e^{j\frac{2\pi d}{\lambda}}}_{\text{direct wave}} + \underbrace{R_{v,h}\cos^3(\psi) e^{j\frac{2\pi d_{ref}}{\lambda}}}_{\text{ground-reflected wave}} + \underbrace{\left(1 - R_{v,h}\right) F_{v,h}\cos^2(\psi) e^{j\frac{2\pi d_{ref}}{\lambda}}}_{\text{surface wave}} \right].$$ (10)

In equation 10, λ denotes the free space wave length, $R_{v,h}$ the plane-wave reflection coefficient of the ground, $F_{v,h}$ the surface-wave attenuation function and d_{ref} the reflected path at angle ψ. Both, the reflection and the attenuation functions depend on the current distribution of the antenna and are given by [9] for vertical and horizontal antenna orientations. The related far field solution is valid for sufficient great distances d and depends in first place on the mathematical description of the surface-wave attenuation function. If an adequate description can be assumed the minimal valid distance, referred by Norton in [8], has to be greater than one free space wave length. An additional limitation is the assumption of locally plane waves for the derivation of the reflection coefficient of the ground. Under the assumption that the surface acts as a perfect mirror the electrical antenna size may be enlarged by the mirror image. Analog to equation 2 the far field distance d_{ff} is defined by:

$$d_{ff} = \frac{2D^2_{max,eff}}{\lambda},$$ (11)

where the modified maximum antenna dimension is denoted by $D_{max,eff}$. For an adequate derivation of the far field distance the quantity $D_{max,eff}$ has to be chosen appropriate under the aspect that the enlargement of the antenna by the ground acts primarily in normal direction to the surface. For distances greater than the minimum far field distance, given by equation 11, the formulation of equation 10 enables the definition of the directivity for on-body scenarios. Analog to equation 3 the directivity D is defined by the normalization of the power density to the far field model:

$$D = \frac{|S|}{|S_{Norton}|}.$$ (12)

The related electromagnetic field is in general a superposition of TE-, TM- and TEM-wave components and therefore a function of the parameters given in equation 8. Despite this fact, the TEM-wave component contributes the most significant part to the power flux density. Even if a strong surface wave is excited, the resulting TM-wave component is comparatively low. This fact enables a simple approximation of the power density by the given expression of electric field:

$$D \approx \frac{|E|^2}{|E_{\text{Norton}}|^2} = \frac{|E|^2}{\left|\frac{U_{\text{n}}}{r}N\right|^2}. \tag{13}$$

Considering the free space and in-body definition of the directivity it is consistent to define the normalization value U_{n} related to a isotropic source which is modified by the function N:

$$D = \frac{|E|^2}{\frac{\eta_0 P_{\text{rad}}}{4\pi d^2}|N|^2}. \tag{14}$$

Due to the fact that the effective area of an antenna is defined for the condition in which the antenna receives a locally plane electromagnetic wave [10] the received power of an antenna inserted in the far field of the transmitting on-body antenna cannot be calculated directly by equation 6. Nevertheless, as shown by a preceding study, see [11], even for body worn antennas a derivation of the received power as a function of the incident power density is possible. The results imply a nearly constant ratio between the received power P_{out} and the incident power density S if the corresponding antennas are farther than the minimum far field distance apart. The constant ratio A'_{eff} can be calculated by

$$A'_{\text{eff}} = \frac{|P_{\text{out}}|}{|S|}. \tag{15}$$

Note, that the ratio defined by equation 15 is a function of the parameters given above and is therefore limited in its applicability to the specific setup. Apart from these aspects the equation enables the opportunity to calculate the received power as a function of arbitrary far field distances.

3.1.1. Validation of the on-body model on the example of an UWB teardrop antenna

The verification of the suggested model has been done for a vertically and horizontally orientated UWB teardrop antenna for the frequency range defined by [3]. The effective height of the antenna has been set to a quarter free space wave length at $f = 3.1$ GHz to avoid intersections between the antenna and the tissue medium. In contrast to the in-body design an encapsulation of the radiating antenna elements is not necessary to obtain an adequate matching of the antenna. The key parameters of the antenna geometry are set to $l_z = 39$ mm and $l_x = 18.5$ mm, see Figure 2, to achieve a return loss below 10 dB for the desired frequency range. The tissue properties have been set to muscle tissue, given by [5], and analog to [7] the geometry has been numerically calculated by the FDTD method presented in [6]. The minimum far field distance d_{ff} has been calculated along equation 11 and is shown in Table 2 for a vertical and horizontal orientated teardrop antenna.

Table 2 also shows the quantities D and A'_{eff} which are calculated for the minimum far field distance $d_{\text{ff,v}} = 0.54$ m and $d_{\text{ff,h}} = 0.43$ m. Analog to the in-body scenario the quality of the suggested model has been verified by the absolute difference ΔS defined by equation 7. The absolute difference shows a sufficient applicability of the suggested on-body model. In contradiction to the in-body scenario it describes a non monotone behavior for the target frequencies due to the inverse frequency dependence of the reflection coefficient of the ground and the surface-wave attenuation function defined by equation 10. As shown in Table 2, the

	$i_{v,h}$	$f = 3.1$ GHz	$f = 6.85$ GHz	$f = 10.6$ GHz
$d_{ff}[m]$	v	0.16	0.35	0.54
	h	0.13	0.28	0.43
$\Delta S[\%]$	v	10.38	4.83	5.56
	h	6.64	3.15	3.48
$D[lin]$	v	1.37	1.07	0.99
	h	1.34	1.23	0.47
$A'_{eff}[m^2]$	v	$8.7 \cdot 10^{-4}$	$1.7 \cdot 10^{-4}$	$4.4 \cdot 10^{-5}$
	h	$10.3 \cdot 10^{-4}$	$1.8 \cdot 10^{-4}$	$3.4 \cdot 10^{-5}$

Table 2. Calculated on-body antenna parameters of the UWB teardrop antenna. The quantities ΔS, G and A'_{eff} are calculated for the maximum far field distance of the considered frequency range with $d_{ff,v} = 0.54$ m and $d_{ff,h} = 0.43$ m.

directivity of the evaluated antennas decreases with increased frequencies. This behavior implies a reduced excitation of surface waves in the upper UWB frequency range due to the greater effective height of the antenna. Note, the derived gain is not directly comparable to the free space or in-body values. Due to the dependency of the channel model to the antenna polarization even the vertical and horizontal quantities are incomparable to each other. Despite this restriction the formulation of the directivity defines a quantity which enables an adequate discussion of various on-body antenna types and to enhance the corresponding design process.

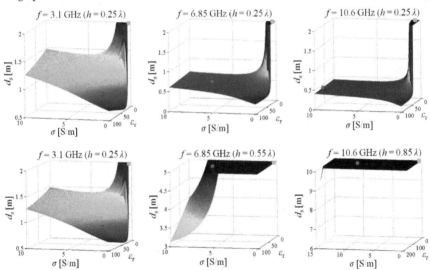

Figure 5. Distance to the antenna where the surface wave exceeds the space wave as function of the frequency and tissue parameters of a vertical polarization; First row: The effective antenna height is set to a quarter wave length of the respective frequency; Second row: Effective antenna height is set to a quarter wave length at $f = 3.1$ GHz; Green square marker: Fat tissue; Red round marker: Muscle tissue.

Figure 5 shows an analysis of the effective antenna height in relation to the surface wave excitation. It discusses the distance d_s where the surface wave exceeds the space wave as function of the tissue parameters. The first row shows the intersect point for effective antenna heights of a quarter wave length of the respective frequency. The second row shows the intersection point for a fixed effective antenna height which has been set to a quarter wave length of the lower UWB edge frequency ($f = 3.1\,\text{GHz}$). The comparison of the results shows that with increasing frequency even at distances greater than 10 m the surface wave component is lower than the corresponding space wave. This fact implies a relatively weak far field and causes a reduction of the directivity at high frequencies. With respect to the design of future UWB on-body antennas this circumstance has to be considered. Additional investigations have also shown that vertical polarized antennas excite a much more dominant surface wave than equivalent horizontal orientated antenna configurations, see [7]. These results are in accordance to the theory given by [8] and should be considered to optimize UWB applications for given propagation scenarios.

3.1.2. Limitations of the on-body model

The validation of the suggested model with respect to the anatomical structure of the human body, with its numerous tissue types and curved surfaces, is done by a path gain calculation of a complete human body voxel model. Basis for this derivation is the numerical IT'IS virtual family Duke model [12]. The selected scenario consists of a transmitting antenna TX which is located at the right shoulder front. The corresponding receiver RX is shifted along the front side of the trunk above the right leg to the right foot. Figure 6 shows the path gain along the chosen path d. In addition, the path gain of the suggested on-body model has been calculated for homogeneous muscle and fat tissues. As seen in Figure 6, the calculated path gain of the voxel model lies between the graphs of the theoretical models.

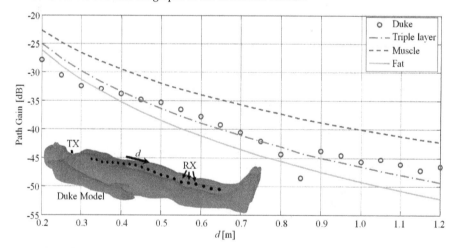

Figure 6. Path gain of the Duke voxel model in comparison to the homogeneous model of fat and muscle; Additional included is a numerical validation graph of a layered model analog to [7].

Analog to [7] a numerical model of a layered plane surface has been implemented to realize a more realistic representation of the human tissue structure. The suggested simulation model consists of 2 mm skin and 5 mm fat tissues which are positioned on an infinite muscle tissue.

As shown in Figure 6 the modified model is capable to give an adequate description of the detailed voxel simulation. In future approaches this fact can be used to enhance the presented on-body far field model by a modification of the surface-wave attenuation function $F_{v,h}$, see equation 10, by an adaption of the numeric distance as function of the surface impedance [13].

An additional effect, which is not included in the model presented, is the propagation in shadowed regions of the human body. While locally small shadowed regions are still covered by the model, large shadowed regions seem not to be described by this theory [7]. Nevertheless, these aspects have also been discussed in the last century with respects to the wave propagation above a spherical earth [9] and may also be transferred to the field of body centric communications.

4. Conclusion

The study has shown that an antenna de-embedding for in- and on-body applications can be realized by the derivation of corresponding far field models with reasonable accuracy for practical applications. Related to this theory, quantities as the directivity and the effective antenna area have been defined to derive good approximations of propagation models. Especially for on-body applications the suggested model gives a detailed insight by the separation of the electromagnetic field in its space and surface wave components.

Moreover, the presented theory enables the calculation of average path gain models of arbitrary antennas which can be reduced to a source of vertical and horizontal orientated current distributions. By this, the numerical calculation space can be reduced to the minimum far field distance of the corresponding model. Additionally, an insulated UWB tear drop antenna design has been presented for in-body communication applications to give an adequate validation example. For the on-body scenario the UWB teardrop antenna has been modified and also been discussed.

In future studies the in-body approach has to be modified to a multipath channel model to include additional propagation effects like surface waves. In addition, the on-body model has to be extended to give a wider applicability with respects to the complex structure of the human body. Moreover, the effective antenna area for on-body applications has to be described as function of the given model. With these improvements a structured combination of the on-, off- and in-body scenarios seems realizable to develop an optimized antenna theory for body centric communications.

Author details

Markus Grimm and Dirk Manteuffel
University of Kiel, Germany

5. References

[1] Grimm, M. & Manteuffel, D. (2011). Characterization of electromagnetic propagation effects in the human head and its application to Deep Brain Implants, *IEEE-APS Topical Conference on Antennas and Propagation in Wireless Communications (APWC)*, pp. 674-677, ISBN 978-1-4577-0046-0

[2] Balanis, C. A. (2005). *Antenna Theory*, John Wiley & Sons., ISBN 978-0-4716-6782-X

[3] Federal Communication Commission (2002). Revision of Part 15 of the communication's rules regarding ultra wideband transmission systems. First report and order, ET Docket 98-153, FCC 02-48

[4] Tai, C. T. & Collin, R. E. (2000). Radiation of a Hertzian Dipole Immersed in a Dissipative Medium, *IEEE Transactions on antennas and propagations*, Vol. 48, No. 10, pp. 1501-1506

[5] Gabriel, C. & Gabriel, S. et al. (2012). An Internet Resource for the Calculation of the Dielectric Properties of Body Tissues, Italian National Research Council website, Available from: http://niremf.ifac.cnr.it/docs/DIELECTRIC/home.html

[6] IMST (2012). EMPIRE XCcelTM, URL: http://www.empire.de

[7] Grimm, M. & Manteuffel, D. (2010). Electromagnetic Wave Propagation on Human Trunk Models excited by Half-Wavelength Dipoles, *Antennas and Propagation Conference (LAPC)*, pp. 493-496, ISBN 978-1-4244-7304-5

[8] Norton, K. A. (1937). The Propagation of Radio Waves over the Surface of the Earth and in the Upper Atmosphere, *Proceedings of the Institute of Radio Engineers*, , Vol. 24, No. 10, pp. 1367-1387

[9] Norton, K. A. (1941). The Calculation of Ground-Wave Field Intensity over a Finitely Conducting Spherical Earth, *Proceedings of the Institute of Radio Engineers*, Vol. 29, No. 12, pp. 623-639

[10] Friis, H. (1946). A Note on a Simple Transmission Formula, *Proceedings of the IRE*, Vol. 34, pp. 254-256

[11] Grimm, M. & Manteuffel, D. (2012). Evaluation of the Norton Equations for the Development of Body-Centric Propagation Models, *European Conference on Antennas and Propagations (EUCAP)*

[12] IT'IS Foundation (2012). Whole-Body Human Models, Enhanced Anatomical Models, URL: http://www.itis.ethz.ch

[13] Wait, J.R. (1996). *Electromagnetic Waves in Stratified Media*, IEEE Press, ISBN 0-7803-1124-8

Cooperative Localization and Object Recognition in Autonomous UWB Sensor Networks

Rudolf Zetik, Honghui Yan, Elke Malz, Snezhana Jovanoska, Guowei Shen, Reiner S. Thomä, Rahmi Salman, Thorsten Schultze, Robert Tobera, Hans-Ingolf Willms, Lars Reichardt, Malgorzata Janson, Thomas Zwick, Werner Wiesbeck, Tobias Deißler and Jörn Thielecke

Additional information is available at the end of the chapter

1. Introduction

Ultra-wideband (UWB) radio sensor networks promise interesting perspectives for emitter and object position localization, object identification and imaging of environments in short range scenarios. Their fundamental advantage comes from the huge bandwidth which could be up to several GHz depending on the national regulation rules. Consequently, UWB technology allows unprecedented spatial resolution in the geo-localization of active UWB radio devices and high resolution in the detection, localization and tracking of passive objects.

With the lower frequencies (< 100 Hz) involved in the UWB spectrum, looking into or through non-metallic materials and objects becomes feasible. This is of major importance for applications like indoor navigation and surveillance, object recognition and imaging, through wall detection and tracking of persons, ground penetrating reconnaissance, wall structure analysis, etc. UWB sensors preserve their advantages -high accuracy and robust operation- even in complicated, multipath rich propagation environments. Compared to optical sensors, UWB radar sensors maintain their capability and performance in situation where optical sensors collapse. They can even produce useful results in non-LOS (non-Line of Sight) situations by taking advantage of multipath.

Despite the excellent range resolution capabilities of UWB radar sensors, detection and localization performance can be significantly improved by the cooperation between spatially distributed nodes of a sensor network. This allows robust localization even in the case of partly obscured links. Moreover, distributed sensor nodes can acquire comprehensive knowledge of the structure of an unknown environment and construct an electromagnetic image which is related to the relative sensor-to-sensor node coordinate system. Distributed observation allows the robust detection and localization of passive objects and the identification of certain features of objects such as shape, material composition, dynamic parameters, and time-variant behavior. This all makes UWB a promising basis for the autonomous navigation of mobile sensor nodes, such as maneuverable robots- in an unknown environment that may arise as a result of an emergency situation.

The objective of the CoLOR project (Cooperative Localization and Object Recognition in Autonomous UWB Sensor Networks) was to develop and demonstrate new principles for localization, navigation and object recognition in distributed sensor networks based on UWB radio technology. The application scenario of the CoLOR project is described by mobile and deployable sensor nodes cooperating in an unknown or even hostile indoor environment without any supporting infrastructure as it may occur in emergency situations such as fire disasters, earthquakes or terror attacks. In this case, UWB can be used to identify hazardous situations such as broken walls, locate buried-alive persons, roughly check the integrity of building constructions, detect and track victims, etc. In this scenario, it is assumed that optical cameras and other sensors cannot be used. Data fusion of optical image information and UWB radar was not in the scope of this project.

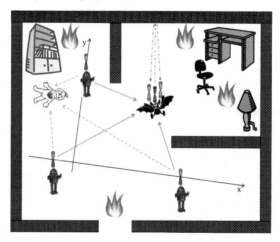

Figure 1. CoLOR scenario.

A possible scenario is shown in Fig. 1. The unknown environment is first explored by autonomous robots that deploy fixed nodes at certain positions. Those nodes being able of transmitting and receiving shall play the role as anchor nodes. They span a local coordinate system and should be placed at "strategic" positions, i.e. they span a large volume and ensure a complete illumination of the environment. Moving nodes will localize themselves relative to this anchor node reference. Moreover, when moving they collect information about the structure of the environment by receiving reflected waves. This way, they build an "electromagnetic image" of the environment and recognize basic building structures ("landmarks"). Step by step a map of the environment is built which can be used as another reference for navigation. This procedure is already well known from autonomous robot navigation as SLAM (Simultaneous Localization and Mapping). However, here we do not consider optical methods but UWB to recognize the feature vector. If there are solitary objects, the moving sensor may scrutinize shape and material composition by circling around. Other objects like humans may walk around and create time-variant reflections that identify their moving trajectory. Humans and animals may also reveal themselves by time variant features resulting from vital functions such as breathing.

The organization of this chapter is as follows. Section 2.1 describes the architecture of the sensor network and basic parameters of UWB sensor nodes that we used to achieve our objectives. Section 2.2 specifies the simulated test scenario that was applied for the development of the localization and imaging algorithms. Simulated data allowed us to develop algorithms in parallel to the demonstrator, which is presented in Section 2.3. The demonstrator was used to assess performance and to evaluate the developed data extraction algorithms in realistic scenarios. The algorithms were developed and evaluated using data obtained from the UWB wave propagation simulator described in Section 3. Within the CoLOR project algorithms were developed for: the cooperative localization of sensor nodes, see Section 4, the evaluation of sensor network topology, see Section 5, simultaneous localization and map building, see Section 6, object parameter estimation, see Section 7, imaging of environment, see Section 8 and the detection and localization of moving objects, see Section 9. Special attention was given to algorithms that promise real-time capability.

2. System architecture

2.1. Sensor network architecture

To accomplish the tasks described above, the UWB sensor nodes used can be heterogeneous in terms of their sensing capabilities and mobility. Most simple nodes may act just as illuminators of the environment. This requires only transmitting operation, but no sensing and processing capability. However, multiple transmit signal access has to be organized. This could be CDMA (Code Division Multiple Access) or TDMA (Time Division Multiple Access). TDMA requires some time frame synchronization. CDMA, on the other hand, would need multiple orthogonal codes which would complicate transmitter circuit design and increase self-interference because of non-perfect orthogonality. We preferred TDMA switching.

With its unprecedented temporal resolution, ToA (Time of Arrival) based localization methods are the natural choice for localization in the case of UWB. ToA, however, requires temporal synchronization between the nodes. This can be achieved by the RTToA (Round Trip Time of Arrival) approach which means that sensors involved must be able to retransmit received signals.

Deployable nodes, placed at verified positions may act as location reference or anchor nodes for the localization of roaming nodes. Other nodes are spread around as static or moving observers. Static observers are well suited to observe moving objects since they can most easily distinguish between desired information and reflections from the static environment (clutter) by exploiting time variance. On the other hand, moving observers (or illuminators, since propagation phenomena are reciprocal) can collect information about static objects and environments (e.g. structure of walls). By applying coherent data fusion methods, moving nodes will create an image of the static propagation environment. This is a full multi-static approach which requires a number of widely distributed cooperating sensor nodes having precise (relative) location information and synchronization at least between subsets of sensors (e.g. between receivers or transmitters or both).

Synchronization issues can be relaxed if we construct a more complex node containing, e.g. one transmitting antenna (Tx) and two receiving (Rx) antennas. Such a sensor already constitutes a small-baseline bi-static radar, which somehow resembles the sensing capability of a bat and allows the estimation of both object range and direction (by using time difference of arrival, TDoA). We will also refer to these nodes as "bat-type sensors". Their advantage is that mutual coherent processing is restricted to one platform. So, if several of those nodes are cooperating, they can directly exchange range and DoA (direction of arrival) information, which we call "non-coherent cooperation" since no exact phase synchronization between nodes is required. There is, of course, also a mixed approach which may consist of non-synchronized illuminators and locally coherent (differential) observers. So, with regard to the capabilities of sensor nodes involved and to their mutual cooperation, we distinguish between three basic structures of sensor networks:

- The multi-static approach, which assumes full coherent cooperation. The position of those nodes is estimated and tracked w.r.t. the anchor nodes.

- The "bat-type" approach consisting of self-contained sensor nodes which are able to detect and recognize characteristic features of the propagation environment on their own. This allows the building up of partial maps of the environment, the investigation of unknown objects in more detail, the identification of object features, etc.

- The mixed approach is characterized by a cooperation of all types of sensors. For example, the multi-static sensors will support the localization of the scout sensors, and these will deliver additional reference information that relates the local sensor coordinate system to the structural details of the environment.

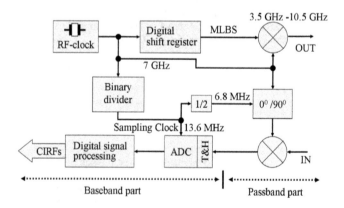

Figure 2. Basic architecture of UWB sensors node.

The basic architecture of a bat-type sensor is described in Fig. 2, see [47], [29], and see Chapter 14. The transmit signal is a periodic M-sequence spread spectrum signal which is generated by a high speed 12-stage digital shift register. It is clocked by a stable RF-clock generator - a dielectric resonance oscillator operating at 7 GHz which allows a usable frequency band up to 3.5 GHz (baseband). The sequence contains 4095 chips corresponding to 585 ns duration and about 175 m of usable distance. In the passband mode the signal is up-/down-converted by the same frequency which results in usable frequency band from 3.5 GHz to 10.5 GHz, which corresponds to the well-known FCC-specification. The receiver applies 512 time subsampling which relaxes ADC (analog-to-digital converter) bandwidth requirements and still allows real-time data recording with a measurement rate of 50 impulse responses per second.

Data measured in real-time by static anchor nodes allow the extraction of information about time-varying objects and people within the inspected scenario. Real-time data measured by moving nodes of the network allow the extraction of information about the geometric structure of the environment. In order to develop data extraction algorithms in parallel with the development of the demonstrator, we simulated a test scenario in terms of electromagnetic waves propagation. The simulated scenario is described in the following Section 2.2.

2.2. Simulated scenario

As we did not have everything ready for the demonstration scenario, e.g. the robot was not available, a scenario was simulated using the ray tracing tool 3, in order to develop and test the algorithms for map building, localization and imaging independently.

For the final demonstration, the scenario described in Section 2.3 was used.

The simulated scenario is shown in Fig. 3. It consists of a room of the size 9x8x4 m with a pillar in its center. There are six different objects (shown in Fig 3 on the right side) distributed in the room. Their positions are marked in the figure. Metal is chosen as material for the walls and for the six objects.

The bat-type sensor follows the track indicated by the dotted line. The sensor is equipped with one transmit and two receive antennas. The transmit antenna is located at the point (0,0,1) of the local Cartesian coordinate system of the robot (with x being the moving direction, y, and z the height) and the receive antennas at (0,0.5,1) and (0,-0.5,1). At 78 positions, indicated by the red circles on the track, the robot stops and takes measurements (the channel is simulated accordingly). Therefore, the transceiver array rotates in the azimuth plane (x-y-plane) with an angle resolution of $3°$, so that at each position 120 simulations are performed. A horn antenna with a 3 dB beam-width of around $60°$ at 9 GHz was used as the radiation pattern for the transmit and receive antennas. The channel was simulated from 4.5 to 13.4 GHz in steps of 6.25 MHz in order to describe the frequency selective behavior realistically.

The inclusion of high-frequency aspects in terms of the transmission channel is particularly important for the accuracy of the simulation results. Otherwise, unrealistic conclusions will result from the simulation. In order to achieve a realistic evaluation of the efficiency of sensing and imaging systems, the channel model must describe the multi-path propagation realistically. A description of the used channel simulator as well as some verification measurements are given in Section 3.

2.3. Demonstrator

To prove robustness and applicability, research results obtained within CoLOR were experimentally validated in an extensive measurement campaign. The main focus was placed on a robust transfer of the algorithms from laboratory conditions to a more realistic indoor propagation scenario. The goals of these extensive multi-disciplinary measurements were

- to verify UWB sensing algorithms in real indoor scenarios,

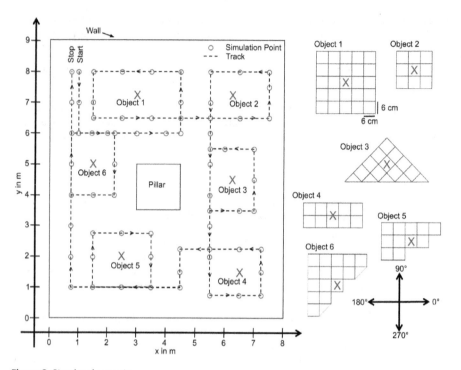

Figure 3. Simulated scenario.

- to perform and evaluate the simultaneous working algorithm for the cooperative approach,

- to determine to what extent previously achieved research results are applicable under these conditions.

With regard to a final experimental validation, the investigations in CoLOR were accompanied by experimental practice throughout the project, i.e. algorithms were always validated with regard to real world conditions.

The measurements made so far were simplified to limited sensor and/or object motions, to ideal movements by means of motorized linear arrays or by reducing the number of simultaneously performing cooperative algorithms. Therewith, only hardware complexity and mechanical effort could be reduced and meaningful validations of the investigated algorithms could be obtained. However, measurement scenarios and instrumental investments were extensively expanded within the scope of a demonstrator to fully meet the demands of a realistic indoor propagating scenario. Hence, an autonomous mobile security robot with professional motion units was fully equipped with UWB-devices, RF components, a power supply unit and a laptop for data acquisition and communication with the data fusion computer. This security robot serves as the previously mentioned mobile bat-type sensor.

Most algorithms of previous investigations could be adopted with manageable complexity to the mobile bat-type scenario. However, some challenges came up unexpectedly, e.g. erroneous robot motion due to uneven floor and slippage of robot tires, potentially more clutter in a realistic scenario, handling and transformation between a global coordinate system of the static environment and an additional local coordinate system of the dynamic robot.

To provide a realistic indoor scenario with corners as well as edges and dimensions like that of a larger office room (56 m²), the fire detection laboratory of the University Duisburg-Essen was modified and used as the location for the measurement campaign. The modification consists of partly installed portable metallic walls to give the room a more complicated shape. The ground plane of the designed indoor scenario is depicted in Fig. 4 .

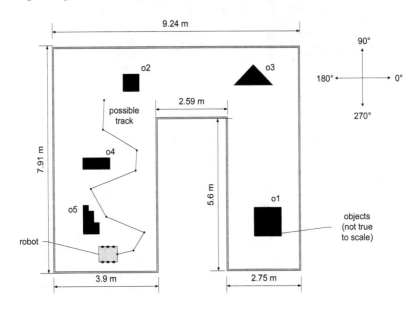

Figure 4. Ground plane of measurement location and robot with possible track.

As previously mentioned, the bat-type sensor is realized by a mobile security robot which is schematically represented in Fig. 5

As super-resolution techniques in short-range UWB sensing have to be performed, certain demands on the motion accuracy result are necessary. However, localization accuracy is predominantly achieved by advanced algorithms described later in chapter 7. Additionally, some assisting robot specifications have also been taken into account to improve accuracy. There are three actuators in the robot, two in the motion unit at the bottom, and one at the top which rotates the antenna array. The actuators are all hollow-shaft servo motors, which offer unique features unsurpassed by conventionally geared drives. Used in highly demanding industrial and medical servo systems, they provide outstanding precision motion

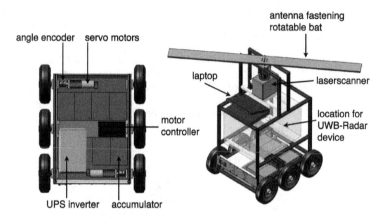

Figure 5. Schematic drawing of the motion unit (left) and of the robot (right).

control in sub-mm range and high torque capacity in a very compact package. The robot has 3 solid rubber tires on both sides which are connected by a chain-drive. Actually, both sides could be used autonomously with different acceleration, deceleration and speed, which results in curved tracks. However, for further accuracy different triggering is avoided so that the servo motors are driven synchronously. Evidently, the robot moves straight forward when the motors drive into the same direction and rotation is performed when the motors drive in the opposite direction. To maintain a more gliding rotation of the robot with reduced positioning errors, the circumference of the middle tire is minimally higher than those of the other ones. The dimensions of the robot as well as the tire position maintain a rotation center in the middle of the robot which also equals the middle of the bat antenna platform at the top. Hence, the movement of the robot was entirely restricted to translations and rotations, strictly avoiding curvature paths. Because of that, a track is split into several straight segments which are separated by a change of orientation. A resulting possible track is shown in Fig. 4. As mentioned previously, to further minimize erroneous robot motions, the bat is equipped with its own rotational unit. The orientation change can be performed by just rotating the bat, which is preferred compared to rotating the whole robot. This is more sensitive to errors due to an uneven or slippery floor. The robot is also equipped with a laser-based indoor navigation system. This highly accurate and well-proven localization system shall provide reference data for subsequent performance analysis. It has neither assisting nor guiding functionality in the localization process of CoLOR. The localization process is first and foremost handled by UWB-Radar technology.

Due to the different localization and imaging applications in this project, the requirements placed on the antenna characteristics differ. Fig. 6 gives an image and a photograph of the antenna array used. It consists of three different antenna types. A broadband monopole antenna (Tx_{2_1}), a dual-polarized broadside radiating antenna ($Tx_{2\_2}, Rx_{2\_1}, Rx_{2\_2}$) and an end-fire radiating antenna ($Tx_{1\_1/1\_2}Rx_{1\_1/1\_2}$). The requirements as well as the design of the antennas itself are described in detail in subsections 7 and 6 .

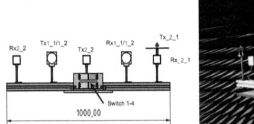 

Figure 6. Schematic diagram (left) and picture (right) of the antenna array used

An array of switches, as shown in Fig. 6, allows the change between the different receiver and transmitter configurations.

3. Hybrid deterministic-stochastic channel simulation

In the framework of this project, a realistic UWB multi-path propagation simulation tool was developed in order to test and compare different algorithms and antenna arrangements for indoor UWB sensing and imaging. Multi-path propagation implies that the transmitted signal does not only arrive over the direct propagation path at the receiver, but also over paths which are dependent on the propagation environment in a complex manner. The received signal is then a combination of a multiplicity of reflected, diffracted and scattered electromagnetic waves. Wave propagation models, in general, can be classified into deterministic and stochastic ones. Deterministic models are based on the physical propagation characteristics of electromagnetic waves in a model of the propagation scenario. In contrast, stochastic models describe the behavior of the channel through stochastic processes.

By now, some statistical channel models have been established for the early design phase and for testing the ideas for possible applications. Statistical models randomly generate channel impulse responses of a channel based on the probability functions, which are usually obtained from measurements. However, if a system has to be tested in a specific environment, deterministic channel models are required, which approximate real physical phenomena.

One of the most popular deterministic channel modeling approaches is ray tracing (RT), based on geometrical optics and the uniform theory of diffraction. In outdoor areas ray tracing simulations emulate the propagation conditions very well [18]. Furthermore, it has been shown that ray tracing can be also easily extended to simulate ultra-wideband channels.

However, comparisons between the measurements and simulations with respect to UWB indoor channels show that the ray tracing results are often underestimated in terms of received power, mean delay and delay spread [26, 34, 42]. This is due to insufficient modeling or the complete neglect of diffuse scattering in the ray tracing model.

Diffuse scattering causes contributions to the power delay profile, which are not resolvable (dense multipath components). Through these contributions the power delay profile is smoother than in the scenarios with reflection and diffraction only.

In the model described in this section, a simple approach is proposed which combines the ray tracing method with statistically distributed scatterers. The approach is inspired by the diffuse scattering model for UWB channels presented in [33], by the spatiotemporal model for urban scenarios presented in [46] and by the geometrically-based stochastic channel model [40]. The scatterer placement and properties are bound to the geometry of the considered scenario. The parameters for the stochastic part of the model are derived from measurements. As only few additional scattering contributions per surface are added, the increase in computational effort is very little. The placement of scatterers ensures that part of the contributions are resolvable for the UWB system. Some preliminary results of this model were previously presented in [23], [24] and [25].

To get an impression of the sources of dense multipath components, the spatial behavior of the channel a stationary office scenario (scenario A), shown in Fig. 7, is analyzed.

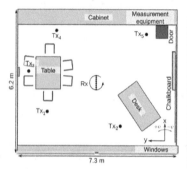
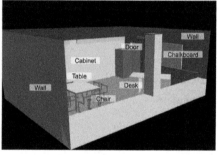

(a) Schematic diagram of the measurement scenario

(b) Simulation environment

Figure 7. Schematic diagram of the measurement (left) and simulation scenario (right, view from outside through the windows) for the DOA analysis of dense multipath components.

The measurement setup used consists of a vector network analyzer, low noise amplifier and of a set of step motor controlled positioners. The frequency range sampled by the analyzer is 2.5 - 12.5 GHz. For the measurements in scenario A, the motor controlled turntable is used to rotate the strongly directive antenna array described in [2] around its z-axis. This antenna is used as a receiver (Rx) and placed approximately in the middle of the room. The transmitter (Tx) is equipped with a UWB omnidirectional monocone antenna and placed at 5 positions marked in Fig. 7 (left).

In the same scenario, ray tracing simulations, as shown in Fig. 7 (right), with up to 5 reflections and up to 3 diffractions have been performed. At this point, no scattering is considered in the simulations. The transmission has not been considered here, earlier radar measurements in comparable rooms, [23], showed that no significant paths are to be expected from their side of the walls. On the other hand, objects inside the wooden cabinets may cause significant dense contributions. The patterns of the antennas used in the measurements have been measured in an anechoic chamber for the considered frequency band, and are considered in the simulations.

In Fig. 8 the measured and simulated power delay profiles (PDP) are depicted for the transmitter position 1, see Fig. 7 (left). For each rotation angle indicated on the x-axis, the time dependent PDP is plotted along the y-axis. As the radiation pattern of the antenna array has a narrow main beam the directions of arrival of the propagation path, impinging on the antenna can be directly identified in the picture.

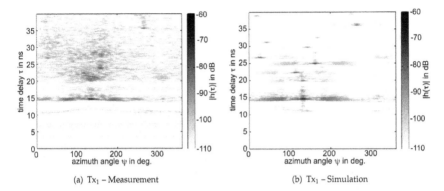

(a) Tx_1 – Measurement   (b) Tx_1 – Simulation

Figure 8. Measured and simulated angle-dependent power delay profiles for the transmitter position Tx_1.

The comparison shows that most of the strong contributions are present in both measurements and simulations. Moreover, the amplitudes of the measured and simulated reflection contributions have been captured with good accuracy.

Nevertheless, a significant amount of power is missing in the simulations. Figure 9 shows the power delay profiles of Fig. 8 averaged over the delay time and over the angle. It can be observed that although some strong contributions such as the direct path and a reflection at τ = 25 ns is at the same level as that from the measurements, for most delay times the simulated power is considerably lower. From the PDP values averaged over the angle we may conclude that this effect is present for all observation angles. The dense components are not distributed evenly over the angle, but create *clusters* around the strong contributions. The observed dense contributions are not an effect of rough surfaces, as all surfaces in the room can be considered to be smooth within the frequency range used. Some contributions may arise from small objects such as doorknobs present in this scenario. Other contributions are most likely due to the scattering from the inhomogeneities within the walls or cabinets. In [44] it was shown that typical inhomogeneous building materials distort the transmitted signal significantly. Such distortions are expected to be present also in the reflected signal and are likely to cause dense components with delay times slightly larger than the delay time of the reflected signal. The amplitudes of such components are decreasing almost exponentially with the delay time.

This means that if the dense contributions are to be modeled, their delay times and angles of arrival should be grouped around the significant contributions.

3.1. Scattering model

To achieve the clustering effect, additional contributions are generated by placing point scatterers around the reflection points calculated by the ray tracing model, see Fig. 10. These scatterers represent small structures on the surface, which have not been considered in the

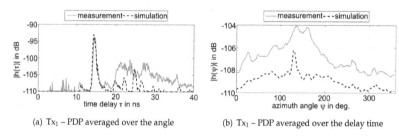

(a) Tx$_1$ – PDP averaged over the angle   (b) Tx$_1$ – PDP averaged over the delay time

Figure 9. Averaged angle-dependent power delay profiles for the transmitter position Tx$_1$.

scenario data so far, as well as interactions with inhomogeneities inside the objects and with objects behind the walls. The delay time of the additional multi-path contributions due to the scattering points is approximately equal to the delay time of the reflected path. Their scattering coefficients are adjusted so that the resulting amplitudes of these multi-path components are slightly below the amplitude of the reflected path.

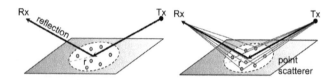

Figure 10. Modeling approach for single reflections.

The number of scatterers $n = 1 \dots N$ is a model parameter and is assumed to be constant for all clusters. The scatterers are distributed uniformly on the objects' surface within a radius r around the reflection point. The scatterers whose position is outside the considered area are discarded.

In order to keep the model as general as possible, each scatterer is characterized by the complex full polarimetric scattering matrix $\underline{S}$. The field scattered from the n-th scatterer $\underline{E}^s$ is described in the frequency domain by:

$$\underline{E}^s = \frac{e^{-jk_0d}}{d} \cdot \underline{S} \cdot \underline{E}^i \tag{1}$$

where $\underline{E}^i$ is the incident field, k_0 is the wave number, and d is the distance between the scatterer and the observation point.

In this subsection, only the vertical co-polarized element $\underline{S}_{vv}$ is considered. The parameterization of other scattering matrix components can be done in the same way. As the frequency band used is very wide, at least some of the contributions can be resolved by the system. Therefore, the scattering contributions in the model are coherently summed at the receiver.

To obtain the amplitudes of scattered contributions in the same order of magnitude as the amplitudes of the reflected path, their scattering factors are related to the reflection coefficient $\underline{\Gamma}$ by a proportionality factor a, which is derived from the measurements as well. The reflection factor $\underline{\Gamma}$ is calculated at the position of the scatterer using the material parameters of the corresponding surface. Depending on the polarization of the impinging wave, the reflection coefficient either for parallel or vertical case is used. Thus, in the case of single reflection paths and assuming vertical polarization, the resulting field at the receiver consists of the reflection contribution and of the sum of scattered contributions $\underline{E}^s_{v,tot}$ given by:

$$\underline{E}^s_{v,tot} = \sum_{n=1}^{N} \frac{e^{-jk_0 d_n}}{d_n} \cdot a\underline{\Gamma}_{v,n} \cdot \underline{E}^i_{v,n} \qquad (2)$$

Due to the single scattering approach, the model covers only the part of the power delay profile with relatively short excess delay times. For the reliable simulation of delay spreads longer multiple reflected propagation paths have to be considered. The intuitive approach would be to place additional scatterers around the higher order reflections points and to use the impinging reflected wave as an excitation, see Fig. 11. This would then have to be incorporated into the reflection path search algorithm of the ray tracing approach, which would require much computational effort. To keep the excess simulation time of the hybrid part of the model as short as possible, the multiple scattering processes are replaced here with "virtual" single bounce scatterers. These scatterers are placed at the point of the multiple reflections. Their scattering factors contain an additional term $e^{-jk_0\delta}$, where δ corresponds to the path length between the point of the first interaction and the considered higher order interaction. This term adds δ/c_0 to the delay corresponding to the distance between the transmitter and the scatterer d_{geom}.

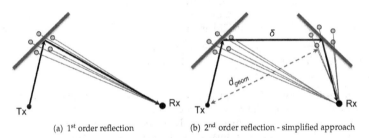

(a) 1st order reflection   (b) 2nd order reflection - simplified approach

Figure 11. Modeling approach for multiple reflections

Thus, the delay time similar to the delay of a propagation path containing one or more reflections can be realized. However, the amplitude of such a path decreases proportionally to $\delta \cdot d_{geom}$ whereas the amplitude of a path containing the reflection would experience a slower decay. To counterbalance this effect, an additional term $p \cdot \delta$ is included in the scattering coefficients of the higher order scatterers resulting in the following expression for all scatterers:

$$\underline{S} = (a + p \cdot \delta) \cdot \underline{\Gamma} \cdot e^{-jk_0\delta} \qquad (3)$$

For the scatterers placed around single reflection, this expression reduces to $\underline{S} = a \cdot \underline{\Gamma}$ since in this case $\delta = 0$. Thus, the resulting model is characterized by 4 parameters:

- N - number of scatterers
- r - placement radius

-   a and p - proportionality factors.

These parameters are estimated from the measurements. The derivation approach is described in the following Sections.

3.2. Derivation of the model parameters

For the derivation of the model parameters, a series of measurements with synthetic arrays at both transmitter (Tx) and receiver (Rx) have been conducted in three different office and lab scenarios in the IHE building of the Karlsruhe Institute of Technology. Two are office scenarios (scenario B and D), where the number of details in both rooms is small. The third scenario (scenario C) is a cluttered lab scenario [25]. Here, a large number of small details such as cables, tools, books etc. is distributed over the tables and shelves. These small objects were neglected in the scenario model. The transmitter is placed within a 0.12 m long linear positioner, and the receiver is moved on a 1.2 m by 0.6 m rectangle. Thus, a linear and a rectangular virtual array are obtained. The spacing between two consecutive antenna positions in both Tx and Rx arrays is 3 cm. With the exception of the antennas, the measurement setup used is identical with the setup described before. The simulation settings are also the same.

For the derivation of the model parameters, the behavior of the channel characteristics (path loss L and delay spread σ_D) in the measurements and the simulations are analyzed and compared [25]. To find adequate model parameters, simulations with different parameter sets are conducted and compared with the measurements. As the test of all possible parameter combinations would be computationally prohibitive, an initial parameter set has been chosen based on previous work findings in [23] and the parameters have been varied one by one.

The scatterer generation is done only once in each realization for Tx position in the middle of the Tx array and for Rx position in the middle of the Rx arrays. For each other Tx/Rx configuration the same scatterers are used.

Due to the statistical nature of the model, some variation of the simulated channel parameters for consecutive simulations with the same model parameter set is to be expected. Hence, for each parameter set 5 realizations are then simulated and the channel parameters derived from them are averaged. This number is small enough to be simulated quickly, and large enough to give approximate mean values for a given parameter set.

To derive the model parameters, their influence on the chosen channel characteristics is analyzed. It can be observed that:

-   the mean relative error in path loss has a minimum at $a = 0.175$, $N = 22$ and $p = 0.075$.
-   the mean error in delay spread rises with rising a and N. It changes also very quickly with p. The error minimum is at $p = 0.025$.
-   for $r <= 0.5$ m all error values rise. For $r > 0.5$ m and $r < 1.5$ m, the error values are stable.

Considering this observation, first p is set to 0.03 because it has the strongest influence on the error. Thus, values of $a = 0.2$ and $N = 16$ are chosen which give a good tradeoff between path loss and delay spread errors. Finally, the scattering radius is set to $r = 1$ m.

Another indirect model parameter is the order of reflection which is considered in the scatterer placement. The influence of the considered reflection order on the delay spread is shown for a single position in the middle of the Rx and Tx array of an office scenario. The measured delay spread for this point is 3.8 ns.

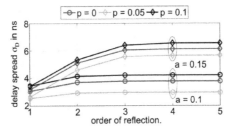

Figure 12. Influence of the considered reflection order on the delay spread in scenario 2.

The curves show that depending on the chosen model parameter (a and p have the strongest influence here) the inclusion of reflections of up to 3rd order influences the delay spread. The same has been observed in other scenarios and for the path loss. Thus, in the following the scatterers will be placed around the reflection points of up to the 3rd order.

3.3. Model performance

To test the parameterized model, it is compared with measurements with respect to path loss, delay spread, azimuth spread, power delay profiles and azimuth spectra at the receiver [25]. For the estimation of the power delay profiles and azimuth spectra, the first Tx position and a rectangular track along the edges of the positioning table at Rx is considered. Each edge of the rectangular track is placed 9 cm (3 Rx positions) away from the edge of the positioning table. The estimation of azimuth angle is done using the sensor-CLEAN algorithm [8] using 4×4 elements with a spacing of 6 cm, with the midpoint at each comparison-track point. In contrast to the measurement the simulations can provide also the angle of arrival of individual paths. However, due to the enormous amount of data obtained if the properties of each individual path are recorded, it is more convenient to apply the estimation also to the simulation data. In this case, only the coherent sum of all paths for a particular Tx/Rx position has to be recorded.

The placement of the comparison points and of the arrays used for calculating the angles of arrival (AoA) at the receiver is shown in Fig. 13 .

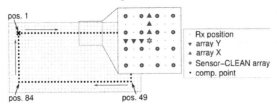

Figure 13. Placement of the comparison points and of the arrays used for the calculation of AoA at the rectangular positioner.

The "array X" configuration is used for positions along the shorter edge, whereas the "array Y" configuration is used for positions along the longer edge of the rectangular positioner. The elevation is neglected here as the measurements with a 2-D array do not allow for resolution of paths impinging from below and above the array.

The analysis of the corresponding PDPs shows that the impulse responses simulated with the hybrid approach bear much more similarity to the measurements. Although the scatterers are generated in a statistical way, their properties are tightly bound to the properties of underlying reflections so that their contributions do not dominate in the channel impulse response but fill the missing dense components of the impulse responses and angular spectra.

In the next step, the mean error μ_e and the standard deviation of the error between the measurement and the hybrid model σ_e is calculated for the path loss L, delay spread σ_D and the angular spread $\sigma_{\psi R}$. For this purpose, all possible Tx/Rx positions as described in Subsection 3.2 are used. These values and the corresponding values of the error between the measurement and conventional ray tracing are shown in Table 1 .

	Scenario B		Scenario C		Scenario D.	
	RT	Hyb.	RT	Hyb.	RT	Hyb.
μ_{e_L} in dB	4.34	1.64	1.98	-0.07	3.27	1.68
σ_{e_L} in dB	0.57	0.77	1.25	1.03	0.73	0.75
$\mu_{e_{\sigma_D}}$ in ns	1.70	0.83	1.23	0.04	1.81	-0.76
$\sigma_{e_{\sigma_D}}$ in ns	0.78	0.59	0.60	0.56	0.66	0.67
$\mu_{e_{\psi R}}$ in deg	22.35	10.26	19.85	7.15	4.33	-1.19
σ_{ψ_D} in deg	9.00	9.86	7.10	8.01	5.16	6.03

Table 1. Mean values and standard deviations of the error between the measurement and ray tracing simulation (RT) and between the measurement and hybrid simulation (Hyb.) .

Except for azimuth spread, the standard deviation values are very small. In the case of azimuth spread, however, additional errors are imposed due to path estimation. In a few cases, an insignificant rise is observed. The mean values are improved simultaneously for all considered channel characteristics.

The spread of the error values of path loss, delay spread and capacity resulting from the statistical nature of the model is analyzed also. For this purpose 40 realizations of the channel with the same parameter set are generated. For each realization, the mean error of each channel parameter is calculated. To describe the spread, the standard deviation over all mean values is adopted.

Finally, the derived model is applied to scenario A from Subsection 6.4 to prove the space-time distribution of the additional contributions. The angle dependent PDP simulated with the hybrid method is shown in Fig. 14 . The comparison with Fig. 8 shows that the additional contributions are properly placed in the azimuth-delay space, thus, depicting better the clustering effects in the scenario.

With this, a simple and effective modeling approach for directional UWB channels is proposed. The ray tracing method is combined with a simple geometric-stochastic model which represents the dense part of the channel.

The parameters of the stochastic model are connected to the properties of reflected paths so that they form a cluster with a certain delay and angle range around the reflected contribution. The stochastic clusters are also implemented around the points of multiple reflections. The

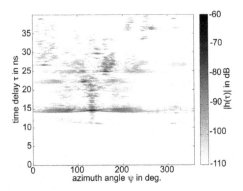

Figure 14. Angle-dependent power delay profile for the transmitter position Tx_1 in Scenario A simulated with the hybrid method.

model delivers very realistic channel impulse responses, azimuth spectra, and resulting channel parameters. The mean error between measurement and simulation is considerably improved in comparison to conventional ray tracing. This includes also the geometrical structure of the channel. Moreover, the deviation of the values of the simulated channel parameters due to the random placement of the scatterers is very small. Thus, a good reproducibility of the results is given.

4. Cooperative localization of mobile sensor nodes

In the application scenario envisaged in the introduction, an unknown environment is inspected by a UWB sensor network. Static anchor nodes of the network are placed at strategic positions. They span a local coordinate system and passively localize people or other moving objects just by electromagnetic waves scattered from them. "Electromagnetic images" of the environment are provided by moving nodes of the network. All data extraction algorithms that evaluate data measured by the sensor network require a priori information about the position of corresponding sensor nodes. In this section, basic principles of the cooperative localization of sensor nodes are described.

UWB localization is usually achieved in two steps, parameter extraction and data fusion, [20, 57]. The parameter extraction estimates parameters of signals received by sensor nodes that are required in the data fusion step. Typical parameters that are used in radio based localization systems are time of arrival (ToA), time difference of arrival (TDoA), angle of arrival (AoA) and/or received signal strength (RSS). The range-based schemes, ToA and TDoA, are shown to yield the best localization accuracy due to the excellent time resolution of UWB signals [19]. The range based ToA approach appears to be the most suitable approach for localization in UWB sensor networks. However, there are still many challenges in developing a real-time ToA based indoor UWB localization system. Due to the number of error sources, such as thermal noise, multipath propagation, direct path (DP) blockage and DP excess delay, the accuracy of the range estimation may get worse. In indoor environments, it is proven that the major sources of errors are multipath components (MPCs) and the NLOS situation [54, 56] that strongly influence the parameter estimation step - the range estimation. The quality of the range estimation is related to the SNR (or distance between Tx and Rx) and the LOS/NLOS situation. It could be improved if suitable a priori information is available. This information is usually obtained from subsequent location estimations. In what follows, we propose a novel UWB localization approach which does not require such a priori information and, instead, is based on the NLOS identification and mitigation.

4.1. UWB localization in realistic environments

The first step in our approach is high precision ToA estimation. Conventionally, ToA estimation for UWB localization is performed via a correlator or equivalently, via a matched filter (MF) receiver, [19]. However, it is difficult and not practical to implement this estimator since the received waveform with many unknown parameters must be estimated. This is almost impossible especially in realistic indoor scenarios. Another approach is the maximum likelihood (ML) based method for joint estimation of path amplitudes and ToAs described e.g. in [31, 67, 76]. This is, however, a computationally extensive method that is not suitable for real-time operations. Although various low-complexity ranging algorithms exist, their performance is not sufficient for high precision ToA estimation. Examples of low-complexity threshold-based methods such as the peak detection method, the fixed threshold method, or the adaptive threshold approach are given e.g. in [11], [17] and [22]. In these approaches, the received signal is compared to an appropriate threshold δ, and the first threshold-exceeding sample index corresponds to the ToA estimate, i.e.,

$$\hat{t}_{ToA} = t_n, \; n = \min \{i | z[i] \geq \delta\}. \tag{4}$$

For the high precision ToA estimation we proposed an adaptive threshold-based ToA estimation algorithm, the maximum probability of detection (MPD) method, in [56]. It aims at improving the robustness in multipath and NLOS situations. The main idea is to compare the probabilities for a number of possible peaks in the obtained CIR of being the ToA estimates. The probability that a certain sample, e.g. the ith sample, is determined as the ToA estimate when its amplitude, $z[i]$, is equal to or greater than the threshold and the samples before are smaller than it, i.e.,

$$P_d(i) = P(\hat{n}_{ToA} = i) = \left[\prod_{n=1}^{i-1} P(z[n] < \delta)\right] \cdot P(z[i] \geq \delta), \tag{5}$$

where, $\hat{n}_{ToA}$ denotes the estimated index, and $i = 1, 2, \ldots, N$ are the sample indices. The one which has the highest probability leads to the final ToA estimate.

The next step in our localization approach is the NLOS identification and mitigation. The advantage of this approach is that, if the identification is correct, the accuracy of the localization can be considerably improved. Several attempts to cope with the NLOS identification problem have been proposed, such as, methods based on the sudden decrease of the SNR or on the multipath channel statistics, or method by comparing statistics of the estimated distances with a threshold in [7, 68]. However, these methods usually need to record a history of channel statistics. The advantage of our approach based on a hypothesis test proposed in [54] is that it could also be applied in cases, when the target node is static or within the halting period of a moving node. The algorithm compares the mean squared error (MSE) of the estimated range estimates with known variance of the LOS range estimates. The two hypotheses are:

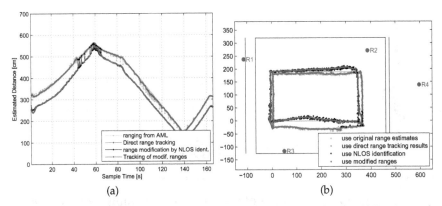

Figure 15. Data processing results of scenario with 2 LOS nodes and 2 NLOS nodes. (a) Ranging results by using different approaches for one NLOS channel; (b) the localization results by using different approaches before. location tracking.

$$\begin{cases} H_0 : M \leqslant \sigma_{LOS}^2, & \text{no NLOS node exists}, \\ H_1 : M > \sigma_{LOS}^2, & \text{NLOS nodes exist}. \end{cases} \tag{6}$$

where M is the MSE of the estimated range estimates.

After the NLOS identification, the location estimation is performed by using the identified LOS nodes only. For the implementation of the location estimation, trilateration systems are widely used. Many range-based location estimation methods with different complexity and restrictions have been proposed in the literature. All of them try to acquire a high precision of the location estimate from the range estimates. Different location estimation algorithms, which aim to find the closest position to the current coordinate of the target node, offer different accuracies and complexities. In [55], performances of a number of location estimation algorithms are compared, such as the least squares method, the Taylor series method and the approximate Maximum likelihood method.

4.2. Measurement-based verification

In order to verify our localization approach described above, a measurement was performed in a radar laboratory environment. Two Rx antennas were situated in one room, another two antennas were situated in the neighboring room and in the corridor. The Tx antenna was mounted on a positioning unit and moved along a predefined rectangular track. The MPD-based algorithm was used for range estimation. The hypothesis test-based NLOS identification and mitigation algorithm, which compares the MSEs of range estimates with the variance of the LOS range estimates, was used for location estimation. For comparison, the approximately Maximum likelihood method was applied for the location estimation, too. For both, range tracking and location tracking, the Kalman filter was applied.

The ranging results obtained for a sensor network containing one NLOS node is shown in Fig. 15(a). The result of the localization is displayed in Fig. 15(b). Both figures illustrate the feasibility of the proposed localization approach and its better performance in most cases compared with a number of other approaches.

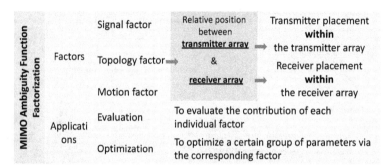

Figure 16. Factorization of MIMO ambiguity function and its potential applications.

5. Evaluation and optimization of the topology via ambiguity function analysis

As discussed in the chapter, sensor network imaging is one of the important applications of UWB sensors. In the UWB sensor network, there are stationary sensors (e.g. anchors), and mobile sensors. In the perspective of radar imaging, the spatial distribution of the stationary sensors would form a "real array", while the movement of the mobile sensors would generate a "virtual array" (i.e. synthetic aperture). The beam patterns of this "real array" and "virtual array" highly depend on their spatial configurations (topologies). In other words, the resolving performance of the real/virtual array highly depends on the topology itself.

Beyond the topologies of the real and virtual arrays, the signal parameters such as the waveform, bandwidth, etc., could also impact the resolving performance of the system. That is, the overall resolving performance of the system is jointly decided by the parameters, including the topologies of the real and virtual arrays, as well as the signal parameters. This makes the analysis of the topology even more challenging.

Sensor networks are designed to be highly accurate for their intended purpose. Always, designers and engineers are required to know the level of resolution expected from a particular sensor configuration. In order to evaluate and optimize the topologies, ambiguity function analysis is introduced in this section. Via ambiguity functions, we could know how the topology of the real/virtual array (i.e. the array formed by stationary/mobile sensors) contribute to the resolving performance of the system, and then further optimize it [30, 69].

Generally, in the far field, the ambiguity function can be factorized into several factors such as the signal related factor, the topology factor (associated with the "real array"), and the motion factor (associated with the "virtual array") [69], as shown in Fig. 16. The combination of these factors results in the overall resolving ability of the system. Theoretically, each individual factor can be used to evaluate a certain aspect of the resolution characteristics or optimize certain parameters instead of using the complicated ambiguity function of the system as a whole.

As described in the scenario, a number of UWB sensors are deployed to image the environment in order to provide necessary information for further applications. As shown in Fig. 17, it can consist of a number of moving transmitters ($T_i \in \{T_1, T_2, ..., T_M\}$) and a number of stationary receivers ($R_j \in \{R_1, R_2, .., R_N\}$) . In this way, a UWB sensor network is constructed. The transmitters can move along predefined tracks (e.g. Track 1 & 2) to probe

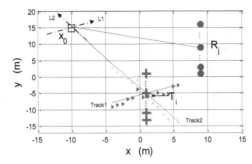

Figure 17. UWB MIMO imaging scenario. "Track 1 & 2" are the transmitter tracks; Triangles: nonlinear tracks.

the environment. The receivers collect the backscattered probing signals to produce an image of the environment. Meanwhile, they could also serve as anchor sensors to support other applications, such as localizing and tracking the position of the moving transmitters [61].

The sensor motion factor is shown in Fig. 18 (a) and (b), with respect to different tracks (Track1 and Track2, as defined in Fig. 17). In the figures, apparently, the ripples are narrower in the direction of L1 compared to the direction of L2. It indicates that the resolving performance in the direction of L1 is better than that of L2, due to the total angular rotation in the direction of L1 is far greater than the angular rotation in the direction of L2 with respect to the reference x_0. For similar reasons, the resolving performance of "Track1" is better than the resolving performance of "Track2" in the corresponding directions. In addition, it is shown in Fig. 18 (a) that a "ghost" object occurs in the direction of L2, due to an insufficient illumination of the object. Generally, it would generate a false object image, and consequently worsen the quality of the image.

In Fig. 18 (a) and (b), the motion factors are given with respect to linear tracks. However, in practice, the sensors are not necessarily moving along linear tracks. There may be more practical irregular tracks as shown in Fig. 17 where the triangles indicate the transmission positions. The irregular movement of the sensors could improve the performance of ghost suppression, since the irregular tracks can provide a more sufficient illumination of the environment compared to the linear tracks.

According to the sensor topology in Fig. 17, the topology factor is given in Fig. 18(c). In the figure, the ghost image is partially suppressed. As shown in the figure, the suppression residuals exist at the ghost image position. However, they are not as strong as the real object. Theoretically, the ghost image can be further suppressed by optimizing the sensor spatial placement.

Figure 18 (a), (b) and (c) indicate the resolution contribution of the sensor motions and the sensor placement topology to the overall resolution. As given in Fig. 17, the overall performance of the system is the combination of all involved individual factors. It implies that we can try to realize a better overall resolving performance by (i) optimizing each individual factor, or (ii) trading-off between related factors. For example, in order to suppress the "ghost" image, on the one hand, we can optimize the movement tracks via the motion factor and the sensor placement topology via the topology factor. On the other hand, a compromise can be made between the motion factor and the topology factor. In this sense, due to the interaction

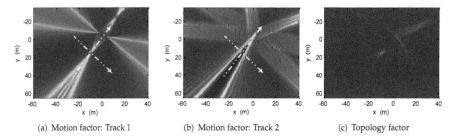

(a) Motion factor: Track 1   (b) Motion factor: Track 2   (c) Topology factor

Figure 18. The motion and topology factors given at a certain frequency f. For the motion factor, $v \cdot PRT = 5c/f$, where v is the sensor speed, PRF is the pulse repetition time, and c is the signal propagation speed. For the topology factor, the sensor element interval is $5c/f$, and the number of sensor elements on each array is 30.

of the factors, the sensor network provides more degrees of freedom for the system designer, compared with a single sensor system.

6. UWB for map building and localization

This section deals with the problem of building a map of the surrounding area using the bat-type scenario introduced in Section 2.1 . This scenario is characterized by the fact that no supporting infrastructure is used and no external information about the location of the mobile, robot-like sensor is needed. Our goal is to build a map of the surrounding for the robot, while at the same time the robot localizes itself relative to the map. In the field of robotics, this problem is known as simultaneous localization and mapping (SLAM).

To solve the complete SLAM problem, many different approaches have been presented e.g. in [62], but there well established sensor technologies like LASERs or optical cameras are used that would not work in the envisaged scenario and cannot make use of the unique capabilities of UWB radar, see Section 2.1. Other solutions are based on WLAN [53], RFID [32] or other external sources of information and, thus, must also be discarded.

There are other approaches for indoor localization and/or map building using UWB technology, but they are restricted to estimate the 2D dimensions of a strictly rectangular room [14] or need a priori information about the positions of walls to calculate virtual anchor nodes [39]. The solution presented here is more general and copes with arbitrary room shapes as long as adjacent walls are straight and orthogonal to each other.

The main advantage of this approach in comparison with the object recognition or the imaging in Section 7 and 8, is the fact that it is able to deliver a solution with a far lower number of measurements.

In the following section, a solution to the SLAM problem using a UWB radar in the bat-type scenario is described. It uses measurement models incorporating three different typical room characteristics: straight walls, corners and edges and a state-space description of the room and the robot. Algorithms for dynamic state estimation are used to calculate the desired states. Data association of measured propagation times and room features is vital here and is dealt with in great detail. Results using simulated and measured data then show the feasibility of the concept. Special requirements for the antennas are also discussed.

6.1. Measurement model

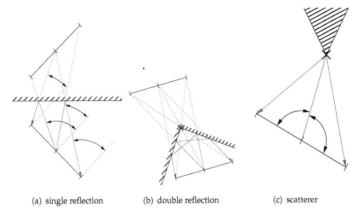

(a) single reflection   (b) double reflection   (c) scatterer

Figure 19. Three room features used in the state-space measurement equation.

To detect and localize the features, the bat-type UWB radar is used to measure the round-trip-times between the transmitting antenna, features of the surroundings, and the receiving antennas, which are extracted from UWB impulse responses and stored in the measurement vector $\mathbf{z}$. A two-dimensional geometrical model of the real world is used in the algorithm, similar to [4]. Walls and corners are represented as single and double reflections, respectively. Edges and small objects are represented as scatterers. Schematic illustrations of the propagation models are shown in Fig. 19 .

Using an estimated initial position and orientation of the antenna array and an estimated initial position of a feature, an expected time-of-flight between transmitter, feature and receiver can be calculated. This can be used for dynamic state estimation, for example in an Extended Kalman filter or a particle filter, to iteratively improve the estimate of the positions. By measuring at different positions or rotating the antenna array, it is possible to distinguish the different features and calculate an initial estimate of their positions. To do this, a state-space description of the room and the robot is needed.

6.2. State-space description of the room and the robot

To solve the SLAM problem, a state-space description is used. The state vector $\mathbf{x}$ to be estimated consists of three different parts.

$$\mathbf{x} = \left[\mathbf{x}_{robot}, \mathbf{x}_{sensor}, \mathbf{x}_{map} \right]^T \tag{7}$$

$\mathbf{x}_{robot}$ contains the information about the robot position in x and y direction, p_x and p_y, as well as the speed of the robot, represented as movement angle p_ϕ and the absolute value of the speed v. All values are in relation to the local coordinate system.

$$\mathbf{x_{robot}} = \left[p_x, p_y, p_\phi, v \right]^T \tag{8}$$

In x_{sensor} relevant information about the sensors like biases, vector $\mathbf{b}$, or the orientation ϕ_{array} of the antenna array relative to the robot is stored.

$$\mathbf{x}_{sensor} = \left[\mathbf{b}, \phi_{array}\right]^T \tag{9}$$

The map consists of the coordinates of recognized features of the surroundings, called landmarks, and are stored in x_{map}.

$$\mathbf{x}_{map} = \left[x_{landmark_1}, y_{landmark_1}, x_{landmark_2}, y_{landmark_2}, \ldots\right]^T \tag{10}$$

This is the largest part of the state vector.

To estimate the current state x_k of the system in time step k, first the a priori state estimate $\hat{x}_k^-$ is calculated using the previous state $\hat{x}_{k-1}$ by using the system transition function g

$$\hat{x}_k^- = g\left(\mathbf{u}, \hat{x}_{k-1}\right) \tag{11}$$

where $\mathbf{u}$ is a control vector used to model external influences like movement commands to the robot.

In a second step, the state estimate $\hat{x}_k$ is updated from $\hat{x}_k^-$ using the measurements from the UWB radar. In what follows, the index k is often discarded to make the text better readable.

The remaining problem is that of data association discussed in the next subsection.

6.3. Data association

A major task in employing the UWB radar for SLAM is data association, in this case the task of assigning the time-of-flight measurements of the radar to corresponding features of the surroundings. The solution presented in this subsection uses two different grouping algorithm, one working in the state space, the other working in the measurement space. In both cases, the estimation of the map is done using a Rao-Blackwellized particle filter, as presented in [13].

The principal challenge we are facing is the fact that in indoor environments there is always an abundance of echoes to deal with. It is not always obvious which particular echoes belong together comparing the two impulse responses from the left and the right receiver channel. It is even harder to identify the feature which caused a particular pair of echoes. In order to improve the current state estimate, the task of determining which pair of impulses belongs to which already identified landmark has to be performed. This is achieved by applying different data association algorithms.

6.3.1. Data association in state space

The first method is a probabilistic method in state space. For a given data association vector $\mathbf{c}$ an importance distribution $\pi_j(i)$ is calculated for all impulses i and landmarks j, using the a priori state estimate $\hat{x}^-$ and the time of flights z_i.

$$\pi_j(i) = p\left(z_i|\hat{x}^-, c_i = j\right) p\left(c_i = j\right) \tag{12}$$

This is possible because the a priori state estimate $\hat{x}^-$ contains the position and orientation of the antenna array as well as the position of the landmarks, so an expected time of flight can be calculated using the measurement functions presented earlier. Here, a Gaussian distribution with known covariance is assumed. $p\,(c_i = j)$ represents the probability of impulse z_i being associated with landmark j.

From the normalized importance distribution π the data association vector c is drawn by means of the Monte Carlo method. In this way, even in the case of false measurements being closer to the predicted measurement than the correct measurement, there is a chance that the right one is chosen. It is important to note that the opposite case, a false measurement chosen over a correct one, is also possible.

So, at first glance, this method has no advantage over a simple Nearest Neighbor method, where only the measurement closest to the prediction is used. The Monte Carlo method tends to produce slightly worse results than simply choosing the association with the highest probability. This method makes sense if not only one state is estimated but many hypotheses of possible states. That is what the particle filter can handle.

The particle filter tracks many hypotheses. These hypotheses are depicted as points, or particles, in the multi-dimensional state space. Each particle s^l is composed of the estimated state $\hat{x}^l$, the covarianz matrix P^l of the state, a data association variable c^l where the association between measurements and landmarks is stored, and a weight w^l.

$$s^l = \left[\hat{x}^l, P^l, c^l, w^l\right] \qquad (13)$$

The index l denotes the lth particle. The weight w^l is an indicator of the likelihood that a certain hypothesis holds true. It is calculated using the weight from the previous iteration. Thus, the weight serves as a memory. In the long run, the hypothesis with the highest weight will be the one that best approximates the real world.

$$w_k^l = w_{k-1}^l\, p\left(z_k|x_k^l, c_k^l\right) \qquad (14)$$

6.3.2. Data association in measurement space

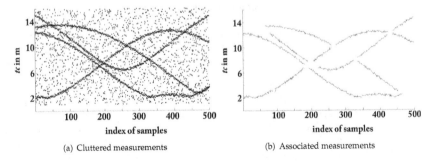

(a) Cluttered measurements   (b) Associated measurements

Figure 20. Left: Cluttered sample measurements over time, right: Associated measurement curves.

One problem that arises in using a particle filter is that due to the probabilistic nature of the algorithm the number of particles can grow considerably high. This is because the number of possible data associations increases with every measurement step. Using more particles or discarding unlikely hypotheses by resampling can only partially solve this problem.

To reduce the necessary number of particles, we use a second approach: Measurements are not directly associated to landmarks. Instead, they are first grouped in the measurement space. To do this, the fact is used that the antenna array moves only in small steps between consecutive measurements, so measurements originating from the same feature also change only slightly. This correlation can be exploited. By employing a simple Kalman Filter in the measurement space, it is possible to predict and group measurements that belong to the same feature. Only whole groups of measurements are passed to the particle filter, which greatly reduces the number of hypotheses needed and therefore the number of particles necessary. Figure 20 shows a simulation of this process. In the left figure, measurements are taken as the robot travels through the environment. Dots indicate extracted time-of-flights. The measurements are cluttered, but almost continuous echoes originating from room features can be made out. The right figure shows the result of the grouping algorithm.

The disadvantage of this procedure is that the grouping introduces a time delay in the system. Moreover, it requires measurements to be made more frequently, and so partly weakens one advantage of the room reconstruction algorithm.

6.4. Simulations

The algorithm was first tested with simulated data. The ray tracing algorithm described in Section 3 was used to calculate the impulse response function of a room. The outline of the room is shown in Fig. 3, alongside with 78 measurement points. At each point, 120 measurements were made by turning the bat-type antenna array in steps of 3 degrees. The simulated environment consisted of a rectangular room with the size of 8 m by 9 m, with a rectangular column the size of 1.5 m by 1.5 m roughly in the middle. Between the walls and the column, six complex objects used to test the object recognition algorithm from Section 7 were placed. Walls, objects and the column were assumed to be of metal. The frequency response function was calculated from 4.5 GHz to 13.4 GHz for an antenna array with three double-ridged horn antennas in the bat-type configuration. The distance between the antennas was set to 0.5 m.

Figure 21 shows the complete radargram for one whole rotation of the antenna array at point 16. In this example, it can be seen that for every angle, there are clear peaks in the impulse response that connect to peaks in the next measurement step, so data association in measurement space is possible. Mapping the whole room just from this position is not possible, because some features simply cannot be seen from there. To map the whole room, measurements from all 78 points were used. The result can be seen in Fig. 23. Walls, shown as solid and dotted lines, and corners, shown as triangles, are mapped at an accuracy of approximately 20 cm. Due to their small size, the placed objects are detected as point scatters which are depicted as stars. Their positions correspond to the object locations shown in Fig. 3. At this point, a separate object recognition algorithm as described in Section 7 could be used to identify and distinguish them. Note that the origin of the coordinate system is set arbitrarily at the point of the first measurements.

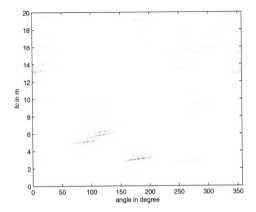

Figure 21. Radargram at position 16 of the simulated room.

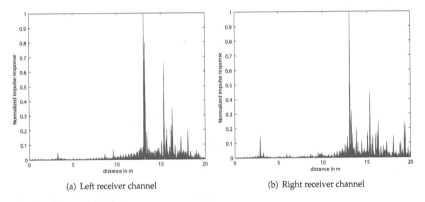

(a) Left receiver channel   (b) Right receiver channel

Figure 22. Impulse responses at position 16 of the simulated room, facing 0 degrees.

Figure 22 shows the impulse responses for the left and right channel at point 16 of the scenario described in Section 2.2. On the x-axis, the time t times the speed of light c indicates the distance the pulse has travelled. Peaks reflected or scattered from different room characteristics can easily be separated.

6.5. Measurements

To further verify the results, measurements were made in a laboratory room. The room included furniture, some metal pipes on the walls, and was filled with assorted laboratory equipment at one end of the room. The sensor array consisted of three double ridged horn antennas 0.46 m apart, similar to the simulation.

The antenna array was placed in the middle of the room and rotated manually. Pictures of the room can be seen in Fig. 25. As in the simulation, no information about the current angle of

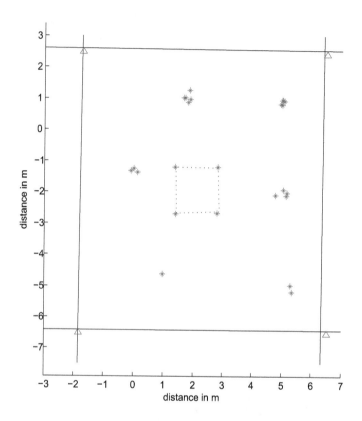

Figure 23. Reconstructed room.

the array was passed to the algorithm. The algorithm only used the UWB measurements to reconstruct the room. Figure 24 shows a sample of a recorded impulse response.

In a first test, the array was rotated only by 180°, illuminating the tidy side of the room. In this case, results similar to those of the simulation could be produced; walls and corners could be mapped with 10-20 cm accuracy. A reconstruction of the whole room was not possible. Many objects on the other, chaotic side of the room produced a large number of echoes and made it impossible to associate the measurements reliably. Here, the algorithm reached its limits.

In a second scenario, measurements were made at 15 positions in an L-shaped corridor, as depicted in Fig. 26. The array was rotated in 3 degree steps at every position, resulting in 1800 measurements.

To test the ability of the algorithm to cope with sparse measurements, only 24 measurements at 5 positions (station 1, 4, 6, 8 and 12 of the scenario) were used, resulting in a total of only 120 measurements. Here, additional information about the position of the robot had to be used, in this case odometry data about the way the robot traveled and the direction the antenna faced. The use of inertial measurement units is also possible.

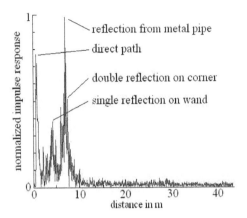

Figure 24. Impulse response showing the features of the room

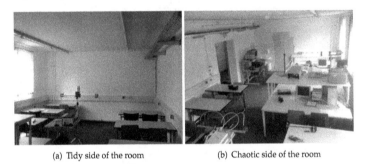

(a) Tidy side of the room   (b) Chaotic side of the room

Figure 25. Room used for measurements

The results in Fig. 27 show that the algorithm is able to recognize the outline of the room using only these few measurements, although a higher number of measurements still improves the quality of the reconstruction. There is also a trade-off between the different data association methods. While grouping in measurement space is only possible if the measurements are taken frequently, data association in state space can cope with few measurements, but rely on additional sensor data.

6.6. Optimized antenna design for SLAM

To further optimize the results of the SLAM algorithm described in this section, antennas with a broader 3 dB beam-width (>60°) than for the object recognition in section 7 are needed.

Apart from the broad frequency band of 3.5 to 10.5 GHz in order to meet further conditions the antenna also has to be dual-orthogonally polarized. The radiation phase center should be constant over frequency, and the two polarizations should have identical radiation conditions. In literature several types of UWB antennas can be found. Most of them are either biconical

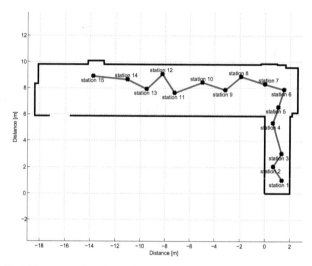

Figure 26. Outline of the measurement scenario.

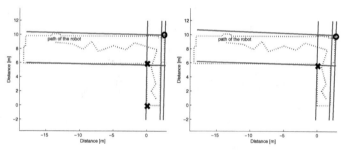

(a) Reconstruction using 1800 measurements (b) Reconstruction using 120 measurements

Figure 27. Reconstructions of an L-shaped corridor using different numbers of measurements. Dotted lines show actual room outline.

structures [5], traveling wave radiators [16], or even a combination of both [6]. The requirement of a common phase center (over frequency) limits the possible solutions. For this purpose a planar solution, a broad-side radiating antenna is chosen and optimized.

The antenna consists of two elliptically shaped dipoles surrounded by a metallic ground plane as shown in Fig. refschematic . The ellipses for each polarization (vertical and horizontal) are orthogonal to each other. Contrary to normal dipoles the feeding is separated and placed between the ground plane and the single ellipses. This is outlined by the arrows in the radiating element shown in Fig. 28. This type of feeding allows a separate feeding for each polarization and helps to keep the current distribution in the radiation zone symmetrical resulting in a constant phase center (of each polarization) exactly in the middle of the elliptical dipoles (two monopoles) [1].

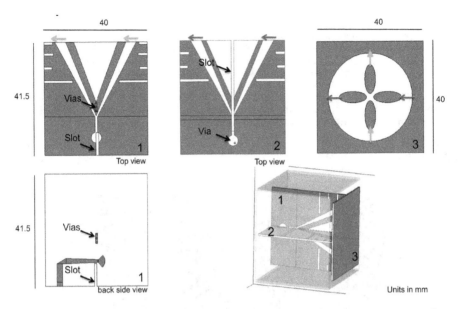

Figure 28. Schematic illustration of the dual polarized antenna element and the feeding networks, all units in mm).

The feeding networks themselves are placed orthogonally to the radiating element, see Fig. 29 (left). Similar to the Vivaldi structure, Fig. 32, a balun is used for microstrip to slotline transition. The slotline is then split up and used to feed the monopoles. The two polarizations are realized by shifting the two orthogonal feeding elements into each other as shown in Fig. 28. Therefore, a slot has to be cut into both (feeding) networks. The gaps in the metallic structures have to be closed again. This is realized by soldering through vias in the respectively orthogonal feeding network.

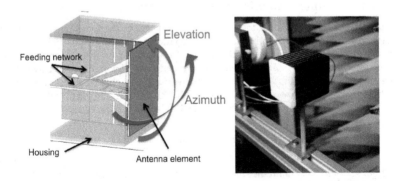

Figure 29. Schematic illustration (left) and photograph (right) of the 4-elliptical antenna

As this type of antenna is radiating broad-side in both directions (forward and backward) and should be used as (mono-) directional radiating antenna, the backward radiation (illuminating towards the feeding network) must be absorbed by a carbon fiber housing as shown in Fig. 29. This results in a reduction of the radiation efficiency. An alternative would be to use a reflector, which, however, would limit the bandwidth of the antenna.

The antenna characteristics are measured with a vector network analyzer in an anechoic chamber. The input impedance matching is around -10 dB between 3.5 and 10.5 GHz and the decoupling of the two ports is approximately 20 dB.

Figure 30 shows the 2D gain over frequency and angle for both planes (E- and H-plane) of one polarization in co-polarization arrangement. The measured gain of the second polarization is very similar and not specifically shown.

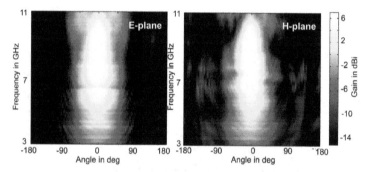

Figure 30. Measured gain over frequency and angle in the E-plane (left) and H-plane (right).

This solution of a planar, dual-polarized UWB antenna covers the frequency range between 3.5 and 10.5 GHz. In this frequency range, the radiation pattern of the antenna remains stable and directive. Both polarizations have the same radiation phase center, which is frequency-independent.

7. Short range super-resolution UWB-radar sensing

In recent years short range UWB radar sensing and imaging has gained steadily increasing interest in research. The demand for a wide absolute bandwidth results from the smallest dimensions to be resolved. However, the request for increased resolution capabilities strove for innovative algorithms, new hardware equipment, and for performance which is not restricted by the bandwidth defined by the hardware. In this chapter novel and pioneering methods, algorithms and antennas are presented which were investigated within CoLOR for UWB radar applications especially in short range UWB radar applications.

7.1. Antenna design and measurement results

As the use of polarization diversity allows further information about the object characteristics to be obtained, for instant about the surface structure as it is shown in this section, the antenna has to be orthogonally polarized. Apart from orthogonal polarizations, further conditions for the (object recognition) antenna design are a high gain and a common phase center (for both polarizations).

A common antenna for such an application is the so-called Vivaldi antenna. The radiation mechanism is based on exponentially tapered slots and the traveling wave principle [3]. This type of antenna has a convenient time domain behavior as shown in [59] and a relatively stable radiation pattern in the whole frequency range. In CoLOR the frequency range for the final demonstrator covers 3.5 to 10.5 GHz. A second band from 4.5 to 13.5 GHz was also used during the development process. Therefore, the objective of the antenna design is to cover the hole frequency range from 3.5 GHz to 13.5 GHz.

A 3D illustration (left) and a photograph (right) of the fabricated antenna is shown in Fig. 31, see also [45]. Combining the integration of two tapered slot line antennas on a single substrate with embedding them into Polytetrafluoroethylene (PTFE) allows the total bandwidth to be covered. The integration of two radiation elements per polarization yields a higher gain and saves space compared with an array of two separated antennas. Furthermore, it is less cost intensive and easier to manufacture. The possibility of varying the tapering of the inner and outer structure, see Fig. 32, can be used to focus the beam.

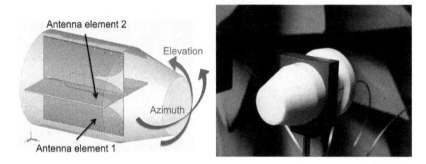

Figure 31. Antenna, embedded in the dielectric.

Both radiating structures are fed by a network shown in Fig. 32 (left), and for the second polarization in Fig. 32 (right), respectively. Starting with microstrip, where the connector is soldered on, an aperture coupling transforms to slotline which finally divides the power and feeds it to the two elements.

As already mentioned, the integration of the antenna elements into a dielectric reduces the effective wavelength. This affects several advantages compared to an antenna in free space. The antenna is capable of radiating a lower frequency, the far-field conditions are fulfilled in a closer distance and the shaping of the dielectric can (also) be used to focus the beam and for sidelobe suppression. For this work PTFE is chosen as dielectric. Its permittivity of $\epsilon_r = 2.1$ is similar to that of the substrate used (Rogers Duroid 5880 with $\epsilon_r = 2.2$). Furthermore, PTFE has low losses and can be easily shaped to adapt to the antenna design. The shaping and the dimensions of the PTFE structure are given in [45]. The conically shaped rod (in radiation direction) allows a smoother transition of the guided wave into free space.

The two polarizations are realized by shifting two orthogonal elements into each other as shown in Fig. 31 left. For doing this, a slot has to be cut into both elements. Thus, the metalized structures are interrupted, see Fig. 32. They have to be galvanically connected again. This is realized by introducing vias in the orthogonal antenna.

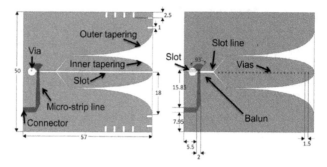

Figure 32. Schematic illustration of the antenna elements 1 (left) and 2 (right), in [mm].

The antennas are manufactured with aid of a circuit board plotter on a Duroid RT5880 substrate of a thickness of 0.79 mm. The measured S-parameter, see Fig. 33 left, show a good impedance matching for both polarizations and antenna elements, respectively, between 3.5 GHz and 13.5 GHz (and even higher). The decoupling (S_{21}, S_{12}) between the two elements is over the biggest portion of the bandwidth better than -25 dB.

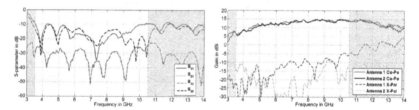

Figure 33. Measured S-parameter (left) and antenna gain in the main beam direction (right).

The gain and the pattern of the antenna were measured in an anechoic chamber. The results for antenna element 2 are presented in Fig. 34. The E-plane corresponds in this case to the azimuth direction, see Fig. 34 left, the H-plane to the elevation one. Antenna element 1 shows a similar characteristic. The maximum gain measured is around 15 dBi at 9 GHz.

To evaluate the polarization properties, the gain for both co-polarizations (Co-Pol) and both cross-polarizations (X-Pol) in the main beam direction was measured, see Fig. 33 right. The difference (between Co-Pol and X-Pol regarding the antenna gain) provides the information about the polarization purity. The cross-polarization suppression is better than 20 dB at the low frequencies up to 10.5 GHz. Starting from 10.5 GHz, the values of the X-Pol of antenna 2 are increasing (deteriorating). This is due to the current distribution of higher modes which cannot be avoided for higher frequencies. Nevertheless the measured performance allows a successful use in polarization diversity systems even above 10.5 GHz, see [45].

7.2. Data pre-processing

The raw radar data provided by the M-sequence radar needs some form of data pre-processing to smooth pulse shape, improve dynamic range, minimize the signal to interference plus noise ratio (SINR) by reducing range sidelobes and finally to enhance the

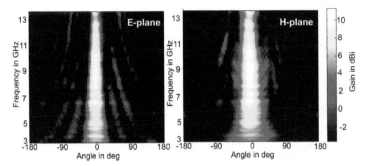

Figure 34. Measured gain [dBi] and pattern of antenna element 2 in the E-plane (left) and H-plane (right).

temporal resolution. The channel impulse response of the radar link can be extracted by deconvoluting with a reference pulse, as we assume the link as an LTI-system. However, it is well known that classical deconvolution by spectral division may drastically distort the result especially at low SNR values. A highly efficient method with low complexity to perform the deconvolution is to apply a simplified Wiener filter with the transfer function

$$H_{wiener}(f) = \frac{1}{H_{ref}(f)} \frac{|H_{ref}(f)|^2}{|H_{ref}(f)|^2 + 1} = \frac{H_{ref}(f)^*}{|H_{ref}(f)|^2 + 1} \tag{15}$$

where $H_{ref}(f)$ is the Fourier transform of a previously measured offline reference pulse. Hence, the estimate of the deconvoluted channel impulse response $h_{deconv}(t)$ is then obtained as $h_{deconv}(t) = h_{wiener}(t) * h_{measured}(t)$, with $h_{measured}(t)$ being the measured impulse response under test. Depending on the power level, the Wiener filter either acts as an inverse or matched filter for the deconvolution. In Fig. 35 an example of the channel impulse extraction is shown. Note that both pulses are normalized to the same power to enable visual comparison in the plot.

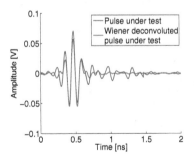

Figure 35. Example of deconvoluted pulses normalized to the same power

7.3. Material characterization

A material characterization in hostile and pathless scenarios requires a remote measurement. Hence, a method known from optics, the ellipsometry has been adapted to the UWB microwave range. The estimation of the dielectric characteristics, especially the permittivity and the emissivity are based on the ratio of the reflected power measured at two orthogonal polarizations. The orientation of the polarization is defined with regard to the plane of incidence. The plane of incidence is orthogonal to the surface of the object and is spanned by the incoming and the reflected ray. In Fig. 36 a drawing of the functional principle is given.

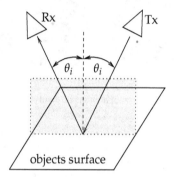

Figure 36. Schematic representation of the functional principle.

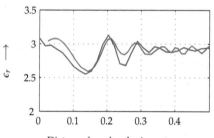

Distance from border in meter $\longrightarrow$

Figure 37. Permittivity (blue) and simulation values (red) as a function from the distance from the border.

The calculation of the permittivity is performed by the inverse application of the Fresnel-formulas. Assuming a material with a relative permeability $\mu_r = 1$, the expressions for the calculation of the relative permittivity ϵ_r and the emissivity e can be written as follows, with E as the received electric field strength for each polarization:

$$\epsilon_r = \left(\frac{\sin^2 \theta_i \left(\frac{E_\perp}{E_\parallel} - 1 \right)}{\cos \theta_i \left(\frac{E_\perp}{E_\parallel} + 1 \right)} \right)^2 + \sin^2 \theta_i \quad \text{and} \quad e = 1 - \left| \frac{1 - \sqrt{\epsilon_r}}{1 + \sqrt{\epsilon_r}} \right|^2 . \tag{16}$$

Here, it is important to note that the given expression for the emissivity e is valid for a straight monitoring of the hot spot. The additional information about the hot spot dimension and distance e.g. to a radiometer can be supplied by the UWB radar. So, an estimation of the hot spot temperature is possible.

The ellipsometry method allows an accurate characterization of plane surfaces with the main restriction to measure at a distance of at least 25 cm to any corner. The estimation of the permittivity of a small object or of objects with a complex shape is significantly influenced by effects of diffraction and scattering. As an example in Fig. 37, the deviation of the estimated relative permittivity of a MDF wall (with relative permittivity of approximately 2.9) is plotted as blue line over the distance from the antennas to the corner of the wall. In order to overcome this restriction, the UWB ellipsometry method is used in a combination with the object recognition process using the imaging methods as described later in this chapter. The distortions of the estimated permittivity values, which arise due to the diffraction effects, are then simulated (red curve in Fig. 37) and the calculated patterns of the permittivity curve are then compared with the corresponding measured pattern. For a first investigation, a simplified simulation algorithm (designed for online measurements) to consider the effects of reflection and diffraction for canonical 2 D objects was implemented. The results show that an accurate estimation of the dielectric characteristics of small objects is possible, with an accuracy of about ±3 % for typical indoor objects (e.g. composed of fiberboards or bricks) with dimensions greater than 10 cm.

The effect of the object surface structure on the material estimation was analyzed by measurements of bulk materials. For slight roughness, i.e. height deviations much smaller than the wavelength, there is almost no influence on the estimation of the permittivity. For surfaces with a roughness in the order of the wavelength, the estimation of the permittivity has an uncertainty of less than 20 %. The surface roughness can be estimated by the analysis of the depolarization, i.e. measuring cross-polarized to the transmitted polarization. For the measured indoor materials with rough surfaces, the cross-polarized power is at least about 15 dB higher than for flat surfaces.

7.4. Pulse separation

The fundamental problem common to all super-resolution approaches is the precise extraction of the round trip time of UWB pulses. While this approach can easily be performed for single reflection measurements, things become challenging when the distance between multiple scattering centers drops below the range resolution. Constructive and destructive interferences are caused, and the shape of the resulting superposed pulses is distorted massively. Common algorithms for this purpose were analyzed, evaluated and extensively tested under various circumstances. In most cases, they can hardly resolve richly interfered pulses which overlap almost the whole pulse width or have vast computational load. Often, to some extent a priori information is necessary (e.g. the number of pulses to be separated), otherwise these algorithms suffer from inflexible termination conditions or need huge post-processing.

Within the CoLOR project a novel wavefront extraction algorithm called Dynamic Correlation Method (DCM) was proposed, [51]. The DCM is based on a correlation search using a set of two differently shifted reference pulses. Thus, the resulting correlation coefficients are no more just a function of one temporal parameter but rather of two parameters which result in a matrix of correlation coefficients. DCM does not ignore the interfering signature of backscattered pulses and provides a pair of pulses taking the interference pattern of them into account. Additionally, it terminates adaptively to different power levels which enables the detection of weak reflections and avoids post-processing. For a further detailed description and a comparison with alternative methods, see [51].

7.5. Short range imaging algorithm

In [28] , an imaging algorithm called Range Point Migration (RPM) was proposed that utilizes fuzzy estimation for the Direction Of Arrival (DoA). It extracts a direct mapping by combining the measured distance of the wavefront with its DoA. It realizes a stable imaging of even complex objects and requires no pre-processing like clustering or connecting discontinuous wavefronts. The 2D RPM was introduced in [28] for a planar sensor track nearby the object which allows either only a limited image of the lateral region of the object or requires huge scan distances. Consistently, the back region of the object is not scanned and, hence, not imaged. In [28] the RPM was extended to 3D where the sensors are placed on a planar surface in front of the 3D object. However, this 2D sensor track, too, is not capable of a full perspective of all stereoscopically distributed voxels (volume pixel) of 3D objects.

In CoLOR, the RPM was extensively analyzed, validated with numerous measurements in different scenarios and further improved [49]. For a full perspective of the object, the sensor track and the antenna alignment need circumnavigation to extract entirely all stereoscopically distributed voxels. In order to reconstruct a full 3D object contour, the scan pattern of the sensors was modified and extended to a spatial scanning including the z-axis. Experimental validations were carried out based on complex test objects with small shape variations relative to the wavelength used (for results see Fig. 40).

The main principle of the improved RPM, which is called Fuzzy imaging henceforth, shall briefly be described on the basis of the following illustration of a simple scenario with 3 measurements.

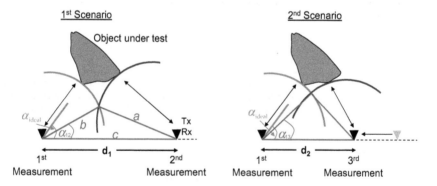

Figure 38. Measurement scenario to illustrate the principle of Fuzzy imaging

In the 1st scenario on the left side of Fig. 38, an object under test is measured at two antenna positions which are separated from each other by a distance d_1 which equals c. Both positions provide the time of flight and the distances a and b, respectively. With a geometric approach (law of cosines), the intersection points of the wavefronts as well as the angle α_{12} are extracted. Afterwards, shown on the right side of Fig. 38 a 3rd measurement is performed at a distance of $d_2 < d_1$ and provides the new angle α_{13}. Evidently, for $d_2 \to 0$, the antenna configuration converges to a mono-static configuration. In that case, $\alpha_{13} \to \alpha_{ideal}$ which would result in an exact image point when combined with the corresponding time of flight measurement.

The set of angles α_{1n} with its n neighboring wavefronts are called crisp set in terms of Fuzzy logic. Each of these scalar values could be regarded as one Dirac delta function at the corresponding angle value and is used for further processing. However, this would only make sense in the case of ideal point scatterers. Once the dimensions of the object are expanded or if the object consists of additional complex structures (e.g. edges and corners), it would result in erroneous image points. Fuzzy technology is applied here to compensate such influences and to still use a kind of convergence of nearby wavefronts. Therefore, each angle of the discrete crisp set is Fuzzyfied by a Gaussian membership function. Hence, the result does not only depend on one scalar but on a Fuzzy set around each scalar.

A scaling/clipping operation of the amplitude of the neighboring reflection point is performed to focus on strong reflections and scattering. Additional weighting is performed which scales the fuzzy sets as a function of the distance between the neighboring position and the one under test. Thus, it is ensured that the influence of sensors being further away is minimized. Afterwards, an accumulation of these differently weighted fuzzy sets is performed. The DoA can be estimated by a maximum defuzzyfication operator.

Within CoLOR it was investigated whether the standard deviation of the Gaussian membership functions are crucial parameters for the image processing. Depending on these parameters, the algorithm either extracts straight parts of the contour of the object, if it is larger than several wavelengths, or it extracts object features, i.e. scattering centers (edges and corners). A strong echo, i.e. a specular reflection is received only when the main lobe of the antenna is aligned to the normal of a smooth surface of the object. However, in the case of a circular track this occurs very rarely, if the cross-section of the object is not a circle. Scattering and diffraction effects overbalance immensely within a circular track, even more if the scan track is spread over a large circular arc. To overcome this problem, weak echoes which spread spherically from the edges can be recorded from any line of sight position. For detailed information, an extensive discussion, as well as the extension to 3D imaging see [49].

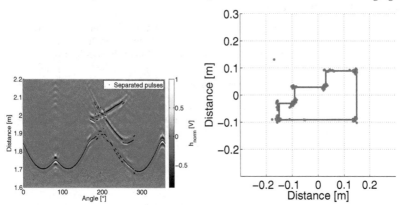

Figure 39. Radargram with extracted wavefronts on the left side and corresponding object on the right with extracted image points.

On the right side of Fig. 39 the object under test is depicted in blue. This object was scanned on a circular track with $1°$ grid resulting in the radargram shown on the left side. The wavefronts which are extracted with the DCM are also plotted in the radargram. With the previously discussed Fuzzy imaging the red image points shown on the right side are extracted.

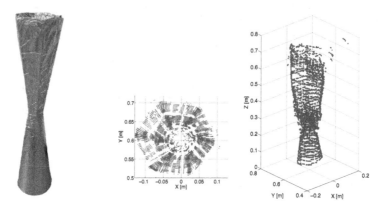

Figure 40. Photograph of an 3D test object on the left. Extracted 3D Radar image with Fuzzy imaging in the middle and top view on the right.

7.6. Exploitation of polarimetric diversity gain

The exploitation of polarization diversity in radar application provides additional information compared to mono-polarized sensing techniques. This polarization diversity gain enhances the efficiency of object classification according to the information contained in the backscattered signal. Hence, additional characteristics of objects such as shape, details of surface structure and orientation are gathered which may remain invisible for mono-polarized systems. However, in the literature polarization characteristics are rarely considered and most approaches use mono-polarized EM-waves. By the exploitation of polarization diversity the performance can be increased significantly.

Unpolarized electromagnetic incident waves on an object are diffracted or scattered in all directions. The spatial distribution of scattered energy depends on the object geometry, material composition, the operating frequency and polarization of the incident wave.

7.6.1. Polarization diversity gain in short-range UWB radar object imaging

Investigations and results shall be demonstrated by a complex edged 2D object with 6 corners. This object is measured on a circular track at a $1°$ grid which results in 360 measurements. The contour of the object can easily be recognized on the right side of Fig. 41. The sensors are the previously introduced crossed Vivaldi antennas embedded in PTFE which allow full polarimetric measurements.

The imaging algorithm used in this work is Kirchhoff Migration (KM). KM relies on some form of coherent summation, which means that a pixel of the radar image is produced by integrating the phase-shifted radar data of the field amplitude measurements at each antenna position. KM image spots of high intensity correspond to the scattering centers of the object. The image contrast is higher with increasing number of recorded impulse responses at different antenna positions. Here, the object is of 1 m height with about 1 m distance to the object in a bi-static configuration with 0.25 m distance between the transmitter and receiver. Actually, the object is a column with no variation in the height. Thus, it has vertical predominant directions causing stronger reflection in co-polarization or VV, respectively (notation: the first index indicates the polarization of the transmitter, the second index the one of the receiver).

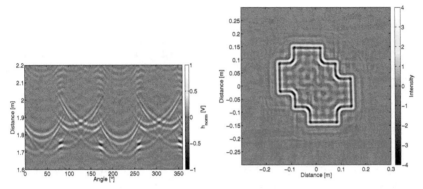

Figure 41. Radargram of the object under test with cross-polarization and 45° rotated antennas on the left and the extracted KM image from this radar data.

However, this object has more or less parts which imitate the scattering and reflection characteristic of flat plates or 0°- dihedrals. In the field of polarization research it is well known that dihedrals have strong polarizing effects. For example, a 0°-dihedral (the angle between the fold line of the dihedral and the vertical axis) has only co-polarized components, whereas a 45°-dihedral has only cross-polarized components. Therefore, dihedrals are especially suitable for calibration in polarimetric measurements.

In order to exploit polarimetric diversity gain a 45° shift is missing in the radar link [52]. As mentioned before, this would actually depolarize the wave. However, by rotating both antennas by 45°, the scattering characteristic of the object edges are comparable to 45°-dihedrals. Hence, in Fig. 42 both sensors are positioned diagonally. Using such a rotated configuration a cross-polarized measurement was performed with cross polarization.

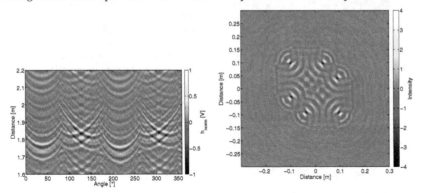

Figure 42. Radargram of the object under test with cross-polarization and 45° rotated antennas on the left and the extracted KM image from this radar data.

In Fig. 42, it can clearly be seen that only the corners are focused in the resulting image but not the flat structures. The reason for this effect is that a flat plate does not depolarize but a 45°-dihedral does.

Thus, the detection capability of a UWB radar system can be improved by exploiting polarization diversity. Under certain circumstances, the radar images detect object features which would have remained invisible in mono polarized radar systems. Supplementary information about the contour, orientation and dimensions of the object can thus be obtained, which upgrades super-resolution UWB radar significantly.

7.6.2. Polarimetric investigation of bulk goods with rough surfaces

One possible discrimination criterion between smooth and rough surfaces is to exploit specific depolarization effects. Rough in this context means that the standard deviation of the outer surface height distribution is in the range of some wavelengths of the operating carrier frequency, i.e. 9 GHz in this case. To obtain results which are independent both of material and shape, extensive measurements were performed.

Four metallic objects with smooth surfaces were used for the investigations, 2 objects with a square cross section of different size, an object with isosceles triangular cross section, and an object with a rectangular cross section. Details are shown in Fig. 44. For comparison with rough surfaces 4 polystyrene bins were built which were filled with bulk goods made of chunky gas concrete, chunky sand-lime brick, medium density fiberboard (MDF) blocks with 0.01 m edge length and M16 × 25 mm screws. Due to the material composition, the polystyrene bin itself has a vanishing radar cross section so that reflections caused by the bin are negligible.

For depolarization investigations, these 8 objects under test were scanned in both co- as well as in both cross-polarized configurations on a circular track with a radius of 1 m at a 1° grid. The depolarization is expressed by the relation of signal powers P_{VH}/P_{VV}.

In Fig. 43 the results for an object with a square cross section are compared with those of bulk good objects. Expectedly, objects with a smooth surface depolarize least, i.e. about -20 dB in the mean of all measurements. This complies with the 20 dB cross-polarization suppression of the antenna characteristic. The depolarization is least when an edge with predominantly vertical orientation is illuminated. Gas concrete and MDF are the bulk goods which depolarize most. The higher the permittivity is the more the bulky material depolarizes, i.e. sand-lime brick with about 8 dB more and the screws which depolarize most with over 10 dB more than the objects with a flat surface.

So UWB radar seems to be capable of discriminating objects of different materials by the roughness of their surface, subject to the condition that the height deviation is not much smaller than the wavelength. These results highlight the superior capabilities of fully polarimetric systems and recommend their use in future radar systems.

7.7. Object recognition for full and restricted circumnavigation

The object recognition (OR) method proposed in this work is part of the previously introduced super-resolution radar imaging system. The investigated objects and the reference alphabet derived from these consist of simple canonical and some polygonal complex objects shown in Fig. 44.

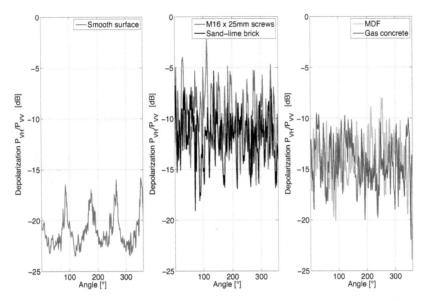

Figure 43. Results of the depolarization investigation of the object with square cross section, with 0.3 m edge length and the bin with same inner dimension, respectively. The depolarization coefficient P_{VH}/P_{VV} of the smooth surface is depicted on the left. The same coefficient is shown of the bulk goods in the middle and on the right side.

As mentioned above, the OR algorithms in [50] and [48] yield very robust results with the method of moment invariants and Fourier descriptors, which was proven for images obtained on complete circular tracks around the objects. However, in many cases such complete tracks are not possible as they lead to unfinished object images. To perform an OR also in such situations the method of Curvature Scale Space (CSS) was applied due its ability to robustly recognize contour parts [35].

7.7.1. Object recognition algorithm with the curvature scale space

The CSS representation is invariant against rotation of objects as this causes a circular shift of the CSS which has no effect on the recognition process, since the CSS of an object under test is compared by a correlation to the CSS of all reference objects. Moreover, CSS is highly robust against noise as most of its influence is compensated for to some degree by a smoothing Gaussian filter. Another property of the CSS is that it retains the local properties of shapes. Each peak of the CSS corresponds to a concavity or a convexity. A local deformation of the shape causes a change just in the corresponding local contour of the CSS image. Thus, a restricted curve can exactly be found in the CSS of the whole curve. Moreover, the absolute value of a CSS peak indicates the curvature radius, and the algebraic sign of the peak indicates whether the curve is concave or convex.

In Fig. 45 a plane curve with 8 convex or concave parts is drawn in blue. The Gaussian smoothed curve is drawn in red. On the right side of Fig. 45 , the corresponding curvature of the smoothed red curve is plotted.

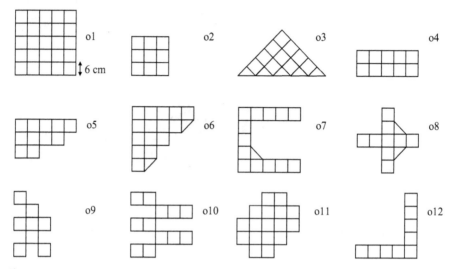

Figure 44. Cross-section of the 12 reference objects. The first 6 objects o1 - o6 were used for experimental OR investigations.

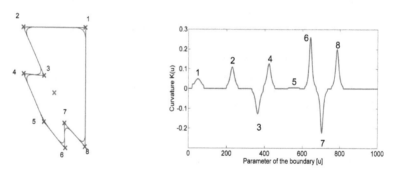

Figure 45. A plane curve in blue with its Gaussian smoothed version on the left, and the curvature of the smoothed curve on the right.

7.7.2. Object recognition performance test

The previously discussed OR algorithm based on the comparison of CSS was applied to Objects o1 - o6 in a vast measurement series. Every object under test was measured for 400 different orientations and different locations with an offset up to 0.2 m for a full circular track, a restricted track of 270° and 180°, respectively.

In the evaluations, every CSS of an object under test was compared with all 12 reference CSS data by a correlation function. The global maximum of the 12 correlation functions then indicated the recognized object.

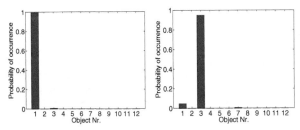

Figure 46. Discrete probability density function of the recognition process of object o1 for 270° track (left) and 180° track (right).

Figure 46 shows that object o1 can clearly be recognized even with the restricted track of 270°. Only for the 180° angle restriction, object o3 is recognized instead of o1 which is not unexpected since o3 equals a part of o1.

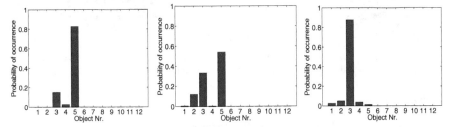

Figure 47. Discrete probability density function of the recognition process of object o5 for a full track (left), 270° track (middle) and 180° track (right).

Figure 47 shows that object o5 can be recognized in about 80% of all cases for a full track. This value drops to about 60% when the track is restricted to 270°. For 180° track o5 cannot be recognized anymore and is confused with o3.

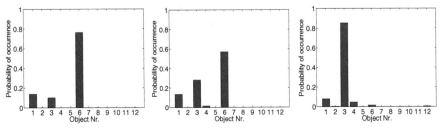

Figure 48. Discrete probability density function of the recognition process of object o6 for a full track (left), 270° track (middle) and 180° track (right).

Object o6 has approximately the same recognition characteristic as o5, as shown in Fig. 48. As both edge lengths of o6 are 0.3 m, o1 is falsely recognized as a certain part of its cross section has the same form as o5 or o6.

7.8. Transfer onto the mobile robot

Research investigations and algorithms had to be transferred onto a new mobile platform to demonstrate the essential results obtained in CoLOR. Many challenges in the area of robotics, localization and robot motion resulted. Additionally, an optimal alignment of the sensors and robust UWB radar sensing conditions had to be taken into account for the motion of the robot. The motion algorithm for the robot when moving around the objects are designed to be contour adaptive, exhibit collision avoidance features, and use the shortest tracks through a room while maintaining optimal antenna alignments.

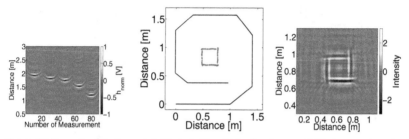

Figure 49. Example of the inspected object o1. The obtained radargram (left), the wavefront-based imaging of the object with robot track (middle) and the same object with a migrated image (right).

8. UWB Imaging

8.1. Imaging in distributed multi-static sensor networks

Although imaging in distributed multi-static sensor networks results in a rough image of the environment, its quality is usually better than the quality of the image obtained by a single bat-type sensor autonomously operating in an unknown environment. Sensor networks offer more diverse information for imaging algorithms. The resulting quality of the image greatly depends on the number and the positions of the illuminating and observing nodes. If there are enough illuminating and observing nodes available simultaneously, then the instantaneous information can be used even for imaging of time variant scenarios with moving objects. For imaging of static environments, one moving observer is enough. The moving observer collects data, and the image is built sequentially and improved gradually by an imaging algorithm. The time domain imaging algorithms are referred to as Kirchhoff migration. They rely on Born's approximation, which presumes undisturbed ray-optical propagation [9]. Their basic principle is depicted in Fig. 50 (a).

The illuminator transmits a signal at the fixed point $[x_T, y_T]$. At the variable position $[x_{Ri}, y_{Ri}]$ the receiver collects the impulse responses. Assuming single bounce reflection and the propagation velocity v, the echo reflected from an object situated at the position $[x_o, y_o]$ can be found at the time delay $\tau_i = (r_T + r_{Ri})/v$ in the measured impulse response $R_i(\tau)$. One observation determines an ellipse around the transmitter-receiver pair. In order to focus the image, observations from different positions must be fused. Conventional migrations propose to fuse the observations by a simple summation

$$o(x_o, y_o) = \frac{1}{N} \sum_{i=1}^{N} R_i(\tau_i) \tag{17}$$

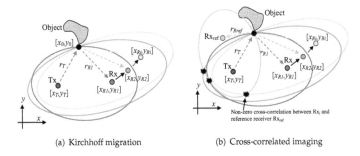

(a) Kirchhoff migration   (b) Cross-correlated imaging

Figure 50. Principle of imaging algorithms.

where N is the number of observations available for the data fusion. A disadvantage of this fusion is indicated in Fig. 50 (a). The signal energy does not only cumulate at the desired pixel. Other local maxima arise at every position where ellipses are crossing. Moreover, the elliptical traces are added to the focused image as well. Both effects decrease the image quality. In [73] we have proposed a data fusion method which reduces these artifacts and enhances the image quality. The algorithm is based on the cross-correlation of multiple observations. The spatial scheme for the cross-correlation algorithm is indicated in Fig. 50 (b). The observation from the reference receiver Rx_{ref} is cross-correlated with each observation of the moving receiver. The cross-correlation results in non-zero values only at positions where the ellipse related to the reference receiver crosses the ellipse related to the moving receiver. The non-zero values which result from the cross-correlation are summed by the data fusion algorithm according to

$$o(x_0, y_0) = \frac{1}{N} \sum_{i=1}^{N} \int_{-T/2}^{T/2} R_i(\tau_i + \zeta) R_{ref}(\tau_{ref} + \zeta) d\zeta \qquad (18)$$

where $\tau_{ref} = (r_T + r_{Rref})/v$ is the time delay which is related to the echo measured by the reference receiver, and T is the pulse duration of the autocorrelation function of the stimulation signal. In order to improve image quality, the number of reference receivers can be increased. The performance of this imaging algorithm strongly depends on the spatial arrangement of the cross-correlated receivers. If the arrangement of the receivers is disadvantageously chosen, the performance of this algorithm tends toward the conventional migration described by (17). If the spatial arrangement is selected in an optimum way the performance of the cross-correlated imaging outperforms the conventional algorithm. In order to illustrate the differences between the conventional and the cross-correlated algorithm, data simulated by the ray tracing algorithm [60] were used for the data fusion. The simulated scenario consisted of a sensor node which moved through a rectangular room of size 8 m by 9 m. The inspected room contained seven objects, namely, one large distributed object (some meters) and six smaller objects (decimeters). The sensor node had one transmitting and two receiving antennas and operated in the frequency band from 4.5 GHz to 13.4 GHz. The antennas were directional, therefore, the moving sensor scanned the room by a full turn at selected positions. Almost 10 thousand impulse responses were simulated for this scenario. Figure 51 (a) shows the result obtained by the conventional migration algorithm. The focused image contains intersecting ellipses. However, as indicated before, the intersections are not exclusively at the target positions. The ellipses also intersect at positions where no target is situated and even traces of these ellipses make the interpretation of the image difficult. Figure 51 (b) illustrates the image of the same scenario obtained by the cross-correlated algorithm. The color scaling of both images in Fig. 51 (a) and (b) is the same. The reduction of image artifacts by means of the cross-correlated algorithm is evident. Elliptical traces were significantly reduced. This helps to identify the inspected environment in this figure more clearly. A more detailed discussion on the cross-correlated imaging with measurement examples is given in [74] and [75].

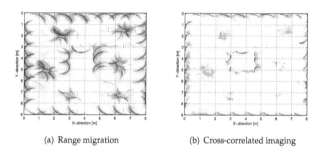

        (a) Range migration   (b) Cross-correlated imaging

Figure 51. Imaging in sensor networks.

8.2. Stationary target detection in an unknown indoor environment

Indoor stationary object detection in an unknown environment is important for many civilian and military applications, such as indoor surveillance, search and rescue operations, logistics, security and so on. Compared with target detection in a known environment, it presents some challenges:

- *The detection takes place in an unknown environment (e.g. an environment after disaster).* Thus, it is not possible to probe the environment before the presence of target. A range of techniques based on *"a priori"* information of the background cannot be utilized (e.g. [41, 64]). Additionally, the statistical distributions of clutter and noise are sometimes unknown, leading to further complications in the detector design.
- *The targets concerned are stationary with respect to the background.* There is no distinct speed difference between the targets and the background. Hence, it prevents the application of the motion-based detection techniques, such as Doppler based approaches, subtraction (or cancellation) between sequence snapshots, etc [58].
- *It is a highly cluttered environment.* The targets (objects of interest) are typically surrounded by clutter (objects, which are not of interest, as shown in Fig. 52). In this case, the response of targets is not always stronger than clutter. In other words, the Signal to Interference and Noise Ratio (SINR) of the system is not always high enough to ensure a reliable detection.

8.2.1. Signal enhancement

Due to the challenges mentioned above, a possible detection scheme is given in Fig. 53. In the scheme, the objective of the first step, "signal enhancement" which is realized by a *"time-shift & accumulation"* operation, is to raise the SINR of system, and to transform the unknown environment into a Gaussian clutter and noise environment so that the detector could be

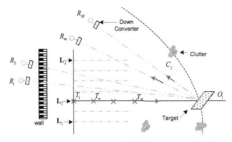

Figure 52. Measurement configuration. $T_1...T_n...T_N$ are different transmission positions on the track $\mathbf{L}_l$. "$R_1...R_M$" are sparsely spaced receivers with different reception angles. The dashed curve is an ellipse segment with the foci R_m and T_n.

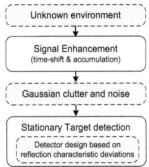

Figure 53. Flow chart of the algorithm .

designed based on Gaussian clutter and noise. Mathematically, the "*time-shift & accumulation*" operation is described as

$$\sum_{n=1}^{N} s_{Rm,Tn}\left(t + t_{Tn,T1}^{diff}\right) \tag{19}$$

where $s_{Rm,Tn}(t)$ is the signal received by the receiver R_m with respect to the transmission position T_n, and $t_{Tn,T1}^{diff} = |\mathbf{T}_n - \mathbf{T}_1| / c$ is the time-delay to be compensated for. $\mathbf{T}_n$ and $\mathbf{T}_1$ are the position vectors of T_n and reference transmission position T_1 , respectively. In the operation,

- the responses of the objects (O_i) in the direction of interest (for example the direction of $\mathbf{L}_l$ in Fig. 52, are time-shifted, aligned and then accumulated. It is a coherent operation in which the parameter $t_{Tn,T1}^{diff}$ is used for time-delay correction. Hence, responses are enhanced by N times compared to the case of single channel data.

- The responses of objects (C_j) out of the direction of interest, are non-uniformly time-shifted, disturbed and then accumulated. It is an incoherent operation. As a consequence, for a certain time-slot, the responses from objects located on different ellipses (or ellipsoids) are non-uniformly time-shifted and accumulated. Hence, these responses are attenuated compared to the output of the coherent operation above, [70].

Figure 54. Measurement environment. R_3 and R_4 are placed behind the wall.

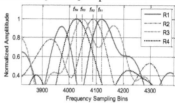

Figure 55. Measured spectrum-shift

8.2.2. The detection principle

For a detection problem, generally, detection algorithms are designed based on the differences (or deviations) between the targets and the background (clutter, noise). The ability to distinguish objects depends on how much their properties (e.g. electromagnetic properties, motion properties) deviate from the properties of the background. For the concerned stationary target which has no speed differences relative to the environment it can be detected based on the reflection characteristic deviations between the target and the background. Here, we take a planar-surface target as an example to illustrate the principle of indoor stationary target detection.

Consider a detection scenario as shown in Fig. 52 and Fig. 54. The transmitter moves along track $\mathbf{L}_l$ to probe the environment, and to enhance the signal. The target is a concrete brick-like object with a square plane-surface ($50cm \times 50cm$). It is located at $x = 24cm$ on the track $\mathbf{L}_l$, with an orientation angle of $60°$. The receivers R_1, R_2, R_3 and R_4 receive the scattered signals from the surroundings. They are located at different directions with respect to the target. The transmitter-target-receiver angles are $13.9°$, $20.3°$, $26.3°$ and $31.7°$, respectively. The antennas of the transmitter and receivers are omnidirectional and horn antennas, respectively. The receivers R_1 and R_2 are in the same room with the target and the transmitter, while the receivers R_3 and R_4 are placed behind a 20cm-thick concrete wall.

Theoretically, it can be proved that if the target is a planar-surface object with diffuse reflections, after passing through the down-converter and certain mathematical transforms, the received signals from different directions would have a spectrum-shift, which is given by $\Omega_{j,i} = \omega_c(\gamma_{j,i} - 1)$ under the illuminating geometry in Fig. 52. Here, the subscripts i and j are the indices of the receivers located at the different directions, ω_c is the angular central frequency, and $\gamma_{j,i}$ is a parameter associated with the reflection angles,

$$\gamma_{j,i} \approx 1 + \left(\sin\theta_{Rj} - \sin\theta_{Ri} \right) / \left(|\mathbf{V}_{bi}| \sin\theta_{bi} \right) \tag{20}$$

where, θ_{Ri} and θ_{Rj} are the reflection angles of the receiver R_i and R_j, respectively. $\mathbf{V}_{bi}$ is the bisector of the angle $\angle R_i X T_1$. θ_{bi} is the reflection angle of $\mathbf{V}_{bi}$. According to this principle, stationary plane-surface objects can be detected.

In practice, (20) is demonstrated in Fig. 55, where the spectra of receivers R_1-$R4$ are measured based on the scenario of Fig. 54. It can be shown that $\frac{f_{Ri}-f_{R1}}{f_{Rj}-f_{R1}} \approx \frac{\gamma_{i,1}-1}{\gamma_{j,1}-1}$, where $f_{Ri}, f_{Rj} \in [f_{R_1}, f_{R_2}, f_{R_3}, f_{R_4}]$ are the frequencies at which the maximum amplitudes are located. The differences between f_{R1}-f_{R4} indicate the spectrum shifts between R_1, R_2, R_3 and R_4, correspondingly.

8.2.3. Detection

According to the principle discussed above, if we take the observation of R_1, $\bar{s}_{R1,Tn}(\tau)$, as a reference signal, the observation of receiver R_j can be given by $\bar{s}_{Rj,Tn}(\tau) \approx \bar{s}_{R1,Tn}(\tau)\exp(-j\gamma_{j,1}\tau)$, where $\gamma_{j,1}$ is determined by the illuminating geometry according to (20). Based on this relationship, if we consider the effects of unwanted contributions due to clutter $\mathbf{c}$ and noise $\mathbf{n}$, the signal model for detection can be given as

$$\mathbf{y} = \bar{\mathbf{s}}^{ref} \otimes \mathbf{K} + \mathbf{c} + \mathbf{n} = \mathbf{x} + \mathbf{c} + \mathbf{n} \tag{21}$$

where $\mathbf{y}$ is an $N_{MN} \times 1$ vector. $N_{MN} = M \times N$, where M and N are the numbers of receivers and transmission positions, respectively. $\mathbf{x} = \bar{\mathbf{s}}^{ref} \otimes \mathbf{K}$, where $\bar{\mathbf{s}}^{ref} = [\bar{s}_{T1}^{ref}, \bar{s}_{Tn}^{ref}, ..., \bar{s}_{TN}^{ref}]^T$, $\bar{s}_{Tn}^{ref} = \bar{s}_{R1,Tn}(\tau)$, and $\mathbf{K} = [1, \exp(-j\gamma_{2,1}\tau), ..., \exp(-j\gamma_{M,1}\tau)]^T$. The symbol $\otimes$ denotes the Kronecker product.

According to the Central Limit Theorem (CLT) of the probability theory, if a large number of clutter from different sources (scattered from different objects) is accumulated, the statistical distribution of the sum will approach a Gaussian distribution. As the scenario concerned takes place in a cluttered indoor environment which has many scatterers, we assume that the clutter $\mathbf{c}$ and the noise $\mathbf{n}$ can approach a Gaussian distribution due to the *"time-delay & accumulation"* operations given by (19). Hence, our detection problem simplifies to searching targets in Gaussian clutter and noise. We assume that the noise $\mathbf{n}$ and the clutter $\mathbf{c}$ are $N_{MN} \times 1$ independent zero-mean complex Gaussians with $N_{MN} \times N_{MN}$ known covariance matrices $\mathbf{M}_{c+n} = E[(\mathbf{c}+\mathbf{n})(\mathbf{c}+\mathbf{n})^H]$. $\mathbf{M}_{c+n}$ is a positive semidefinite and Hermitian symmetric matrix [63]. The superscript H indicates conjugate transpose of a matrix. Based on the signal model above, a matched filter detector is given as

$$\mathbf{h} = \mathbf{M}_{c+n}^{-1}\mathbf{x} = \mathbf{M}_{c+n}^{-1}(\bar{\mathbf{s}}^{ref} \otimes \mathbf{K}) \tag{22}$$

The probability of detection can be given as $P_d = 1 - \Phi(\frac{k_{th}-\mathbf{h}^H\mathbf{x}}{D})$, and the probability of false alarms can be given as $P_f = 1 - \Phi(\frac{k_{th}}{D})$. Here, k_{th} is the threshold for a likelihood test and $\Phi(x) = \frac{1}{\sqrt{2\pi}}\int_{-\infty}^{x}e^{-t^2/2}dt$. The parameter D is defined as $D^2 = \mathbf{x}^H\mathbf{M}_{c+n}^{-1}\mathbf{x} = N_{MN}\frac{x_{av}^2}{E[(\mathbf{y}-\mathbf{x})^H(\mathbf{y}-\mathbf{x})]}$, where x_{av}^2 is the average signal power of $\mathbf{x}$. Generally, $\frac{x_{av}^2}{E[(\mathbf{y}-\mathbf{x})^H(\mathbf{y}-\mathbf{x})]}$ can be regarded as the SINR at the output of the detector [43].

In terms of the Neyman-Pearson criterion, the threshold k_{th} could be set as $k_{th} = D\Phi(1 - P_f)$, and the detection probability can be given as $P_d = 1 - \Phi(\Phi^{-1}(1 - P_f) - D)$.

9. Detection and localization of moving objects

In this section we describe methods for the detection and localization of several moving persons who do not have a tag or device attached to them. This is useful for applications where the targets being tracked are not expected to cooperate with the system.

A single UWB sensor suffers from narrow-band interferences as well as shadowing when detecting multiple targets. Often, the closest target can be observed best. Due to shadowing caused by the closest target, the other targets are usually invisible, although they lie within the coverage of the sensor. The closest target attenuates the electromagnetic waves that propagate toward the targets located behind it. Thus, the shadowed targets are almost impossible to detect by a single sensor node. Using a distributed network of UWB radars, the estimated target positions are refined by fusing the information available from the multiple sensors which are able to detect the targets. It also allows for the observation of the targets from different viewing angles.

A network of multiple static bat-type sensors distributed around the inspected area is used. Each bat-type sensor node is capable of autonomously detecting and localizing the present targets. The weak echoes of the targets are first detected in the backscattered signal, after which the detections are correctly assigned to the targets and finally, the target information from all sensors is fused together resulting in location estimates of the targets in the scenario. Here we describe two methods that can cope with this challenging task. The first one uses simplification assumptions of one target detection per sensor to combat the data association problem. It is described in Section 9.1. The second one is based on the probabilistic hypothesis density (PHD) filter for range estimation and position tracking and is described in more detail in Section 9.2.

In both methods, background subtraction is used for target detection [9, 21, 38, 71]. The echoes evoked by the moving targets are usually weak and must be detected in the presence of other strong multipath signals. The disturbing signals are usually time invariant and overlay echoes from the target. Therefore, one received impulse response is insufficient to separate them. However, due to their time-invariance these background signals can be estimated by a suitable estimation technique from a sequence of received impulse responses. The subtraction of the estimated background from the measured data leads to a signal where the weak target echo can be recognized (see Fig. 56(b)) and its range can be estimated more easily.

For the verification of the methods, a sensor network constellation as in Fig. 56(a) is used. Five UWB bat-type sensors are placed around a large foyer, and one bat-type sensor is placed behind one of the walls, directed towards the area of interest. Directional horn antennas with different size and quality are used, resulting in variable target detection performance between the sensors. A scenario with three moving persons was used. All three persons move in the area of interest during the whole measurement time. They move in a straight line from one wall to the opposite and back. The starting position of the targets is shown by a star in Fig. 56(a). The arrow signifies the direction of movement at the start of the measurement.

9.1. Low complexity method for the localization of multiple targets in sensor networks

The method presented in this section was proposed as a computationally efficient method for the localization of multiple targets in dense sensor networks. Each target can be usually observed by at least one sensor node. As can be seen in Fig. 56(b), the closest target

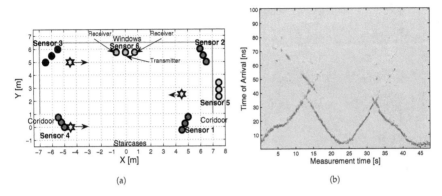

(a)   (b)

Figure 56. a) The verification scenario and b) the received impulse responses after background subtraction

to the sensor evokes the strongest echo, and almost completely shadows the other two targets when it is very close to the sensor. In order to simplify the localization algorithm, we make the assumption that each sensor node can observe only the closest target. This assumption allows the algorithms for multi-target detection such as the constant false alarm rate (CFAR) techniques [15], algorithms for data association, or multi-hypotheses tests to be skipped. This makes the proposed algorithm computationally efficient and suitable for real-time applications. On the other hand, this simplification restricts the number of targets that can be localized by the approach. A sensor network with N sensor nodes can localize up to N targets in real time. A more detailed description of this method is presented in [72].

9.1.1. Single target range estimation

After the time-invariant background signals have been removed from the received impulse response, the targets' ranges need to be estimated. The distance from the transmitter to the target and back to the receiver is considered as target range. The target range of the closest target can be estimated by using one of the various threshold-based approaches where the leading edge of the received signal is detected [10–12, 22, 56]. Threshold-based approaches offer simple techniques which detect the leading edge of a received signal by comparing the received signal magnitude or energy with a predefined threshold. The choice of an appropriate threshold is mandatory for a good performance of this estimator.

9.1.2. Localization

As in our scenario each sensor has two receivers, from each sensor node we have the estimated range of the closest target by the two receivers. The estimated ranges determine two ellipses whose focal points are determined by the locus of the transmitting antenna and the locus of the respective receiving antenna. The target position is determined as the intersection of the two ellipses within the area of detection of the respective sensor. The location estimates with respect to each of the six sensor nodes is shown in Fig. 57(a). Here, the estimates are represented by the same color as the respective sensor node.

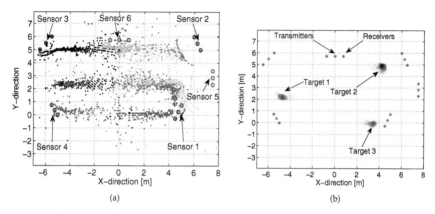

Figure 57. The target position estimates a) by each sensor and b) after data fusion

9.1.3. Data fusion

The location estimates from all sensor nodes must be fused together by an appropriate algorithm. We propose an imaging algorithm that does not require any data association. It creates a sampled image of the inspected area at a given time i which is stored in a matrix $\mathbf{P_i}$. Each element of the matrix corresponds to the spatial coordinates within the inspected area. The image is updated according to

$$\mathbf{P_i} = \alpha \mathbf{P_{i-1}} + (1 - \alpha)\mathbf{U_i} \tag{23}$$

where $\mathbf{U_i}$ is the innovation matrix and α is the forgetting factor. Thus, the new image estimate $\mathbf{P_i}$ takes a fraction of the previous estimate $\mathbf{P_{i-1}}$ and a fraction from the innovation matrix $\mathbf{U_i}$. The innovation matrix $\mathbf{U_i}$ maps the location estimates of all sensor nodes. The matrix $\mathbf{P_i}$ indicates moving targets in the environment as hot spots within the image.

An example of the data fusion is given in Fig. 57(b). It depicts one snapshot of the inspected area computed for a certain measurement time. Due to the exponential averaging each target appears in the snapshot like a "comet" with a tail. This tail indicates previous locations of the objects. Its length can be adjusted by adjusting the weighting factor α in Eqn. 23.

9.2. Localization and tracking of multiple targets in sensor networks based on the PHD filter

Unlike the method presented in Section 9.1, this method uses the range information related to all targets detected by each sensor. Traditionally, multiple targets are tracked using two-step approaches: data association of the observations to targets followed by single target tracking on the associated data. The traditional approaches are generally computationally expensive, thus, we use a less computationally expensive alternative based on random finite sets (RFS) [36, 37]. It allows for tracking of time-variant number of unknown moving targets in the presence of false alarms, miss-detections and clutter. An approximation of the Bayesian multi-target tracking represented by RFSs, by its first order moment leads to the PHD filter [27]. There are two implementations of the PHD filter, one based on a sequential Monte Carlo approach [65], and the other based on Gaussian mixtures [66]. Here, we use the

Gaussian mixture (GM) approach. We first use it for estimating the target ranges with respect to each sensor, and later for fusion of the target location estimates by all sensors. The method is explained in greater detail in [27].

9.2.1. Multiple target range estimation

In Fig. 58 we describe the processing done on a measured impulse response from raw measurement to range estimation. The impulse response presented is from a scenario with three moving targets. First the measured impulse response is shown in dark blue. As can be seen, the target echoes are non-detectable. After subtracting the time-invariant background (resulting signal shown in green) the echoes of the moving targets are detectable more easily. A Gaussian adaptive threshold constant false alarm rate (CFAR) method as in [15] is used for extracting the ranges from the resulting signal. The adaptive threshold calculated is shown in cyan. A Neyman-Pearson detector is used to discriminate between the noise and the target echo. The ranges extracted using the CFAR approach are shown as points in magenta.

The CFAR detector is not immune to clutter noise and false detections. As we can see in Fig. 58, four targets are detected by the CFAR method although there are only three targets in the scenario. In addition, there are also multiple detections per target, making it difficult to decide if there are multiple targets in the vicinity of each other, or it is only a single target. By using a GM-PHD filter we improve the target range estimates. The ranges extracted by the CFAR method are used as observations for the PHD filter, and a linear Gaussian model is assumed. The target detection and survival probability are assumed to be state independent. To estimate the number of targets and their states, first the Gaussian terms with low weights are pruned and the Gaussian terms that are within a small distance of each other are merged together. The number of targets is estimated by the number of Gaussian terms with a weight above a predefined threshold, and their state is represented by the mean terms of these Gaussian mixtures. The surviving Gaussian terms for the impulse response given in Fig. 58 are shown in black.

9.2.2. Target localization

As in Section 9.1.2, the target locations are computed using the ranges estimated by the two receivers of the sensor. As multiple targets are detected by each receiver, the range estimates from both receivers corresponding to the same target need to be associated. An intersection threshold T_s is defined for each sensor. The ranges of a target in the inspected area with respect to both receivers of a sensor s are first calculated as $r_k^{s,1}$ and $r_k^{s,2}$. The intersection threshold is then defined as the maximum possible absolute difference between these ranges such that the target lies in the area of interest.

$$T_s = \max_{k \in A} |r_k^{s,1} - r_k^{s,2}| \tag{24}$$

As the size of the inspected area is known or approximated by the sensor detection range, the intersection threshold is calculated prior to scanning the environment. The range association is done such that for each range estimate from the first receiver, we associate the range estimate from the second receivers which satisfies

$$|r_k^{s,1} - r_k^{s,2}| < T_s \tag{25}$$

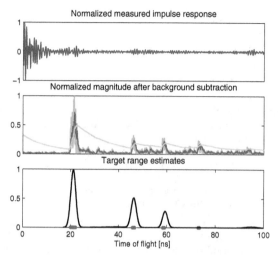

Figure 58. Target echo detection - measured impulse response (blue), normalized signal magnitude after background subtraction (green), CFAR test statistic (red), CFAR adaptive threshold (cyan), indices of detected targets by CFAR (magenta) and Gaussian mixtures representing the estimated target ranges (black) are shown

as a range estimate that corresponds to the same target. When multiple range estimates comply with this rule, the range estimate which results in the smallest absolute difference is chosen.

The target location is analytically calculated as the intersection of the ellipses defined by the associated range estimates. The location estimates with respect to each of the six sensor nodes is shown in Fig. 59(a), where the estimates are represented by the same color as the respective sensor node.

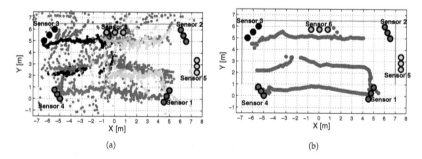

Figure 59. a)The target position estimates by each sensor and b) the target tracks after data fusion

9.2.3. Data fusion and target tracking

The location estimates from all sensors are fused together resulting in a single target location per target. The estimated target locations are not in a track form and contain a significant amount of noise due to small errors in the range estimation. A simplified GM-PHD filter is applied for fusing the target location estimates from all sensors. The location estimates are used to form the observation RFS for the GM-PHD filter. The target state is defined by the 2D target coordinates and velocity vector, $\mathbf{x} = [x\ \dot{x}\ y\ \dot{y}]^T$. The targets are identified using nearest neighbor association, and the results for the three-target scenario can be seen in Fig. 59(b).

Both methods presented in this section can be used for the localization of multiple non-cooperative targets using distributed UWB sensor network with static sensors as in the scenario described in Section 2.1 .

10. Conclusion

The CoLOR project was devoted to the recognition of unknown environments using UWB technology. This topic encompasses a number of partial challenges. In order to obtain a complex picture of some catastrophic scenario, like the detection of victims after some natural disasters, their location in collapsed buildings, the geometrical information and the status of the buildings, we derived new detection, localization and imaging algorithms. Their performance was analyzed on simulated data and data measured in realistic scenarios using UWB sensors. These UWB sensors are capable of real-time operation in MIMO configuration. This allows us to analyze the application of UWB sensor networks and cooperative approaches for the localization of sensor nodes within the network, for the localization of people, for the detection and monitoring of their live signs, and for the imaging of their surroundings.

It was shown that by using a mobile UWB radar with multiple antennas, it is possible in an efficient way to reconstruct the basic layout of rooms and the position of freestanding objects. The detected features are added to a map while at the same time the own position is estimated (SLAM). To minimize the computational cost and the number of measurements needed, simplified models for wave propagation and stochastic, dynamic state space estimators where enhanced. The method of data association proved to be most critical regarding the precision and reliability of the map.

Using this map as a-priori information, the detection, localization and the imaging of the objects within an indoor scenario can be performed using the developed localization and imaging algorithms. By knowing the location of the individual objects, the potential of UWB radar was fully tapped by obtaining super-resolved local information about 2D as well as 3D complex objects (concerning the outer contour). The interior of objects was gathered by novel algorithms which are based on exact radiation patterns depending on the permittivity of the medium while showing low computational load. The obtained radar images are post-processed by means of object recognition algorithms designed for full, fragmented or restricted illumination to provide recognition of the object under test from a finite alphabet. By the adaption of classical ellipsometry to the UWB-range, an estimation of dielectric surface properties can robustly be performed even for small dimensioned objects with a size of a couple of wavelengths. In addition polarimetric measurements as well as polarimetric data processing were taken into account to obtain object features which may remain invisible in mono polarized systems.

In order to test and compare different algorithms and antenna arrangements for indoor UWB sensing and imaging, a realistic UWB multi-path propagation simulation tool was developed. The propagation model is based on a hybrid approach which combines the deterministic ray tracing method (based on geometrical optics and the uniform theory of diffraction) with statistically distributed scatterers. Verification measurements show that the new model delivers very realistic channel parameters like channel impulse response, azimuth spectra, and path loss. Thus, it is suitable for an application in UWB system simulations.

Further within CoLOR a flexible and dual-polarized UWB antenna array has been developed. The major challenges beside the huge bandwidth itself were to design antenna elements which are able to meet the requirements regarding size, pattern, beam-width, polarization and the location of the phase center (over frequency). Via switches it is possible to select and control the single elements of this array, which allows its adaption to the different localization, imaging and object recognition algorithms and applications in this project. Due to the dual polarized antenna elements, the possibility to take advantage of polarization diversity is given and demonstrated.

Our results show that the UWB technology and especially the cooperative approach that fuses diverse information from multiple sensors provide a big potential for safety, security and emergency applications.

Author details

Rudolf Zetik, Honghui Yan, Elke Malz, Snezhana Jovanoska, Guowei Shen and Reiner S. Thomä
Ilmenau University of Technology, Germany

Rahmi Salman, Thorsten Schultze, Robert Tobera and Hans-Ingolf Willms
University Essen-Duisburg, Germany

Lars Reichardt, Malgorzata Janson, Thomas Zwick and Werner Wiesbeck
Karlsruhe Institute of Technology (KIT), Germany

Tobias Deißler and Jörn Thielecke
Friedrich-Alexander University Erlangen-Nürnberg, Germany

11. References

[1] Adamiuk, G., Beer, S., Wiesbeck, W. & Zwick, T. [2009]. Dual-orthogonal polarized antenna for UWB-IR technology, *Antennas and Wireless Propagation Letters, IEEE* 8: 981 –984.

[2] Adamiuk, G., Janson, M., Wiesbeck, W. & Zwick, T. [2009]. Dual-polarized UWB antenna array, *IEEE International Conference on Ultra-Wideband, ICUWB'09*, Vancouver, Canada, pp. 159–163.

[3] Adamiuk, G., Zwick, T. & Wiesbeck, W. [2008]. Dual-orthogonal polarized vivaldi antenna for ultra wideband applications, *17th International Conference on Microwaves, Radar and Wireless Communications, 2008. MIKON.*

[4] Araujo, E. G. & Grupen, R. [1998]. Feature detection and identification using a sonar-array, *In Proceedings of the IEEE International Conference on Robotics and Automation*, IEEE Computer Society Press, pp. 1584–1589.

[5] Balanis, C. A. [1996]. *Antenna Theory: Analysis and Design*, New York: Wiley.

[6] Blech, M. & Eibert, T. [2007]. A dipole excited ultrawideband dielectric rod antenna with reflector, *Antennas and Propagation, IEEE Transactions on* 55(7): 1948 –1954.

[7] Borras, J., Hatrack, P. & Mandayam, N. B. [1998]. Decision theoretic framework for NLOS identification, *Proc. 48th IEEE Vehicular Technology Conf. VTC 98*, Vol. 2, pp. 1583–1587.

[8] Cramer, R.-M., Scholtz, R. & Win, M. [2002]. Evaluation of an ultra-wide-band propagation channel, *IEEE Transactions on Antennas and Propagation* 50(5): 561–570.

[9] Daniels, D. J. (ed.) [2004]. *Ground penetrating radar (2nd Edition)*, Vol. 15 of *IEE radar, sonar, navigation, and avionics series*, Institution of Electrical Engineers, London, UK.

[10] Dardari, D., Chong, C.-C. & Win, M. Z. [2006]. Analysis of threshold-based TOA estimator in UWB channels, *Proc. Eur. Signal Process. Conf. (EUSIPCO)*, Florence, Italy.

[11] Dardari, D., Chong, C.-C. & Win, M. Z. [2008]. Threshold-based time-of-arrival estimators in UWB dense multipath channels, *IEEE Transactions on Communications* 56(8): 1366–1378.

[12] Dardari, D. & Win, M. Z. [2006]. Threshold-based time-of-arrival estimators in UWB dense multipath channels, *Proc. IEEE International Conference on Communications*, Vol. 10, pp. 4723–4728.

[13] Deissler, T., Salman, R., Schultze, T., Thielecke, J. & Willms, I. [2010]. UWB radar object recognition for slam, *Proceedings of the 11th International Radar Symposium*, pp. 378–381.

[14] des Noes, M. & Denis, B. [2012]. Benefits from cooperation in simultaneous anchor-less tracking and room mapping based on impulse radio - ultra wideband devices, *19th International Conference on Systems, Signals and Image Processing, IWSSIP 2012*.

[15] Dutta, P., Arora, A. & Bibyk, S. [2006]. Towards radar-enabled sensor networks, *Information Processing in Sensor Networks, 2006. IPSN 2006. The Fifth International Conference on*, pp. 467 –474.

[16] Elsherbini, A., Zhang, C., Lin, S., Kuhn, M., Kamel, A., Fathy, A. & Elhennawy, H. [2007]. UWB antipodal vivaldi antennas with protruded dielectric rods for higher gain, symmetric patterns and minimal phase center variations, *Antennas and Propagation Society International Symposium, 2007 IEEE*, pp. 1973 –1976.

[17] Falsi, C., Dardari, D., Mucchi, L. & Win, M. Z. [2006]. Time of arrival estimation for UWB localizers in realistic environments, *EURASIP Journal on Applied Signal Processing* 2006: 1–13.
URL: *http://www.hindawi.com/journals/asp/2006/032082.abs.html*

[18] Fügen, T., Maurer, J., Kayser, T. & Wiesbeck, W. [2006]. Capability of 3-D ray tracing for defining parameter sets for the specification of future mobile communications systems, *IEEE Transactions on Antennas and Propagation* 54(11): 3125–3137.

[19] Gezici, S. [2008]. A survey on wireless position estimation, *Wireless Personal Communications* 44: 263–282.
URL: *http://www.springerlink.com/content/w1481x761266m417*

[20] Gezici, S. & Poor, H. [2009]. Position estimation via ultra-wide-band signals, *Proceedings of the IEEE* 97(2): 386 –403.

[21] Gezici, S., Tian, Z., Giannakis, G., Kobayashi, H., Molisch, A., Poor, H. & Sahinoglu, Z. [2005]. Localization via ultra-wideband radios: a look at positioning aspects for future sensor networks, 22(4): 70 – 84.

[22] Guvenc, I. & Sahinoglu, Z. [2005]. Threshold-based TOA estimation for impulse radio UWB systems, *Proc. IEEE International Conference on Ultra-Wideband ICU 2005*, pp. 420–425.

[23] Janson, M., Fügen, T., Zwick, T. & Wiesbeck, W. [2009]. Directional channel model for ultra-wideband indoor applications, *IEEE International Conference on Ultra-Wideband, ICUWB'09*, Vancouver, Canada.

[24] Janson, M., P., J., Zwick, T. & Wiesbeck, W. [2010]. Directional hybrid channel model for ultra-wideband MIMO systems, *4th European Conference on Antennas and Propagation, EuCAP'10*.

[25] Janson, M., Pontes, J., Fügen, T. & Zwick, T. [2012]. A hybrid deterministic-stochastic propagation model for short-range MIMO-UWB communication systems, *accepted for publication in Frequenz, Journal of RF-Engineering and Telecommunications* .

[26] Jemai, J., Eggers, P., Pedersen, G. & Kurner, T. [2009]. Calibration of a UWB sub-band channel model using simulated annealing, *IEEE Transactions on Antennas and Propagation* 57(10): 3439 –3443.

[27] Jovanoska, S. & Thomä, R. [2012]. Multiple target tracking by a distributed UWB sensor network based on the PHD filter, *Proc. 15th Int. Conf. Information Fusion (FUSION)*, Singapore, Singapore. submitted for publication.

[28] Kidera, S., Sakamoto, T. & Sato, T. [2010]. Accurate UWB radar three-dimensional imaging algorithm for a complex boundary without range point connections, *Geoscience and Remote Sensing, IEEE Transactions on* 48(4): 1993 –2004.

[29] Kmec, M., Sachs, J., Peyerl, P., Rauschenbach, P., Thomä, R. & Zetik, R. [2005]. A novel Ultra-wideband real-time MIMO channel sounder architecture, *XXVIIIth URSI General Assembly*, New Delhi, India.

[30] Li, J. & Stoica, P. (eds) [2008]. *MIMO radar signal processing*, Hoboken, NJ: Wiley.

[31] Li, X. & Pahlavan, K. [2004]. Super-resolution TOA estimation with diversity for indoor geolocation, *IEEE Transactions on Wireless Communications* 3(1): 224–234.

[32] Loeffler, A., Wissendheit, U., Gerhaeuser, H. & Kuznetsova, D. [2010]. A multi-purpose rfid reader supporting indoor navigation systems, *IEEE International Conference on RFID-Technology and Applications (RFID-TA)* 1: 43–48.

[33] Lostanlen, Y. & Gougeon, G. [2007]. Introduction of diffuse scattering to enhance ray-tracing methods for the analysis of deterministic indoor UWB radio channels, *International Conference on Electromagnetics in Advanced Applications, ICEAA'07*, pp. 903–906.

[34] Lostanlen, Y., Gougeon, G., Bories, S. & Sibille, A. [2006]. A deterministic indoor UWB space-variant multipath radio channel modelling compared to measurements on basic configurations, *First European Conference on Antennas and Propagation, EuCAP'06*, pp. 1–8.

[35] Mackworth, S. & Mokhtarian, F. [1988]. The renormalized curvature scale space and the evolution properties of planar curves, *Computer Vision and Pattern Recognition, 1988. Proceedings CVPR '88., Computer Society Conference on*, pp. 318 –326.

[36] Mahler, R. [2003]. Multitarget Bayes filtering via first-order multitarget moments, 39(4): 1152–1178.

[37] Mahler, R. P. S. [2001]. *Random set theory for target tracking and identification*, Handbook of Mutlisensor Data Fusion, CRC Press, London, p. 14.

[38] McIvor, A. M., Zang, Q. & Klette, R. [2001]. The background subtraction problem for video surveillance systems, *RobVis*, pp. 176–183.

[39] Meissner, P. & Witrisal, K. [2012]. Multipath-assisted single-anchor indoor localization in an office environment, *19th International Conference on Systems, Signals and Image Processing, IWSSIP 2012*.

[40] Molisch, A., Kuchar, A., Laurila, J., Hugl, K. & Bonek, E. [1999]. Efficient implementation of a geometry-based directional model for mobile radio channels, *IEEE VTS 50th Vehicular Technology Conference, 1999. VTC 1999 - Fall* 3: 1449–1453.

[41] Moura, J. M. F. & Jin, Y. [2008]. Time reversal imaging by adaptive interference canceling, *IEEE Transactions on Signal Processing* 56(1): 233–247.

[42] Nasr, K. [2008]. Hybrid channel modelling for ultra-wideband portable multimedia applications, *Microwaves, Antennas & Propagation, IET* 2(3): 229–235.

[43] Poor, H. V. [1994]. *An introduction to signal detection and estimation, Chapter III*, 2nd edn, Springer-Verlag.

[44] Porebska, M., Adamiuk, G., Sturm, C. & Wiesbeck, W. [2007]. Accuracy of algorithms for UWB localization in NLOS scenarios containing arbitrary walls, *2nd European Conference on Antennas and Propagation, EuCAP'07*, Edinburgh, UK.

[45] Reichardt, L., Kowalevski, J., Zwirello, L. & Zwick, T. [2012]. Compact, teflon embedded, dual-polarized ultra wideband (UWB) antenna, *accepted for pulicatiob on IEEE, Antennas and Propagation Society International Symposium.*

[46] Ren, Y.-J. & Tarng, J.-H. [2007]. A hybrid spatio-temporal model for radio propagation in urban environment, *2nd European Conference on Antennas and Propagation, EuCAP'07*, Edinburgh, UK.

[47] Sachs, J., Peyerl, P., Zetik, R. & Crabbe, S. [2003]. M-sequence ultra-wideband-radar: state of development and applications, *International Radar Conference*, Adelaide, Australia, pp. 224 – 229.

[48] Salman, R. & Willms, I. [2010]. A novel UWB radar super-resolution object recognition approach for complex edged objects, *Ultra-Wideband (ICUWB), 2010 IEEE International Conference on*, Vol. 2, pp. 1 –4.

[49] Salman, R. & Willms, I. [2011a]. 3d UWB radar super-resolution imaging for complex objects with discontinous wavefronts, *Ultra-Wideband (ICUWB), 2011 IEEE International Conference on*, pp. 346 –350.

[50] Salman, R. & Willms, I. [2011b]. Super-resolution object recognition approach for complex edged objects by UWB radar, *Object Registration*, InTech.

[51] Salman, R. & Willms, I. [2012]. Joint efficiency and performance enhancement of wavefront extraction algorithms for short-range super-resolution UWB radar, *Microwave Conference (GeMiC), 2012 The 7th German*, pp. 1 –4.

[52] Salman, R., Willms, I., Reichardt, L., Zwick, T. & Wiesbeck, W. [2012]. On polarization diversity gain in short range UWB-radar object imaging, *2012 IEEE International Conference on Ultra-Wideband (ICUWB).*

[53] Seitz, J., Vaupel, T., Jahn, J., Meyer, S., Boronat, J. G. & Thielecke, J. [2010]. A hidden markov model for urban navigation based on fingerprinting and pedestrian dead reckoning, *Information Fusion (FUSION), 2010 13th Conference on*, IEEE, pp. 1–8.

[54] Shen, G., Zetik, R., Hirsch, O. & Thomä, R. S. [2010]. Range-based localization for UWB sensor networks in realistic environments, *EURASIP Journal on Wireless Communications and Networking* pp. 1–9.
URL: *http://www.hindawi.com/journals/wcn/2010/476598.html*

[55] Shen, G., Zetik, R. & Thomä, R. S. [2008]. Performance comparison of TOA and TDOA based location estimation algorithms in LOS environment, *Proc. 5th Workshop on Positioning, Navigation and Communication (WPNC 2008)*, Hannover, pp. 71–78.

[56] Shen, G., Zetik, R., Yan, H., Hirsch, O. & Thomä, R. S. [2010]. Time of arrival estimation for range-based localization in UWB sensor networks, *2010 IEEE International Conference on Ultra-Wideband (ICUWB 2010)*, Nanjing, China.

[57] Shen, G., Zetik, R., Yan, H., Jovanoska, S. & Thomä, R. S. [2011]. Localization of active UWB sensor nodes in multipath and NLOS environments, *the 5th European Conference on Antennas and Propagation (EUCAP 2011)*, Rome, Italy.

[58] Skolnik, M. (ed.) [2002]. *Introduction to Radar Systems*, 3rd ed edn, Mc Graw-Hill.

[59] Sörgel, W., Waldschmidt, C. & Wiesbeck, W. [2003]. Transient responses of a vivaldi antenna and a logarithmic periodic dipole array for ultra wideband communication, *IEEE, Antennas and Propagation Society International Symposium.*

[60] Sturm, C., Sörgel, W., Knorzer, S. & Wiesbeck, W. [2006]. An adequate channel model for ultra wideband localization applications, *Antennas and Propagation Society International Symposium 2006, IEEE*, pp. 2140 –2143.

[61] Thomä, R. S., Hirsch, O., Sachs, J. & Zetik, R. [2007]. UWB sensor networks for position location and imaging of objects and environments, *Proc. Second European Conf. Antennas and Propagation EuCAP 2007*, pp. 1–9.

[62] Thrun, S., Burgard, W. & Fox, D. [2001]. *Probabilistic Robotics (Intelligent Robotics and Autonomous Agents)*, The Mit Press.

[63] van den Bos, A. [1995]. The multivariate complex normal distribution-a generalization, *IEEE Transactions on Information Theory* 41(2): 537–539.

[64] Varslot, T., Yazici, B., Yarman, C.-E., Cheney, M. & Scharf, L. [2007]. Time-reversal waveform preconditioning for clutter rejection, *Proc. Int. Waveform Diversity and Design Conf*, pp. 330–334.

[65] Vo, B.-N., Singh, S. & Doucet, A. [2003]. Sequential Monte Carlo Implementation of the PHD filter for Multi-target Tracking, *Proc. 6th Int. Conf. Information Fusion (FUSION)*, Vol. 2, Cairns, Australia, pp. 792–799.

[66] Vo, B.-N. V. & Ma, W.-K. [2006]. The Gaussian Mixture Probability Hypothesis Density Filter, 54(11): 4091–4104.

[67] Win, M. & Scholtz, R. [2002]. Characterization of ultra-wide bandwidth wireless indoor channels: a communication-theoretic view, *Selected Areas in Communications, IEEE Journal on* 20(9): 1613 – 1627.

[68] Wylie, M. P. & Holtzman, J. [1996]. The non-line of sight problem in mobile location estimation, *Proc. 5th IEEE Int Universal Personal Communications Record. Conf*, Vol. 2, pp. 827–831.

[69] Yan, H., Shen, G., Zetik, R., Hirsch, O. & Thomä, R. S. [2012]. Ultra-wideband MIMO ambiguity function and its factorability. IEEE Trans. Geosci. Remote Sensing, to be published. DOI: 10.1109/TGRS.2012.2201486.

[70] Yan, H., Shen, G., Zetik, R., Malz, E., Jovanoska, S. & Thomä, R. S. [2011]. Stationary symmetric object detection in unknown indoor environments, *Loughborough Antennas & Propagation Conference 2011, LAPC 2011).*

[71] Yarovoy, A. G., van Genderen, P. & Ligthart, L. P. [2000]. Ground penetrating impulse radar for land mine detection, *in* D. A. Noon, G. F. Stickley, & D. Longstaff (ed.), *Society of Photo-Optical Instrumentation Engineers (SPIE) Conference Series*, Vol. 4084 of *Society of Photo-Optical Instrumentation Engineers (SPIE) Conference Series*, pp. 856–860.

[72] Zetik, R., Jovanoska, S. & Thomä, R. [2011]. Simple method for localisation of multiple tag-free targets using UWB sensor network, *2011 IEEE Int. Conf. Ultra-Wideband (ICUWB)*, pp. 268 –272.

[73] Zetik, R., Sachs, J. & Thomä, R. [2005]. Modified cross-correlation back projection for UWB imaging: numerical examples, *Ultra-Wideband, 2005. IEEE International Conference on*, p. 5.

[74] Zetik, R. & Thomä, R. [2008]. Monostatic imaging of small objects in UWB sensor networks, *Ultra-Wideband, 2008. IEEE International Conference on*, Vol. 2, pp. 191 –194.

[75] Zetik, R. & Thomä, R. S. [2010]. Imaging of Distributed Objects in UWB Sensor Networks, *in* Sabath, F., Giri, D. V., Rachidi, F., & Kaelin, A. (ed.), *Ultra-Wideband, Short Pulse Electromagnetics 9*, Springer New York, pp. 97–104.

[76] Zhan, H., Ayadi, J., Farserotu, J. & Le Boudec, J.-Y. [2009]. Impulse radio ultra-wideband ranging based on maximum likelihood estimation, *IEEE Transactions on Wireless Communications* 8(12): 5852–5861.

Chapter 10

Pedestrian Recognition Based on 24 GHz Radar Sensors

Steffen Heuel and Hermann Rohling

Additional information is available at the end of the chapter

1. Introduction

Radar sensors offer in general the capability to measure extremely accurately target range, radial velocity, and azimuth angle for all objects inside the observation area. These target parameters can be measured simultaneously even in multiple target situations, which is a technical challenge for the waveform design and signal processing procedure. Furthermore, radar systems fulfil these requirements in all weather conditions, even in rain and fog, which is important for all automotive applications, [1], [2]. Advanced driver assistant systems (ADAS) are currently under investigation to increase comfort and safety in general. For Adaptive Cruise Control (ACC) applications a single 77 GHz radar sensor is used, which has a maximum range of 200 m and covers a narrow azimuth angle area of 15 degree for example. Many other and additional automotive applications, like Stop & Go, Pre-Crash or Parking Aid, consider a completely different observation area [3]. In this case a maximum range of 50 m, but a wide azimuth angle area of 120 degrees is required. For these applications 24 GHz radar sensors are used. Besides the range and velocity parameters, additional information concerning the target type are of great interest, as one of the main objectives of future safety systems will be the increased protection of all pedestrians and other vulnerable road users.

By extending the radar signal processing part of a 24 GHz radar sensor with a pedestrian recognition scheme, the same radar sensor which is used for the mentioned applications can be applied additionally for pedestrian recognition and allows the design of pedestrian safety systems. Therefore, the radar signal processing part has to be adapted to the assumption of extended targets with a characteristic range profile and a velocity profile (e.g. based on the Doppler Spectrum) in general [4]. The detailed analysis of the resulting range profile and target's velocities is possible and can be used to recognize pedestrians in urban areas with conventional 24 GHz radar sensors.

2. Radar sensor and measurements

Several proposals for pedestrian recognition schemes have been described, which are based on video cameras and computer vision systems [7], [8]. But automotive radar sensors in the

24 and 77 GHz band are also strong candidates for automotive safety systems. Compared with vision systems, they have some additional important advantages of robustness in all weather conditions, simultaneous target range and radial velocity measurement and a high update rate. These properties are especially important for pedestrian recognition, as the object classification should be available immediately and at any time.

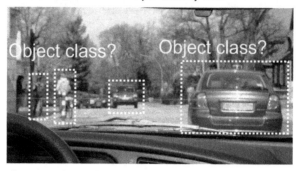

Figure 1. Daily traffic situation in an urban area with an oncoming vehicle and pedestrians walking on the sidewalk.

This chapter presents the modulation scheme of an automotive radar sensor and explains the features of pedestrians and vehicles by which a robust classification is possible in an urban area from a moving vehicle with a mounted 24 GHz radar sensor, see Figure 1.

2.1. Modulation scheme

The automotive 24 GHz radar sensor allows a simultaneous and unambiguous measurement of target range R and radial velocity v_r even in multiple target situations. This is achieved by combining the advantages of the Frequency Shift Keying (FSK) waveform and the Frequency Modulated Continuous Waveform (FMCW) in a so called Multi Frequency Shift Keying (MFSK) waveform [19], which is already used in commercial automotive radar sensors to enable Adaptive Cruise Control (ACC) or Blindspot Detection (BSD) [20], [21]. Applying an FSK waveform, the target range R and radial velocity v_r can be measured. However, there is no range resolution. Multiple objects measured at the same spectral line in the Doppler spectrum result in an unusable range information, as the determination procedure assumes a single target. To mitigate this drawback, the FMCW waveform resolves targets in range R and velocity v_r. Limitations will occur in this case in multi target situations due to ambiguous measurements. The specific MFSK waveform is applied in the 24 GHz Radar sensor for a range and Doppler frequency measurement even in multi target situations with a bandwidth of $f_{sweep} = 150$ MHz and a resulting range resolution of $\Delta R = 1.0$ m. It is a classical step-wise frequency modulated signal with a second linear frequency modulated signal in the same slope but with a certain frequency shift f_{step} integrated into this waveform in an intertwined way. The chirp duration is denoted by $T_{CPI} = 39$ ms which results in a velocity resolution of $\Delta v = 0.6$ km/h.

It is important to notice that this waveform is not processed by a matched filter or analyzed by an ambiguity function. Instead it is processed in a non-matched filter form to get an unambiguous and simultaneous target range and Doppler frequency measurement with high resolution and accuracy. The echo signal of the stepwise and intertwined waveform is downconverted by the corresponding instantaneous transmit frequency into baseband and sampled at the end of each short frequency step. This time discrete signal is Fourier transformed separately for the two intertwined signals to measure the beat frequency f_B which is simultaneously influenced by the target range R and radial velocity v_r.

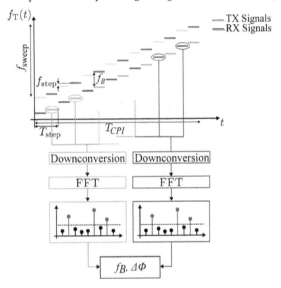

Figure 2. MFSK waveform principle with two intertwined transmit signals.

$$f_B = -\frac{2v_r}{\lambda} - \frac{2R \cdot f_{sweep}}{c} \cdot \frac{1}{T_{CPI}} \tag{1}$$

In any case, a single target will be measured and will be detected on the same spectral line at position f_B for the two intertwined signals. Therefore, after the detection procedure the phase difference $\Delta\Phi$ between the two complex-valued signals on the spectral line f_B will be calculated. The step frequency f_{step} between the intertwined transmit signals determines the unambiguous phase measurement $\Delta\Phi$ in the interval $[-\pi; \pi)$. This phase difference $\Delta\Phi$ again is influenced by the target range R and radial velocity v_r described in Equation (2).

$$\Delta\Phi = -\frac{2\pi}{f_{sample}} \cdot \frac{2v_r}{\lambda} - \frac{4\pi R \cdot f_{step}}{c} \tag{2}$$

The target range R and radial velocity v_r can be determined by solving the linear equation described in Equation (1) and (2) in an unambiguous way. In this case, ghost targets are completely avoided since this waveform and signal processing combines the benefits of linear FMCW and FSK technology. The system design and the sensor parameters can be determined like in a linear FMCW radar system. The range and velocity resolution ΔR and Δv are determined by the bandwidth f_{sweep} of the radar sensor and the chirp duration T_{CPI} as described in Equation (3) and (4), respectively.

$$\Delta R = \frac{c}{2} \cdot \frac{1}{f_{sweep}} \tag{3}$$

$$\Delta v = -\frac{\lambda}{2} \cdot \frac{1}{T_{CPI}} \tag{4}$$

The table below shows the system parameters of the automotive radar sensor in detail.

Carrier Frequency	$f_T = 24\,\text{GHz}$
Sweep Bandwdith	$f_{sweep} = 150\,\text{MHz}$
Maximum Range	$R_{max} = 200\,\text{m}$
Range Resolution	$\Delta R = 1\,\text{m}$
Chirp Length	$T_{CPI} = 39\,\text{ms}$
Maximum Velocity	$v_{max} = 250\,\text{km/h}$
Velocity Resolution	$\Delta v = 0.6\,\text{km/h}$

Table 1. 24 GHz Radar Sensor Parameters.

Classical UWB-Radar Sensors have a sweep bandwidth of $f_{sweep} = 2\,\text{GHz}$. Using such a bandwidth, a high range resolution is determined, which allows also pedestrian classification. The technical challenge in this chapter is to realize pedestrian recognition based on a 24 GHz radar sensor with a bandwidth of only 150 MHz. This sensor is used in automotive applications, therefore an extension of the signal processing in terms of pedestrian classification is desirable.

2.2. Radar echo signal measurements

The possibility to recognize pedestrians with a static radar sensor using the Doppler effect has been shown in [15]. A moving vehicle is equipped with an automotive radar sensor with a built-in feature extraction and classification to recognize pedestrians. The feature extraction in the backscattered radar echo signals resulting from superposition of the reflection points of an object is done automatically in the radar sensor signal processing. Detected targets are therefore tracked in the environment and an additional feature extraction and classification is performed.

To distinguish between the echo signal characteristics of pedestrians and vehicles, a target recognition model is described which is based on the specific **velocity profile** and **range profile** for each object separately [4]. The velocity profile describes the extension of the different velocities of an object measured by the radar sensor, while the range profile shows the physical expansion of a target.

In case of a longitudinally moving pedestrian, different reflection points at the trunk, arms and legs with different velocities are characteristic in radar propagation. Therefore an *extended* velocity profile will be observed in a single radar measurement of a pedestrian as the velocity resolution Δv of the radar sensor is higher than the occuring velocities. Carrying out several measurements with a time duration of 39 ms each, a sinusoidal spreading and contraction of the velocity profile can be observed in the case of a pedestrian, due to the movement of arms and legs for example in the swing and stand-phase of the legs. For a laterally moving pedestrian, the velocity profile is less extended due to the moving direction of the pedestrian. Furthermore, the extension depends mainly on the azimuth angle under which the pedestrian is measured. In contrast, the radar echo signal in case of a vehicle shows a very narrow (*point shaped*) velocity profile due to a uniform motion.

Additionally, a *point shaped* range profile will occur in the case of a longitudinally or laterally moving pedestrian as the physical expansion is small compared to the range resolution of $\Delta R = 1.0\,\mathrm{m}$. In contrast, a vehicle shows an *extended* range profile, due to several reflection points spaced in several range cells. The measurement result of a single observation is shown in the range Doppler diagram in Figure 4.

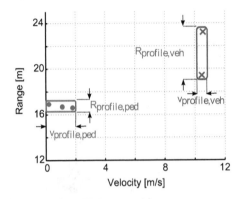

Figure 3. Range profile and velocity profile of a single measurement.

Under the use of an MFSK modulation signal, a range profile and the velocity profile can be extracted from a series of received signals as shown in Figure 4. As an example, four consecutive range and velocity measurements are depicted in a range Doppler diagram. The red dots show a longitudinally walking pedestrian, the blue crosses an in front moving vehicle. The figures depicted are based on radar measurements taken in an urban area with an ego speed of 50 km/h. It can be observed that neither velocity profile nor a range profile can be seen in the first measurement, consequently, those feature values are zero. In the second measurement, however, several range and velocity measurements allow to calculate an extended range profile for the vehicle and an extended velocity profile for the pedestrian.

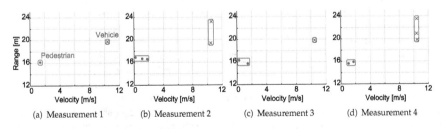

Figure 4. Sequence of range and velocity measurements.

The range profile and velocity profile features do not depend on the modulation signal. Solely, the range and velocity resolution must be smaller than the expected extension. For example, in the case of a continous wave modulation signal, the range profile can be read directly from the Fourier transformed radar echo signal and the velocity profile can be evaluated from the Doppler spectrum. Also, instead of these spectra or the frequency spectrum and phase difference analysis, it is possible to calculate the extension of an object in range and velocity on the basis of target lists by applying a detection algorithm. On this basis, an extended range profile with a point-shaped velocity profile can also be measured for a vehicle. For a pedestrian, the profiles remain vice versa. Figure 5 depicts this context.

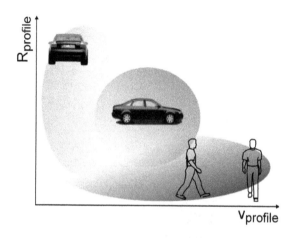

Figure 5. Range profile and velocity profile of a pedestrian and a vehicle.

The longitudinally and laterally moving pedestrians are classified as **pedestrians**, longitudinally and laterally moving vehicles as **vehicles**, all other signals received from objects such as parked cars, poles, trees and traffic signs are classified as **other** objects.

3. Classification

Target recognition is a challenge for each radar engineer. A reliable feature extraction and classification process has to be implemented. To describe the characteristics of pedestrians and vehicles, the velocity profile and range profile signal features have been introduced. These are the basis in the feature extraction and target recognition system based on a single radar measurement (single look of 39 ms duration) as described in subsection 3.1. An extended extraction based on the spreading and contraction of the spectra by observing several measurements is considered in subsection 3.2. Finally in subsection 3.3, a tracker feedback is calculated where additional features based on the Kalman gain and innovation are extracted. In the next step, the classification process is performed, which maps the extracted features into classes. An evaluation of the classification results is shown by means of a confusion matrix for the case of a single measurement- and multiple measurements-feature extraction.

3.1. Classification based on a single radar measurement

Radar sensors provide continuously available measurement results in an interval of a few milliseconds. This interval is determined by the duration of the transmission signal $T_{CPI} = 39$ ms in which a single MFSK signal is transmitted. The echo signal is downconverted and Fourier-transformed, which allows the described features to be extracted continuously. Rather than examinating a received sequence of radar echoes, this subchapter will initially focus on a single radar measurement of $T_{CPI} = 39$ ms.

3.1.1. Feature extraction

Automotive radar sensors are an important source of information for security and comfort systems. The information is measured in terms of range, radial velocity and signal level. However, information about the object types do not exist. To fill this gap, features from the available information are extracted, which describe the object types and allow a decision of the related class on the basis of measured sensor data. To describe a detected object, this signal processing step calculates a number of features, which are discriminant for measurements containing different object types and match for objects from the same type. Thereby, moderately separated features achieve even in a perfect classification algorithm only moderate or even poor results ([16], [17]). An ideal feature extractor on the other hand shows good classification performance by using simple linear classifiers. This is why the feature extraction is so important. For a distinct classification, transformation-invariant features are sought. Still, there is no recipe to determine a feature set and since each sensor type describes an object specifically and each task is different, the feature set for pedestrian recognition based on an automotive radar sensor is explained shortly. Figure 6 shows the feature extraction with the specific object description in the context of the signal processing chain.

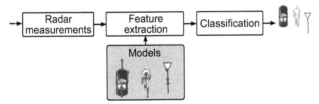

Figure 6. Context of the feature extraction in the signal processing chain and the object description using the range profile and velocity profile.

The basis for feature extraction are the velocity profile and the range profile of a detected object which has been described previously. The term of the profile describes the physical and kinematic dimensions of an object in the distance, angle and velocity. This can be measured in the case of multiple reflection points with different velocities greater than zero for any object. It can involve an extended or a point-shaped profile for the range and velocity depending on the type of expansion. On this basis, a number n of features can be calculated which describe the object in terms of a radar measurement. All n calculated real valued features $x_1, ..., x_n$ are saved in a feature vector $\vec{x}$ and build the basis for further signal processing steps.

$$\vec{x} = (x_1, x_2, ..., x_n)\ n \in \mathbf{N}, x_i \in \mathbf{R} \tag{5}$$

Exemplarily, the calculation of the range profile R_{profile} is given in Equation (6). Analogously, the velocity profile v_{profile} of the spectrum can be calculated, Equation (7).

$$x_1 = R_{\text{profile}} = R_{\max} - R_{\min} \tag{6}$$

$$x_5 = v_{\text{profile}} = v_{\max} - v_{\min} \tag{7}$$

The approach in feature extraction, using stochastic features, assumes that the measured data are random variables with independent and identical distribution. From this data within a single measurement cycle the variance and the standard deviation is estimated. To support the classification process, the number of scatterers is extracted, which describes the number of detected reflection points of an object. This approach allows a classification of the object type within a single measurement. The entire feature set for $n = 8$ features is shown in Table 2 below.

Feature	Annotation	Description
x_1	R_{profile}	Extension in range
x_2	$std(R)$	Standard deviation in range
x_3	$var(R)$	Variance in range
x_4	v_r	Radial Velocity
x_5	v_{profile}	Extension in velocity
x_6	$std(v_r)$	Standard deviation in velocity
x_7	$var(v_r)$	Variance in velocity
x_8	scatterer	Number of scatterers

Table 2. Feature Set of each object in a single measurement.

To determine the quality of a feature, the common area index (CAI) of two histograms is considered. While a common area index of 0 describes a complete overlapping of the feature space, a CAI of 1 describes an absolutely separable feature.

Several urban measurement scenarios of longitudinally moving vehicles and pedestrians were measured with an automotive radar sensor. From the detections of each single measurement cycle the features are extracted. Exemplarily, the velocity profile of a vehicle and a pedestrian is depicted in Figure 7 as a histogram. It shows a strong overlap of the area with a point shaped extension. This results directly from the model. A pedestrian is not extended at all times, because the arms and legs move sinusoidally. In addition, the echo signal fluctuates which causes fewer detections in a measurement. The vehicle equipped with the radar sensor moves also. The quality of the feature is calculated to CAI $= 0.57$.

3.1.2. Classification

The assignment of a measured object to a class is performed by a subjective decision algorithm based on the extracted characteristics. This process is called classification. The features therefore have been described previously, and are extracted within a single radar measurement of $T_{\text{CPI}} = 39$ ms. In supervised classifiers, the model of the classifier is generated in a training phase by using a training data set. The verification is performed in an evaluation phase with a test data set. The training data and test data consist of randomly selected feature vectors $\vec{x}$ of the radar measurements and corresponding assigned class labels. In the training and evaluation phase the classification result can be compared to the class labels and make a

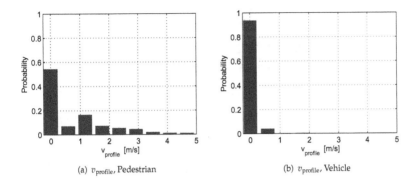

(a) $v_{profile}$, Pedestrian   (b) $v_{profile}$, Vehicle

Figure 7. Feature histogram of the velocity profile using single radar measurements as a basis for feature extraction. The common area index is calculated to 0.57.

statement about the performance of the algorithm and designed model. Figure 8 depicts this process.

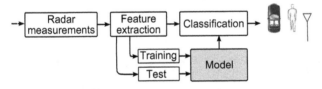

Figure 8. Signal flow graph of the classification process. Using a training data-set a model can be evaluated which performance is measured by a labelled test data set. This model is used for the classification process.

A classifier based on statistical learning theory is the support vector machine (SVM), introduced by Boser et al. in 1992 [24]. The SVM became very famous as studies about classification algorithms show good performance [25]. The classification process has low complexity and is very effective for high dimensional feature vectors. An SVM separates a set of training data by calculating a hyperplane $h(x)$ with maximum margin between the two classes ± 1 in a higher dimensional space in order to find the best classification function.

In this classification process, the SVM is able to map the extracted feature set into three different classes by using a majority voting algorithm. The verification of the previous training is conducted with the help of test data sets recorded from real urban measurements.

Table 3 shows the classification results. A trained and tested SVM was applied to different data sets of the extracted features from single measurements. All measurements were taken in an urban area with an ego velocity of 50 km/h. Applying new test data results in 71.32% true positive for a vehicle and 45.20% true positive for a pedestrian.

These quantitative results show already a possibility to distinguish between vehicles, pedestrians and other objects. However the performance is not good enough. Therefore

	Vehicle	Pedestrian	Other
Vehicle	71.32	5.87	22.81
Pedestrian	10.29	45.20	44.52
Other	23.56	26.65	49.78

Table 3. Confusion matrix: classification applied to a single measurement test data set containing 8000 data samples.

multiple radar measurements are considered to extract a more significant features set for the classification process.

3.2. Classification based on multiple radar measurements

From the continuously available radar measurements, a single measurement can be used to extract a feature set on the basis of range profile and velocity profile and estimated stochastic features. In this section, several range and velocity measurements are buffered and build the basis for the additional extraction process. From these buffered measurements a second **multiple measurement feature set** is extracted. This extends the classification process, which was previously based on a single measurement only.

3.2.1. Feature extraction

To gain performance, the choice of the measurement buffer dimension is crucial. A long measurement buffer builds the basis for a more successful feature extraction, however results in a long classification time, as the classifier has to wait for the buffer to be filled. A short buffer, on the other hand is not always able to build the basis for separable features, as shown in the previous section where a single measurement is used. In this section, the dimension of the buffer is explained by means of "probability of maximum velocity profile" and fast availability deduced from the step frequency of a moving pedestrian.

An ideal measurement of a moving pedestrian allows to extract the step frequency from the spreading and contraction of the velocity profile [9]. This step frequency of $f_{ped} = 1.4 - 1.8\,\text{Hz}$ can be used to determine a necessary buffer dimension. Every $1.4\,\text{Hz}$ the maximum velocity profile can be observed considering a moving pedestrian, which allows to extract the maximum velocity profile and range profile. At all other times, the expansion in velocity is lower or even zero. To detect at least one expansion in velocity, the number of measurements should therefore span a period of $T_{ped} = \frac{1}{f_{ped}}$. However, it can be assumed that all measurements are independent of each another, due to the ego motion of the radar, vibrations and fewer detections. Applying the feature extraction process using a single measurement, the probability P to extract an extended velocity is then given by:

$$P_{v_{profile},single} = \frac{T_{CPI}}{1/f_{ped}} \qquad (8)$$

Assuming an equal distribution, multiple measurements increase the probability to extract a velocity profile by the factor $\frac{T_{Buffer}}{T_{CPI}}$.

$$P_{v_{profile},multiple} = \frac{T_{Buffer}}{T_{CPI}} \cdot \frac{T_{CPI}}{1/f_{ped}} \qquad (9)$$

It can be seen that a long measurement buffer increases the probability to detect the maximum velocity profile. But even a smaller velocity profile can be detected and fulfils the requirements. In these measurements, a buffer of $T_{\text{Buffer}} = 150\,\text{ms}$ is applied.

Using current and time delayed range measurements leads to incorrect range profiles, due to the movement of the objects during the elapsed buffer time. To cope this, the corresponding range measurements inside the buffer must therefore be predicted in range. Each stored measurement is predicted in range by the elapsed time ΔT_{Buffer} and the velocity $\vec{v} = (v_x, v_y)$ during the measurements to compensate the movement. A new estimated range $\hat{R} = |\vec{R}^*|$ can be calculated by a cartesian representation using velocity and elapsed time:

$$\vec{R}^* = \vec{R} + \Delta T_{\text{Buffer}} \cdot \vec{v} \text{ with } \vec{R} = (X, Y) \text{ and } \vec{v} = (v_x, v_y) \tag{10}$$

The multiple measurement feature set is shown in Table 4. It consists of the same characteristics as the single measurement feature set, but is calculated from a basis of several buffered velocity and predicted range measurements.

Feature	Annotation	Description
x_9	$\hat{R}_{\text{profile,buf}}$	Extension in range
x_{10}	$std(\hat{R},\text{buf})$	Standard deviation in range
x_{11}	$var(\hat{R},\text{buf})$	Variance in range
x_{12}	$v_{r,\text{buf}}$	Radial Velocity
x_{13}	$v_{\text{profile,buf}}$	Extension in velocity
x_{14}	$std(v_{r,\text{buf}})$	Standard deviation in velocity
x_{15}	$var(v_{r,\text{buf}})$	Variance in velocity
x_{16}	scatterer,buf	Number of scatterers

Table 4. An additional feature set extracted from multiple measurements. To cover multiple measurements, each single measurement R, v_r is stored in a buffer of several milliseconds. This ensures a quick availability of an additional feature set for pedestrian classification.

In the previous section the characteristic velocity profile of a vehicle and a pedestrian was depicted as a histogram. On the basis of a single radar measurement a quality of CAI = 0.57 was determined. Using multiple radar measurements the feature extraction is based on a larger number of measurement values. This leads to a higher separability of the features as shown exemplarily in Figures 9(a), 9(b). The common area index has a value of CAI = 0.88 and thus increases by 31% compared to the single measurement feature extraction.

3.2.2. Classification

In the single measurement it is described how range and velocity measurements build the basis for the feature vector and the classification process. In the multiple measurement, the basis is an extended feature vector based on several range and velocity measurements of an object stored inside a buffer. Instead of using 8 features from a single measurement, additional 8 features are available for the first time with a filled buffer. For a successful classification using a SVM, a new model is built, which is also trained/tested with in total 16 features and is a basis for the following classification process. As shown in the confusion matrix in Table 5 by the additional features, the correct classification and the overall performance achieved, increases by using a total number of 16 features from single and multiple measurements.

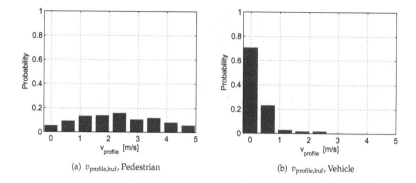

(a) $v_{profile,buf}$, Pedestrian   (b) $v_{profile,buf}$, Vehicle

Figure 9. Feature histogram of the velocity profile using multiple radar measurements as a basis for feature extraction. The common area index is calculated to 0.88.

Additionally, classifying feature vectors from single and multiple measurements results in fewer false positives.

	Vehicle	Pedestrian	Other
Vehicle	90.71	0.58	8.71
Pedestrian	4.94	53.94	41.12
Other	16.28	16.64	67.08

Table 5. Confusion matrix: classification applied to a single and multiple measurement test data set containing 8000 data samples.

Due to additionally multiple measurements as a basis for feature extraction, an improvement in the classification result is shown. Especially in terms of correct classification of a pedestrian and false alarms in which a pedestrian was classified as a vehicle, a significant enhancement is seen.

3.3. Classification based on the tracker feedback

A single radar measurement and multiple radar measurements were considered for feature extraction. These features are already a good basis for the classification process. In this subchapter, a third, additional feature set is described. This set can be extracted from the tracking algorithm. In this adaptive algorithm, different process noise of pedestrian radar measurements and vehicle radar measurements result in different gains for the track. On this basis, the process noise Q and the calculated gain K are additional features and are added to single and multiple radar measurement features in the classification process.

3.3.1. Feature extraction

Tracking is defined by a state estimation of moving targets. This state estimation is determined from the state parameters such as position, velocity and acceleration from a detected target. Known tracking methods are for example the alpha-beta filter or Kalman filter [26] which estimate a new state using a well-known prior state (e.g. position, velocity, acceleration). This reduces false alarms and smoothes movements of the objects.

The Kalman filter is a linear, recursive filter, whose goal is to determine an optimal estimate of the state parameters. The optimal estimate is based on available measurements and the models which describe the observed objects. In the equation of the motion model and observation model, the measurement noise is assumed to be average free, white Gaussian noise with the known covariance Q_{k-1} and R_k respectively. Under the given conditions, i.e., linear models and Gaussian statistics, the Kalman filter provides the optimal solution for the estimation of the state in the sense of minimizing the mean squared error, as described in [26].

The tracking for the object described by the motion model of the Kalman filter works fine as long as the motion models fit to the object. Pedestrians, vehicles, and static objects have different motion, which makes the tracking more difficult. Instead of creating a different motion and observation model for each object, it is proposed to determine the covariance of process noise Q_k and the measurement noise R_k adaptively. The process noise considers a non-modeled behavior in the motion model, while the measurement noise consideres uncertainty in the measurement. The original Kalman filter is not adaptive, which is why deviations from the model can not be handled. The gain matrix K, which is calculated from the process and measurement noise, reaches a stable condition after a short measurement time. An increase in the covariance Q_k leads to a larger value for K, so that the measured values are weighted more strongly, a decrease in Q_k relies more on the estimation.

In addition to an improved tracking effect, additional features can be extracted from the adaptive adjustment of the process noise, as pedestrian measurements in range and velocity differ from those of vehicles. Next to the process noise $Q_{k,v}$ of the velocity, the Kalman gain K_v (velocity component of the matrix K) is a good feature as measurements show. Anyhow, in an adaptive adjustment of the process noise, a compromise between the compensation of non-modeled movements and the filtering effect to reduce noise must be found, even though features are extracted.

The process noise matrix Q describes object-specific measurement properties that are initially set and are readjusted during operation of the tracker. For example, the readjustment of a single coefficient $Q_{k,v}$ in the Q matrix at the measurement k in respect to the velocity v is based on the actual target range R, the velocity v, the parameters a and b in an alpha-beta filter. Equation (11) shows the relation. The velocity of a pedestrian deviates between consecutive measurements, while the velocity deviation of a car within consecutive measurements is small or even zero. Consequently, the velocity v can be used to update the matrix Q.

$$Q_{k,v} = (1 - \beta)Q_{k-1,v} + \beta \frac{|z_{k,v} - v_{k|k-1}|}{R_{k|k} \cdot a + b} \tag{11}$$

The predicted covariance matrix $P_{k|k-1}$ in the tracking process depends on the motion model F_{k-1}, the currently measured covariance $P_{k-1|k-1}$ and the process noise Q_{k-1} as shown in Equation (12).

$$P_{k|k-1} = F_{k-1}P_{k-1|k-1}F_{k-1}^T Q_{k-1} \tag{12}$$

Under the use of this covariance matrix $P_{k|k-1}$ and the innovation covariance S_k, the gain K_k can be calculated.

$$K_k = P_{k|k-1} S_k^{-1} \qquad (13)$$

Both, matrix Q_k and gain K_k are used as additional features in the classification process. The separation of the feature space is depicted in Figure 10. These histograms show that a vehicle (Figure 10(c), 10(d)) has lower process noise Q_v and thus results in a smaller gain K compared to a pedestrian (Figure 10(a), 10(b)). This is due to a mostly linear trajectory of a vehicle with one main reflection point. A pedestrian echo fluctuates, which is reflected in a greater process noise and thus larger gain K.

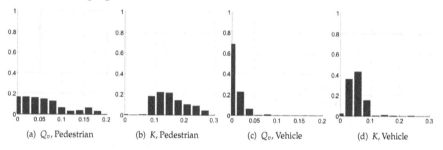

(a) Q_v, Pedestrian   (b) K, Pedestrian   (c) Q_v, Vehicle   (d) K, Vehicle

Figure 10. Feature Histogram of Process Noise Q_v and Gain K.

The additional features Q and K are available for a created and active track. For each detected object a track is created, which is however not activated until several measurements can be associated with the track. A track must be confirmed by subsequent measurements, otherwise the track remains semi-active. In any case, additional features are available for classification.

Feature	Annotation	Description
x_{17}	Q_v	Velocity component of the process noise matrix Q
x_{18}	K_v	Veleocity component of the Kalman gain matrix K

Table 6. Additional Feature Set extracted from the tracker using an adaptive process noise.

3.3.2. Classification

An additional feature extraction based on the process noise and Kalman gain has been described. These features are added to the prior feature set. Table 7 shows the results of the classifier using all proposed features, single measurement, multiple measurements and tracker feedback. Even in an urban area with a high density of static targets pedestrians could be detected, tracked and classified with a true positive rate of 61.22% in the test data set. These results outperform prior outcomes using single- and the multiple measurements as a feature basis.

The classification results show an increasing performance in terms of correct classification and misclassification of pedestrians.

	Vehicle	Pedestrian	Other
Vehicle	92.84	0.50	6.66
Pedestrian	5.71	61.22	33.07
Other	10.06	13.30	76.64

Table 7. Confusion matrix: classification applied to a single, multiple and tracker feedback measurement test data set containing 8000 data samples.

4. Summary and conclusions

This chapter described a pedestrian classification algorithm for automotive applications using an automotive 24 GHz radar sensor with a bandwidth of 150 MHz as a measuring device. Three different systems for pedestrian recognition have been considered. The first system was based on a single radar measurement. The second system extracted a feature set on the basis of multiple radar measurements. Finally a tracking procedure was adapted to extract an additional feature set. The results show an increasing performance in the classification accuracy by using single-, multiple- and tracker feedback features. It is also pointed out that is not necessary to equip radar sensors with large bandwidths in order to classify pedestrians in urban areas.

Author details

Hermann Rohling and Steffen Heuel
Department of Telecommunications, Hamburg University of Technology, Hamburg, Germany

5. References

[1] Foelster, F.; Oprisan, D.; Rohling, H., (2004) Detection and Tracking of extended targets for a 24GHz automotive radar network, International Radar Symposium IRS 2004, Warsaw, Poland; pp 75-80

[2] Meinecke, M.-M.; Obojski, M.; Toens, M.; et.al, (2003) Approach For Protection Of Vulnerable Road Users Using Sensor Fusion Techniques, International Radar Symposium IRS 2003, Dresden, Germany, pp 125-130

[3] Schiementz, M.; Foelster, F. (2003) Angle Estimation Techniques for different 24 GHz Radar Networks, International Radar Symposium IRS 2003, Dresden, Germany, 2003

[4] Rohling, H.; Heuel, S.; Ritter, H. (2010) Pedestrian Detection Procedure integrated into an 24 GHz Automotive Radar, IEEE RADAR 2010, Washington D.C.

[5] Schmid, V.; Lauer, W.; Rollmann, G. (2003) Ultra Wide Band 24 GHz Radar Sensors for Innovative Automotive Applications, International Radar Symposium IRS 2003, Dresden, Germany, pp 119-124

[6] Oprisan, D.; Rohling, H.,(2002) Tracking Systems for Automotive Radar Networks, IEE Radar 2002, Edinburgh

[7] Philomin, V.; Duraiswani, R.; Davis, L. (2000) Pedestrian tracking from a moving vehicle, Proceedings of the IEEE Intelligent Vehicle Symposium 2000, Dearborn (MI), USA

[8] Ran, Y. et al (2005) Pedestrian classification from moving platforms using cyclic motion pattern, IEEE International Conference on Image Processing 2005, Volume 2, pages: II 854-7

[9]   Van Dorp, P.; Groen, F. (2003) Real-Time Human Walking Estimation with Radar, International Radar Symposium IRS 2003, Dresden, Germany, pp 645-650

[10] Lei, J.; Lu, C. (2005) Target Classification Based on Micro-Doppler Signatures, IEEE International Radar Conference 2005; Arlington, USA; pp 179-183

[11] Klotz, M. (2002) An Automotive Short Range High Resolution Pulse Radar Network, Phd. Thesis, Hamburg University of Technology, Hamburg, Germany

[12] Luebbert, U. (2005) Target Position Estimation with a Continuous Wave Radar Network, Phd. Thesis, Hamburg University of Technology, Hamburg, Germany

[13] Foelster, F., Rohling, H. and Meinecke, M. M. (2005) Pedestrian recognition based on automotive radar sensors, 5th European Congress on Intelligent Transportation Systems and Services 2005, Hannover, Germany

[14] Heuel, S.; Rohling, H. (2011) Two-Stage Pedestrian Classification in Automotive Radar Systems, International Radar Symposium IRS 2011, Leipzig, Germany

[15] Nalecz, M.; Rytel-Andrianik, R.; Wojtkiewicz, A. (2003) Micro-doppler analysis of signals received by FMCW radar, Proceedings of International Radar Symposium 2003, Dresden, Germany, pp. 651-656

[16] Duda, R. O.; Hart, P. E.; Stork, D. G. (2001) Pattern classification, Wiley, New York

[17] Schuermann, J. (1996) Pattern Classification - A Unified View of Statistical and Neural Approaches, Wiley, New York

[18] Schiementz, M. (2005) Postprocessing Architecture for an Automotive Radar Network, Phd. Thesis, Hamburg University of Technology, Hamburg, Germany

[19] Meinecke, M.-M.; Rohling, H. (2000) Combination of LFMCW and FSK modulation principles for automotive, German Radar Symposium GRS 2000, Berlin, Germany

[20] Winner, H., Hakuli, S., Wolf, G., (2009), Handbuch Fahrerassistenzsysteme, Vieweg+Teubner Verlag / GWV Fachverlage GmbH, Wiesbaden, Germany

[21] Smart Microwave Sensors GmbH, (2012), UMRR | LCA BSD Technical Information Sheet, available at
      http://www.smartmicro.de/images/stories/contentimage/automotive/LCA   and
      BSD Technical Information.pdf, Accessed: 30 August 2012

[22] Perry, J. (1992) Gait Analysis - Normal and Pathological Function, SLACK Incorporated, ISBN 1556421923

[23] Foelster, F.; Ritter, H., Rohling, H. (2007) Lateral Velocity Estimation for Automotive Radar Applications, The IET International Conference on Radar Systems, Edinburgh, Great Britain

[24] Boser, B. E.; Guyon, I. M.; Vapnik, V. N. (1992) A training algorithm for optimal margin classifiers, In D. Haussler, editor, 5th Annual ACM Workshop on COLT, Pittsburgh, PA, pp. 144-152

[25] Wu, X.; Kumar, V.; Ross Quinlan, J.; Ghosh, J.; Yang, Q.; Motoda, H.; McLachlan, G. J.; Ng, A.; Liu, B., Yu, P. S. and others (2008) Top 10 algorithms in data mining, Knowledge and Information Systems, Vol. 14, No. 1., pp. 1-37

[26] Branko, R.; Sanjeev, A.; Nei, G. (2004) Beyond the Kalman Filter: Particle Filters For Tracking Applications, Artech House Inc.

Permissions

The contributors of this book come from diverse backgrounds, making this book a truly international effort. This book will bring forth new frontiers with its revolutionizing research information and detailed analysis of the nascent developments around the world.

We would like to thank Reiner Thomä, Reinhard Knöchel, Jürgen Sachs, Ingolf Willms and Thomas Zwick, for lending their expertise to make the book truly unique. They have played a crucial role in the development of this book. Without their invaluable contribution this book wouldn't have been possible. They have made vital efforts to compile up to date information on the varied aspects of this subject to make this book a valuable addition to the collection of many professionals and students.

This book was conceptualized with the vision of imparting up-to-date information and advanced data in this field. To ensure the same, a matchless editorial board was set up. Every individual on the board went through rigorous rounds of assessment to prove their worth. After which they invested a large part of their time researching and compiling the most relevant data for our readers. Conferences and sessions were held from time to time between the editorial board and the contributing authors to present the data in the most comprehensible form. The editorial team has worked tirelessly to provide valuable and valid information to help people across the globe.

Every chapter published in this book has been scrutinized by our experts. Their significance has been extensively debated. The topics covered herein carry significant findings which will fuel the growth of the discipline. They may even be implemented as practical applications or may be referred to as a beginning point for another development. Chapters in this book were first published by InTech; hereby published with permission under the Creative Commons Attribution License or equivalent.

The editorial board has been involved in producing this book since its inception. They have spent rigorous hours researching and exploring the diverse topics which have resulted in the successful publishing of this book. They have passed on their knowledge of decades through this book. To expedite this challenging task, the publisher supported the team at every step. A small team of assistant editors was also appointed to further simplify the editing procedure and attain best results for the readers.

Our editorial team has been hand-picked from every corner of the world. Their multi-ethnicity adds dynamic inputs to the discussions which result in innovative outcomes. These outcomes are then further discussed with the researchers and contributors who give their valuable feedback and opinion regarding the same. The feedback is then collaborated with the researches and they are edited in a comprehensive manner to aid the understanding of the subject.

Apart from the editorial board, the designing team has also invested a significant amount of their time in understanding the subject and creating the most relevant covers. They scrutinized every image to scout for the most suitable representation of the subject and create an appropriate cover for the book.

The publishing team has been involved in this book since its early stages. They were actively engaged in every process, be it collecting the data, connecting with the contributors or procuring relevant information. The team has been an ardent support to the editorial, designing and production team. Their endless efforts to recruit the best for this project, has resulted in the accomplishment of this book. They are a veteran in the field of academics and their pool of knowledge is as vast as their experience in printing. Their expertise and guidance has proved useful at every step. Their uncompromising quality standards have made this book an exceptional effort. Their encouragement from time to time has been an inspiration for everyone.

The publisher and the editorial board hope that this book will prove to be a valuable piece of knowledge for researchers, students, practitioners and scholars across the globe.

List of Contributors

Rainer Moorfeld and Adolf Finger
Communications Laboratory, Dresden University of Technology, Germany

Nossek Josef A., Mezghani Amine and Michel T. Ivrlač
Institute for Circuit Theory and Signal Processing, Technische Universität München, Germany

Russer Peter, Mukhtar Farooq, Russer Johannes A. and Yordanov Hristomir
Institute for Nanoelectronics, Technische Universität München, Germany

Noll Tobias and Korb Matthias
Chair of Electrical Engineering and Computer Systems, RWTH Aachen University, Germany

Nuan Song, MikeWolf and Martin Haardt
Ilmenau University of Technology, Germany

Pérez Guirao María Dolores
Institute of Communications Technology (IKT), Leibniz Universitaet Hannover (LUH), Hanover, Germany

Andreas Schenk
Lehrstuhl für Informationsübertragung, Friedrich-Alexander-Universität Erlangen-Nürnberg, Germany

Robert F.H. Fischer
Institut für Nachrichtentechnik, Universität Ulm, Germany

Mohamed El-Hadidy, Mohammed El-Absi and Thomas Kaiser
Duisburg-Essen University, Institute of Digital Signal Processing (DSV), Germany

Yoke Leen Sit and Thomas Zwick
Karlsruhe Institute of Technology, Institut für Hochfrequenztechnik und Elektronik (IHE), Germany

Markus Kock and Holger Blume
Leibniz Universität Hannover, Institute of Microelectronic Systems (IMS), Germany

Gholamreza Alirezaei, Rudolf Mathar and Daniel Bielefeld
Institute for Theoretical Information Technology, RWTH Aachen University, D-52056 Aachen, Germany

Markus Grimm and Dirk Manteuffel
University of Kiel, Germany

Rudolf Zetik, Honghui Yan, Elke Malz, Snezhana Jovanoska, Guowei Shen and Reiner S. Thomä
Ilmenau University of Technology, Germany

Rahmi Salman, Thorsten Schultze, Robert Tobera and Hans-Ingolf Willms
University Essen-Duisburg, Germany

Lars Reichardt, Malgorzata Janson, Thomas Zwick and Werner Wiesbeck
Karlsruhe Institute of Technology (KIT), Germany

Tobias Deißler and Jörn Thielecke
Friedrich-Alexander University Erlangen-Nürnberg, Germany

Hermann Rohling and Steffen Heuel
Department of Telecommunications, Hamburg University of Technology, Hamburg, Germany

Printed in the USA
CPSIA information can be obtained
at www.ICGtesting.com
LVHW020517280823
756436LV00005B/191

9 781632 405050